TIME AND THE METAPHYSICS OF RELATIVITY

PHILOSOPHICAL STUDIES SERIES

VOLUME 84

Founded by Wilfrid S. Sellars and Keith Lehrer

Editor

Keith Lehrer, *University of Arizona, Tucson*

Associate Editor

Stewart Cohen, *Arizona State University, Tempe*

Board of Consulting Editors

Lynne Rudder Baker, *University of Massachusetts at Amherst*
Radu Bogdan, *Tulane University, New Orleans*
Allan Gibbard, *University of Michigan*
Denise Meyerson, *University of Cape Town*
Franois Recanati, *École Polytechnique, Paris*
Stuart Silvers, *Clemson University*
Nicholas D. Smith, *Michigan State University*

The titles published in this series are listed at the end of this volume.

TIME AND THE METAPHYSICS OF RELATIVITY

Edited by

WILLIAM LANE CRAIG

*Talbot School of Theology,
Marietta, GA, U.S.A.*

KLUWER ACADEMIC PUBLISHERS
DORDRECHT / BOSTON / LONDON

A C.I.P. Catalogue record for this book is available from the Library of Congress.

ISBN 0-7923-6668-9

Published by Kluwer Academic Publishers,
P.O. Box 17, 3300 AA Dordrecht, The Netherlands.

Sold and distributed in North, Central and South America
by Kluwer Academic Publishers,
101 Philip Drive, Norwell, MA 02061, U.S.A.

In all other countries, sold and distributed
by Kluwer Academic Publishers,
P.O. Box 322, 3300 AH Dordrecht, The Netherlands.

Printed on acid-free paper

All Rights Reserved
© 2001 Kluwer Academic Publishers
No part of the material protected by this copyright notice may be reproduced or
utilized in any form or by any means, electronic or mechanical,
including photocopying, recording or by any information storage and
retrieval system, without written permission from
the copyright owner

Printed in the Netherlands.

IN MEMORIAM

H. A. LORENTZ
H. E. IVES
GEOFFERY BUILDER
SIMON J. PROKHOVNIK

The Trail of Light

TABLE OF CONTENTS

Preface ix

    Chapter 1   The Historical Background of Special Relativity   1

    Chapter 2   Einstein's Special Theory   21

    Chapter 3   Time Dilation and Length Contraction   47

    Chapter 4   Empirical Confirmation of Special Relativity   65

    Chapter 5   Two Relativistic Interpretations   69

    Chapter 6   The Classical Concept of Time   105

    Chapter 7   The Positivistic Foundations of Relativity Theory   122

    Chapter 8   The Elimination of Absolute Time   149

    Chapter 9   Absolute Time and Relativistic Time   171

    Chapter 10   God's Time and General Relativty   195

    Chapter 11   Conclusion   242

Bibliography 243

Subject Index 269

Proper Name Index 273

PREFACE

The present volume is part of a larger project, which is the attempt to draft a coherent doctrine of divine eternity and God's relationship to time. In my *God, Time, and Eternity*,[1] I argued that whether one construes divine eternity in terms of timelessness or of omnitemporality will depend crucially upon one's views about the objectivity of tensed facts and temporal becoming. If one adopts a tensed, or in McTaggart's terminology, an A-Theory of time, then a coherent doctrine of divine eternity requires that one construe God, at least since the moment of creation, to exist temporally, which implies that divine timelessness can be successfully maintained only if a tenseless or B-Theory of time is correct. Accordingly in my companion volumes *The Tensed Theory of Time: a Critical Examination* and *The Tenseless Theory of Time: a Critical Examination* I set for myself the task of adjudicating the A- vs. B-Theory of time.[2] In the former volume, I examine arguments for and against the A-Theory of time, and in the latter I turn to an examination of arguments for and against the B-Theory. This inquiry took me into a study of relativity theory, its presuppositions and implications. The paucity of integrative literature dealing with the concept of God and relativity theory is striking. I am convinced that this lack is largely due to the fact that theologians and philosophers of religion generally do not understand Einstein's theories and so are reduced to merely parroting what they read in scientific popularizations. The result is a superficial and uncritical interaction. Moreover, due to the influence of verificationism, the philosophy of time and space during the past century has been largely reduced to philosophical reflection on spatio-temporal concepts given by physics. With the collapse of positivism and the rejuvenation of metaphysics, the metaphysical foundations of relativity theory deserve renewed scrutiny. There is thus, I believe, a need for an accessible, largely non-mathematical, and philosophically-informed introduction to relativity theory focusing on the concepts of time contained therein.

I must confess that I was surprised at the conclusions to which this study drove me. I never suspected that Newton would emerge as one of the two heroes, as it were, of this book. Newton's poorly understood views on absolute time and space have been the object of such widespread disdain that I just assumed that they were indefensible and obsolete. I now recall with chagrin my attitude when, as a student,

---

[1] *God, Time and Eternity* (Dordrecht: Kluwer Academic Publishers, forthcoming). The project is a natural extension of my previous work on the *kalam* cosmological argument for a personal Creator of the universe. See William Lane Craig, *The Kalam Cosmological Argument*, Library of Philosophy and Religion (London: Macmillan, 1979); William Lane Craig and Quentin Smith, *Theism, Atheism, and Big Bang Cosmology* (Oxford: Clarendon Press, 1993). The second *kalam* argument I presented presupposes the correctness of the A-Theory of time. The defense of the A-Theory is therefore vital to the argument. The coherence of the conclusion of the *kalam* cosmological argument, that there exists a Personal Creator of the beginning of the universe, requires that a coherent doctrine of divine eternity and God's relationship to time be worked out, thus leading to the concerns of the present project.
[2] *The Tensed Theory of Time: a Critical Examination*, Synthèse Library (Dordrecht: Kluwer Academic Publishers, forthcoming); *The Tenseless Theory of Time: a Critical Examination*, Synthèse Library (Dordrecht: Kluwer Academic Publishers, forthcoming).

I first read Richard Swinburne's defense of a basically Newtonian view of time in his *Space and Time*[3]: I was *embarrassed* for him. I never dreamed that someday I myself would be driven into Newton's arms. I now see Newton's fundamental insights into the nature of time as brilliant and enduring in their validity. Of course, Newton made mistakes; but as John Earman has written, they are "the mistakes of a genius engaged in a great struggle."[4] Of Newton it may truly be said that his failures have a greater luster than even the successes of his critics.[5]

The other hero of this book, also unexpected, is the great Dutch scientist H. A. Lorentz, a giant of nineteenth century physics, who was awarded the Nobel Prize in 1901 and is usually remembered as the forerunner to Einstein. As I have become increasingly familiar with Lorentz's life and works, so have my admiration and affection for him grown. A true gentleman and scholar, he was Einstein's hero, too; "in scientific matters I dare revere you as my teacher," Einstein wrote to Lorentz in 1923; "to follow in your footsteps formed the greatest motive of my life."[6] Lorentz continued to study, lecture, and converse with Einstein about both the Special and the General Theory of Relativity, and though he disagreed with Einstein's interpretation of the mathematical formalism of relativity theory, he always graciously expressed his appreciation of Einstein's view. Lorentz saw clearly that the central issue dividing Einstein's and Lorentz's divergent interpretations of the mathematical core of relativity theory was epistemological in nature; moreover, he was convinced that Einstein's verificationism was untenable and told him as much. In the course of my research I struggled mightily to avoid taking Lorentz's part in this debate over the interpretation of relativity theory, but it was in vain: I could find no plausible escape. I am convinced that Lorentz was correct, and I can only beg the reader's indulgence to consider with an open mind whether that conclusion does not follow from my premises.

In this volume I have reproduced a considerable number of figures from textbooks and discussions of relativity theory. For their permission to reproduce such figures, I gratefully acknowledge Oxford University Press, Insight Press, MIT Press, W. H. Freeman and Company, Akademiai Kiado, and HarperCollins UK.

---

[3] Richard Swinburne, *Space and Time* (New York: St Martin's Press, 1968).
[4] John Earman, "Who's Afraid of Absolute Space?" *Australasian Journal of Philosophy* 48 (1970): 317. Earman comments,
"What I find especially disturbing about such condemnations of Newton is not the injustice they do to Newton but rather the fact that they are possible only after an abdication of philosophical responsibility. In all the philosophical literature with which I am acquainted, there is precious little attempt to give reasonably clear and precise answers to the questions which are central to the cluster of philosophical issues which revolve around Newton's conception of space and time" (Ibid., p. 288).
Cf. the patronizing distinctions made by W.-H. Newton-Smith between "good Newton" and "bad Newton" (W.-H. Newton-Smith, "Space, Time, and Spacetime: A Philosopher's View," in *The Nature of Time*, ed. Raymond Flood and Michael Lockwood (Oxford: Basil Blackwell, 1986), pp. 27-29.
[5] So Julian B. Barbour, *Absolute or Relative Motion?*, vol. 1: *The Discovery of Dynamics* (Cambridge: Cambridge University Press, 1989), p. 629.
[6] Albert Einstein to H. A. Lorentz, 15 July 1923, Albert Einstein Archive, Hebrew National and University Library, Jerusalem, no. 16-552, cited in Jozsef Illy, "Einstein Teaches Lorentz, Lorentz Teaches Einstein. Their Collaboration in General Relativity, 1913-1920," *Archive for History of Exact Sciences* 39 (1989): 281.

The persons who have contributed to the writing of this book by way of discussion, criticism, and suggestion are too numerous to recall, but I should especially like to acknowledge my indebtedness to the late Simon J. Prokhovnik and to Quentin Smith for their many comments on the work as it evolved. I am also deeply grateful to my wife Jan for her faithful labor in production of the typescript and to my research assistants Ryan Takenaga and Mike Austin. I should also like to thank Edward White and the Day Foundation for their generous grant which helped to fund the production of the camera-ready copy and to Mark Jensen and Jennifer Jensen for meticulously bringing this book into its final form.

Atlanta, Georgia   William Lane Craig

CHAPTER 1

THE HISTORICAL BACKGROUND OF SPECIAL RELATIVITY

INTRODUCTION

A noted philosopher of time recently confessed that "most of us philosophers don't understand" the Special Theory of Relativity (SR).[1] The same goes without saying for theologians, who often find themselves obliged to address issues related to time and eternity. I should also say that most physicists who do understand relativity theory scientifically nevertheless evince a remarkable naiveté concerning the epistemological and metaphysical foundations of Einstein's theory. There is need, therefore, of an exposition of the theory which is accessible to philosophers and theologians and which explores the presuppositions and implications of the theory. It is toward this end that this book has been written.

In undertaking such a task, we need to make a terminological clarification right from the start if later misunderstanding is to be averted. It is typically asserted that relativity theory abrogated the concepts of absolute time and space. But the word "absolute" in discussions of time and space has various, quite distinct meanings, which, if not kept clear, will be a source of serious confusion. Friedman has isolated three senses in which the term "absolute" is used in spacetime theories:[2]

> (i) Absolute-relational: this distinction arises in the dispute, made famous by the Clarke-Leibniz correspondence, concerning whether theories about spacetime structure are merely theories about the spatio-temporal relations between physical objects or whether they describe independently existing entities, such as space, time, or spacetime, in which physical objects are contained.

> (ii) Absolute-relative: This distinction concerns those elements of spatio-temporal structure which are either well-defined independently of any reference frame or well-defined only with regard to some specified reference frame. In Newtonian theory, for example, an event's temporal location is absolute, while in SR its temporal location is relative, though its location in spacetime is absolute.

> (iii) Absolute-dynamical: This distinction contrasts geometrical structure that is fixed independently of the processes and events occurring within spacetime and geometrical structure which is shaped by those processes and events. The General Theory of Relativity (GR) proposes a spacetime geometry which is not absolute in this sense, even if it is absolute in sense (i) above.

---

[1] Ned Markosian, "How Fast Does Time Pass?" *Philosophy and Phenomenological Research* 53 (1993): 829.
[2] Michael Friedman, *Foundations of Spacetime Theories* (Princeton: Princeton University Press, 1983), pp. 62-68.

2 CHAPTER 1

At least three more distinctions come to mind in which the term "absolute" is used:[3]

> (iv) Absolute-measured: This distinction was made by Newton with respect to the difference between the true or actual intervals in time and space and our empirical measures thereof, which record with varying degrees of accuracy the actual lapse of time or distance in space. Time and space are ontologically distinct from our physical measures thereof.
>
> (v) Absolute-local: This distinction comes into play in late nineteenth century aether theories, like those of Lorentz and Poincaré, who held that temporal location, while not absolute in sense (ii), is nonetheless absolute in that temporal location relative to the rest frame of the aether is privileged over all other local times reckoned according to frames in motion relative to the aether.
>
> (vi) Absolute-conventional: This distinction governs the dispute between those who hold that even within a single reference frame SR denies the existence of non-arbitrary, objective simultaneity relations and those who hold that such relations are not mere conventions. The same distinction contrasts the positions of those who, like Newton, hold the metrical properties of time and space to be intrinsic and unique versus thinkers who, like Poincaré, regard the metric of time and space as purely conventional.

In our ensuing discussion, I shall take pains to be as clear as possible in which sense I am using the term "absolute" when the occasion arises.

GALILEAN RELATIVITY

In order to understand the philosophical significance of Einstein's theory, it will be helpful if we first cast a glance backwards at classical physics.[4] Relativity already governed Newtonian mechanics (that branch of physics dealing with the motion of ponderable bodies and the action of forces upon them); that is to say, according to Newtonian theory it is impossible for a hypothetical observer associated with a frame of reference traveling at a uniform velocity (also called an inertial frame[5]) to perform mechanical experiments which, by yielding results different from those that would be obtained were he at rest, would disclose to him that he was, in fact, in motion. This principle of relativity was already realized by Galileo, who gave a delightful illustration of it:

---

[3] See discussions in John Earman, "Who's Afraid of Absolute Space?" *Australasian Journal of Philosophy* 48 (1970): 288-293; J. L. Mackie, "Three Steps toward Absolutism," in *Space, Time and Causality*, ed. Richard Swinburne, Synthèse Library 157 (Dordrecht: D. Reidel, 1983), pp. 3-5; Michel Ghins, *L'inertie et l'espace-temps absolu de Newton à Einstein*, Mémoires de la Classe des Lettres 69/2 (Brussels: Palais des Académies, 1990), pp. 16-20.

[4] For a thorough discussion, see the fine work of Arthur I. Miller, *Albert Einstein's Special Theory of Relativity* (Reading, Mass.: Addison-Wesley, 1981).

[5] A reference frame is a conventional standard of rest relative to which measurements can be made and experiments described. "A reference frame is said to be inertial in a certain region of space and time when, throughout that region of spacetime, and within some specified accuracy, every test particle that is initially at rest remains at rest and every test particle that is initially in motion continues that motion without change in speed or direction" (Edwin F. Taylor and John Archibald Wheeler, *Spacetime Physics* [San Francisco: W. H. Freeman, 1966], pp. 9-10). From this definition, it follows that inertial frames are local, not universal. But that does not mean that they are small. The test particles of an inertial frame moving in tandem could be scattered widely throughout space, so that this locally moving frame could be huge relative to a single observer (J. G. Taylor, *Special Relativity*, Oxford Physics Series [Oxford: Clarendon Press, 1975], p. 13).

> For a final indication of the nullity of the experiments brought forth, this seems to me the place to show you a way to test them all very easily. Shut yourself up with some friend in the main cabin below decks on some large ship, and have with you there some flies, butterflies, and other small flying animals. Have a large bowl of water with some fish in it; hang up a bottle that empties drop by drop into a wide vessel beneath it. With the ship standing still, observe carefully how the little animals fly with equal speed to all sides of the cabin. The fish swim indifferently in all directions; the drops fall into the vessel beneath; and, in throwing something to your friend, you need throw it no more strongly in one direction than another, the distances being equal; jumping with your feet together, you pass equal spaces in every direction. When you have observed all these things carefully (though there is no doubt that when the ship is standing still everything must happen in this way), have the ship proceed with any speed you like, so long as the motion is uniform and not fluctuating this way and that. You will discover not the least change in all the effects named, nor could you tell from any of them whether the ship was moving or standing still. In jumping, you will pass on the floor the same spaces as before, nor will you make larger jumps toward the stern than toward the prow, even though the ship is moving quite rapidly, despite the fact that during the time that you are in the air the floor under you will be going in a direction opposite to your jump. In throwing something to your companion, you will need no more force to get it to him whether he is in the direction of the bow or the stern, with yourself situated opposite. The droplets will fall as before into the vessel beneath without dropping toward the stern, although while the drops are in the air the ship runs many spans. The fish in their water will swim toward the front of their bowl with no more effort than toward the back, and will go with equal ease to bait placed anywhere around the edges of the bowl. Finally the butterflies and flies will continue their flights indifferently toward every side, nor will it ever happen that they are concentrated toward the stern, as if tired out from keeping up with the course of the ship, from which they will have been separated during long intervals by keeping themselves in the air. And if smoke is made by burning some incense, it will be seen going up in the form of a little cloud, remaining still and moving no more toward one side than the other. The cause of all these correspondences of effects is the fact that the ship's motion is common to all the things contained in it, and to the air also.[6]

Of course, if Galileo's vessel were suddenly to run aground, then the passenger would certainly know that he was or had been in motion, since he and everything else in his cabin which was not fastened down would suddenly lurch forward with appreciable force. But the relativity principle governing Newtonian mechanics and preserved in relativistic mechanics concerns uniform motion only (sometimes called translatory motion or translation); motions involving acceleration, deceleration, or rotation were not relativized (one thinks of Newton's famous *Gedankenexperiment* of the rotating bucket; the concave shape of the surface of the swirling water showed that it was rotating, not merely relatively with respect to the bucket, whose motion could be abruptly arrested, but absolutely in space itself independently of any reference frame[7]). It was not until Einstein proposed his General Theory of Relativity (GR) in 1915 that an attempt at the relativization of non-uniform motion was made. Galilean relativity leads to a set of equations called the Galilean transformation, which allows one to calculate on the basis of the coordinates which

---

[6] Galileo Galilei, *Dialogue Concerning the Two Chief World Systems—Ptolemaic and Copernican* [1632], trans. Stillman Drake (Berkeley: University of California Press, 1962), pp. 186-188. Galileo's example is not quite inertial because up and down, north and south are still distinguishable. His ship needed to be a spaceship.

[7] Isaac Newton, *The Principia*, trans. I. Bernard Cohen and Anne Whitman, with a Guide by I. Bernard Cohen (Berkeley: University of California Press, 1999), pp. 412-413.

an event has in one's own reference frame the coordinates assigned to that event relative to some other reference frame. Let us imagine two inertial frames S and S' which share a common x axis and that S' is in motion relative to S with velocity v (Figure 1.1).

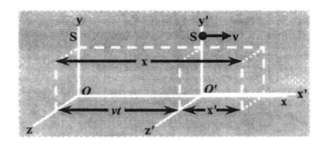

Figure 1.1. Two inertial frames with a common x-x' axis and with the y-y' and z-z' axes parallel. As seen from frame S, frame S' is moving in the positive x-direction at speed v. Similarly, as seen from frame S', frame S is moving in the negative x'-direction at this same speed. The origins O and O' coincide at time $t=0$, $t'=0$.

Since Galilean relativity did not concern time, both S and S' exist in the same absolute time, and we assign $t = 0$ when O and O' coincide. How are the coordinates of an event relative to S related to the coordinates of that same event relative to S'? The answer is

$$x' = x - vt \tag{1}$$

$$y' = y \tag{2}$$

$$z' = z \tag{3}$$

$$t' = t \tag{4}$$

These transformation equations governed mechanics up until nearly the end of the nineteenth century.

THE CHALLENGE OF ELECTRODYNAMICS

The principle of relativity governing Newtonian mechanics is usually stated in terms of the preservation of the validity of the laws of mechanics in any inertial frame. In other words, a hypothetical observer will employ the same mechanical laws of physics regardless of whether he is at rest or in uniform motion. There was a loophole, however, in Galilean relativity that became apparent during the nineteenth century. Electrodynamics, which deals with electric currents and electromagnetic radiation in general, had not been relativized, which in fact threatened to undo Galilean relativity.

Again, a bit of background: Newton's contemporaries vacillated between a corpuscular and a wave theory of light. According to the corpuscular theory, the propagation of light is like the motion of launched balls (a ballistic theory of light propagation). On the wave theory, light consists of vibrations in an invisible medium called the luminiferous aether (wave theory of light propagation). This aether is not a mobile gas, but came to be conceived as a rigid fluid like glass, though completely intangible, and thus at utter rest in space.[8] The discovery of the aberration of starlight in 1725 and its explanation by James Bradley in 1727 eventually tipped the scales in favor of the wave theory. In the phenomenon of stellar aberration, the attitude of a star varies with the position of the earth in its orbit around the sun, the greatest displacement of the star occurring when the earth is moving directly toward or away from it. The displacement is due to the fact that the earth is moving relative to the star: if the telescope were pointed at the actual position of the star in the sky, it would not be seen because in the time it takes for the starlight to pass from the lens of the telescope to the eyepiece, the earth has slightly moved. Therefore, the telescope must be slightly angled so that by the time the light strikes the eyepiece, the latter has moved forward to receive the light (see Figure 1.2).

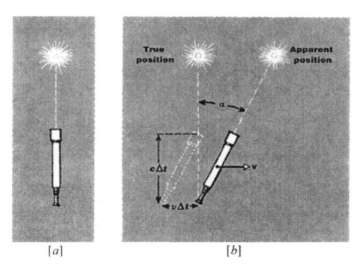

Figure 1.2. Stellar Aberration. [a] The star and telescope have no relative motion (*i.e.*, both are at rest in the aether); the star is directly overhead. [b]

---

[8] The wave theory of light held the oscillations to be *transverse*, not *longitudinal* motions in the aether, so that it could not be conceived as a gas, (which would support longitudinal waves), but rather as an elastic solid.

The telescope now moves to the right at speed *v* through the aether; it must be tilted at an angle α (greatly exaggerated in the drawing) from the vertical in order for one to see the star, whose apparent position now differs from its true position.

Bradley's discovery was eventually to prove problematic for the corpuscular theory because the velocity of light should depend, according to a ballistic model of propagation, on the motion of its emitting source and its measuring receiver, whereas calculations based on stellar aberration seemed indifferent to the star's motion. On a wave theory, stellar aberration could be accounted for if the aether remains virtually completely unaffected by the earth's motion through it. In 1804 Thomas Young proposed a wave theory according to which the luminiferous aether penetrates and pervades all material bodies with little or no resistance, "as freely perhaps as the wind passes through a grove of trees." By the 1830's the wave theory of light had become the predominant model.

James Clerk Maxwell, who wrote the mathematical equations governing electrodynamics (1864), introduced the concept of the electromagnetic field into physics.[9] According to Maxwell, waves in this field propagate through the aether with the constant velocity c, the speed of light. It became evident that light is itself a visible form of electromagnetic radiation (Figure 1.3) and is therefore described by

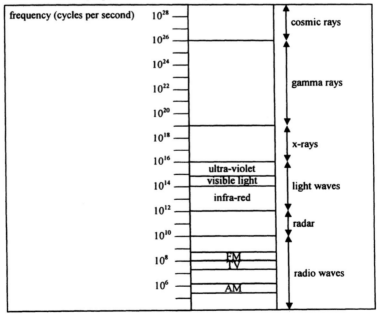

Figure 1.3. The electromagnetic spectrum

---

[9] For an account of Maxwell's equations, see Ghins, *L'inertie et l'espace-temps absolu*, pp. 106-109.

Maxwell's equations. Maxwell had succeeded in unifying electromagnetism and optics, and the aether came to be conceived as the medium of optical, electric, and magnetic phenomena.

But now it seemed possible to subvert Galilean relativity via Maxwell's electrodynamics. For since waves of any sort (sound waves, water waves, electromagnetic waves) propagate through their respective media at a velocity which is independent of any motion of their source, it followed that by measuring the speed at which light waves in the luminiferous aether pass by one's measuring apparatus, one could determine what mechanics alone could not discover: whether the inertial system in which the experiment was performed was at rest with respect to the aether (in which case one's apparatus would measure light as passing by at exactly c) or whether it was in motion (in which case, according to the addition of velocities law in classical physics, the light would pass at $c + v$, if one were moving against the light, that is, directly toward its source, or at $c - v$, if one were moving with the light, that is, directly away from its source, where v is the velocity of one's own inertial frame). Since the speed of light was a known quantity, any significant divergence from c registered by the apparatus would disclose one's velocity through the aether. Since the aether was conceived to be at rest with respect to all of space, any motion through it as disclosed by this divergent quantity (nicknamed the "aether wind") would be absolute, rather than merely relative, motion, thus overcoming Galilean relativity.

The difficulty in carrying out such an experiment was that other factors could be invoked to explain any discrepancy from precisely c in the measurement of the velocity of light relative to one's apparatus. For example, Fresnel (1818) had suggested that bodies in motion through the aether partially drag along the aether within them, and his "dragging coefficient" was invoked to account for why light did not strictly obey the Newtonian law for addition of velocities. To an accuracy of order v/c it was impossible to demonstrate any motion of the earth relative to the aether.

It was not until a young scientist named A. A. Michelson designed an apparatus called an interferometer which was accurate to second order $(v/c)^2$ that an experiment to measure the "aether drift" became feasible. Michelson's device, which was accurate to one part in ten billion, consisted of two arms of equal length at right angles to each other. At the junction of the two arms was a half-silvered mirror that split a beam from a light source fixed to the apparatus into two beams, each of which traveled along one arm to a mirror at its extremity, where the beam was reflected back to combine with the other beam, producing a pattern of interference fringes (see Figure 1.4).

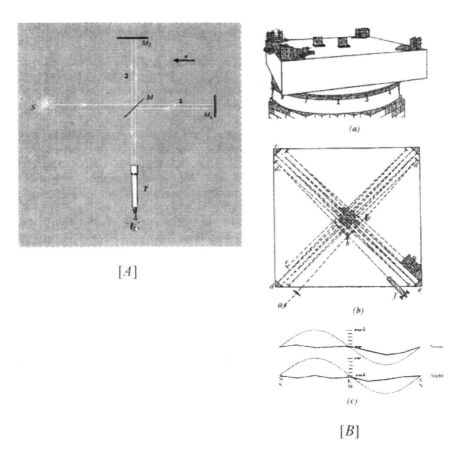

Figure 1.4. Michelson-Morley experiment. [*A*] A simplified version of the Michelson interferometer showing how the beam from the source *S* is split into two beams by the partial silvered mirror *M*. The beams are reflected by mirrors 1 and 2, returning to the partially silvered mirror. The beams are then transmitted to the telescope *T* where they interfere, giving rise to a fringe pattern. In this figure, *v* is the velocity of the aether with respect to the interferometer. [*B*] (*a*) Mounting of the Michelson-Morley apparatus. (*b*) Plan view. (*c*) Observed results. The solid lines show the observed fringe shift in the Michelson-Morley experiment as a function of the angle of rotation of the interferometer. The smooth dashed curves—which should be multiplied by a factor of 8 to bring it to the proper scale—show the fringe shift predicted by the aether hypothesis.

If the device were stationary in the aether, then no matter in which direction the arms were pointed, the interference fringes would be identical, since the light from the source would travel the two arms with equal velocity, namely, *c*. But if the earth

HISTORICAL BACKGROUND OF SPECIAL RELATIVITY

were moving through the aether, then the distances traversed by the light beams along each arm would be unequal. The light beam traveling along the arm parallel to the earth's motion through the aether would have to traverse more than twice the length of the arm because during the beam's out and return journey the arm is moving, so that either in its outward or return leg of the trip the beam will traverse the length of the arm plus the distance the apparatus had moved through the aether. (Of course, on one of the legs of the journey the distance traversed will be less than the length of the arm, but that does not fully compensate for the increased distance of the other leg.) The light beam traveling along the arm perpendicular to the direction of motion must also travel more than twice the length of its arm, too, since relative to the aether the beam on its return leg does not retrace its identical path, but travels at an angle to it. However, the distance traveled by the beam along the parallel arm is greater than the distance traveled by its counterpart along the perpendicular arm by a factor of

$$\sqrt{1 - v^2/c^2} \tag{5}$$

Because the parallel beam thus travels farther, it takes longer than the perpendicular beam to return to the point where they interfere. The time of the round trip journey of the parallel beam is

$$\frac{\ell}{c-v} + \frac{\ell}{c+v} \tag{6}$$

or

$$\frac{2\ell}{c}\left(\frac{1}{1-v^2/c^2}\right) \tag{7}$$

The time of the round trip journey for the perpendicular beam, on the other hand, is

$$\frac{2\ell'}{\sqrt{c^2 - v^2}} \tag{8}$$

or

$$\frac{2\ell'}{c}\left(\frac{1}{\sqrt{1-v^2/c^2}}\right) \tag{9}$$

The difference between the two times is

$$\frac{2}{c}\left[\frac{\ell'}{\sqrt{1-v^2/c^2}} - \frac{\ell}{1-v^2/c^2}\right] \tag{10}$$

What this implies is that if the machine were pointed in a different direction, say, turned at a 90° angle, the observed interference fringes should be different, for now the light source is no longer pointing in a direction parallel to the instrument's motion, but perpendicular to it. In this new position the difference in the beams' travel times should be

$$\frac{2}{c}\left[\frac{\ell' + \ell}{1 - v^2/c^2} - \frac{\ell' + \ell}{\sqrt{1 - v^2/c^2}}\right] \qquad (11)$$

By monitoring the interference patterns for any changes, one could use Michelson's ingenious device to detect the earth's motion through the aether to an accuracy inexplicable by mere first order phenomena.

Michelson carried out his experiment for the first time in 1881, but his results were vitiated by various errors and oversights. Then in 1887 with the cooperation of E. W. Morley, he conducted one of the most famous experiments in the history of science, the Michelson-Morley measurement of the earth's motion through the aether. They mounted the interferometer on a marble slab and floated the slab on mercury so that it could be smoothly rotated around a central pin. Mirrors were arranged to reflect the light beams back and forth through eight round trips, thus making the total distance traveled 11 meters and thereby increasing the accuracy of their measurements. To Michelson's chagrin, the experiment, contrary to expectation, yielded no significant change in the interference fringes whatsoever. Taken literally, what the results seemed to indicate was that the earth was perfectly at rest in the aether! But that seemed clearly impossible: for even if, by some extravagant coincidence, the motion of the earth in its orbit in one direction were exactly counterbalanced by, say, the motion of the solar system in the opposite direction, so that the earth's net motion through the aether was precisely zero, still the earth was itself rotating and yet the readings of the interferometer were constant whether recorded day or night. Moreover, readings taken at different times of the year, for example, six months apart, when the earth is moving in its orbit in exactly opposite directions, remained invariant. Earth could not be at rest in the aether. So why was its motion undetectable?

THE LORENTZIAN *VERSUCH*

Michelson and Morley's results bewildered the giant of nineteenth century physics, the Dutch physicist H. A. Lorentz. Other physicists had speculated that the Michelson-Morley results could be explained by the hypothesis of "aether drag": that the earth carried along with it a sort of aether-atmosphere in which it is at rest. But Lorentz saw that such a suggestion was untenable, since the phenomenon of stellar aberration required that the earth not drag the external aether along with it. Confronted with what Friedman has characterized the "highly counter-intuitive, even paradoxical result that something having velocity c in one reference frame has

c with respect to a second inertial frame in motion relative to the first,"[10] Lorentz hit upon an audacious and ingenious solution: that measuring rods in motion *shrink* in the direction of their motion due to electromagnetic effects at the molecular level caused by the translatory motion.[11] Thus, the reason the Michelson-Morley experiment failed to detect any change in the interference fringes was because whichever arm of the interferometer was parallel to the direction of motion was contracted by a factor of

$$\sqrt{1 - v^2/c^2} \tag{12}$$

This hypothesis had been independently suggested by the Irish physicist George FitzGerald and so came to be known as the Lorentz-FitzGerald Contraction.[12]

Although many would agree with the remark of one commentator that such a move on Lorentz's part was "a physics of desperation,"[13] others have suggested that the hypothesis was not implausible. Eddington, for example, asserted, "There is really nothing mysterious about the FitzGerald contraction...it is an entirely natural

---

[10] Friedman, *Foundations*, p. 15. Tests have continued to uphold the invariance of light's velocity; particles moving at 99.97% the speed of light have been measured as emitting light at the same value for stationary sources to one part in 10^4 (see Taylor, *Special Relativity*, pp. 2-5).

[11] In a classic text, he wrote,

"Surprising as this hypothesis may appear at first sight, yet we shall have to admit that it is by no means far-fetched, as soon as we assume that molecular forces are also transmitted through the ether, like the electric and magnetic forces of which we are able at the present time to make this assertion definitely. If they are so transmitted, the translation will very probably affect the action between two molecules or atoms in a manner resembling the attraction or repulsion between charged particles. Now since the form and dimensions of a solid body are ultimately conditioned by the intensity of molecular actions, there cannot fail to be a change of dimensions as well. From the theoretical side, therefore, there would be no objection to the hypothesis" (H. A. Lorentz, *Versuch einer Theorie der elektrischen und optischen Erscheinungen in bewegten Körpern* [Leiden: E. J. Brill, 1895]; reprinted in *The Principle of Relativity*, with Notes by A. Sommerfeld, trans. W. Perrett & G. B. Jeffery [1923; rep. ed.: New York: Dover Publications, 1952], pp. 5-6). For a good account of Lorentz's electron theory, see Stanley Goldberg "The Lorentz Theory of Electrons and Einstein's Theory of Relativity," *American Journal of Physics* 37 (1969): 982-994.

[12] An interesting story, which has only recently come to light, lies behind FitzGerald's priority for this hypothesis. (See G. Bush, "Note on the History of the FitzGerald-Lorentz Contraction," *Isis* 58 [1967]: 230-232.) FitzGerald had proposed the contraction in a communication to the journal *Science*. But the journal informed FitzGerald that it was being discontinued and so could not publish his paper. Since FitzGerald did not resubmit it elsewhere, Lorentz has been credited with the first published proposal of the contraction hypothesis, though he graciously deferred to FitzGerald's being the first to suggest it, once he learned of the situation. Ironically, however, it turns out that unbeknownst to FitzGerald, *Science* did not close its doors and FitzGerald's paper was published as "The Ether and the Earth's Atmosphere," *Science* 13 (1889): 390, three years before Lorentz's "The Relative Motion of the Earth and the Ether," *Verslagen Koninklijke Akademie van Wetenschappen Amsterdam* 1 (1892): 74!

[13] Miller, *Einstein's Special Theory of Relativity*, p. 30; cf. idem, *Imagery in Scientific Thought: Creating 20th Century Physics* (Cambridge, Mass.: MIT Press, 1986), p. 64. Cf. Minkowski's remark: "This hypothesis sounds extremely fantastical, for the contraction is not to be looked upon as a consequence of resistances in the ether, or anything of that kind, but simply as a gift from above,—as an accompanying circumstance of the circumstance of motion (H. Minkowski, "Space and Time," in *The Principle of Relativity*, by A. Einstein, *et al.*, trans. W. Perrett and G. B. Jeffery (rep. ed.: New York: Dover Publications, 1952), p. 81.

property of a swarm of particles held in delicate balance by electromagnetic forces...a rod in motion is subjected to a new magnetic stress, arising not from unfair outside tampering but as a necessary consequence of its own electrical constitution; and under this stress the contraction occurs."[14] In 1896 Morton and Searle showed that a configuration of electrically charged particles will change shape when placed in motion, and Lorentz showed that the new configuration corresponds exactly to the Lorentz-FitzGerald contraction.[15] The difficulty with Lorentz's hypothesis was that as his theory developed it became necessary to posit a contraction in the elementary particles themselves, called "electrons" by Lorentz, and this flattening of individual particles could not be explained via molecular forces among ensembles of particles.[16] Since atomic theory was in its nascence, Lorentz was swimming in uncharted waters, which could not be navigated until the development of quantum theory. Thus, he wrote, "Our assumption about the contraction of the electrons cannot in itself be pronounced to be either plausible or inadmissible. What we know about the nature of electrons is very little and the only means of pushing our way farther will be to test such hypotheses as I have made here."[17] In his analysis of why Einstein's research programme superseded that of Lorentz, Elie Zahar muses that Lorentz was just unlucky in his choice of problems, for he was immediately involved in problems which were to eventually defeat Einstein at a later date, namely, elementary particle physics.[18]

---

[14] Sir Arthur Eddington, *The Nature of the Physical World*, with an Introductory Note by Sir Edmund Whittaker, Everyman's Library (1928; rep. ed.: London: I. M. Dent & Sons, 1964), p. 18. Elsewhere he wrote that while FitzGerald's hypothesis seems "strange and arbitrary," nevertheless it was rendered "very plausible" by Lorentz and Larmor. For the molecular binding forces are electrical in nature, and so it cannot be a matter of indifference to those forces how the electromagnetic field is flowing with respect to those molecules (Idem, *Space, Time and Gravitation*, Cambridge Science Classics [1920: rep. ed.: Cambridge: Cambridge University Press, 1987], p. 19). Cf. Herbert Dingle, *The Special Theory of Relativity* (London: Methuen, 1940), p. 18: "It is quite consistent with the electrical theory of matter, electrical and optical phenomena being correlated through the appearance of c as the velocity of light and also as the ratio of electrical units." Similarly, C. Møller, *The Theory of Relativity*, 2d ed. (Oxford: Clarendon Press, 1972), p. 28 thinks that Lorentz gave "a plausible explanation" for the contraction.

[15] See Paul S. Epstein, "The Time Concept in Restricted Relativity," *American Journal of Physics* 10 (1942): 4, who comments, "the physicist possesses a complete dynamical explanation of the mechanism of the Lorentz contraction, which is thus *real* in the ordinary sense of the word" (Ibid., p. 5).

[16] See comments of Max Born, *Einstein's Theory of Relativity*, trans. Henry L. Brose (New York: E. P. Dutton, n. d.), p. 187; Møller, *Theory of Relativity*, p. 28.

[17] H. A. Lorentz, "Electromagnetic Phenomena in a System moving with any Velocity Smaller than that of Light," *Proceedings of the Royal Academy of Amsterdam* 6 (1904): 809ff, rep. in H. A. Lorentz, *Collected Papers*, 9 vols., ed. P. Zeeman and A. D. Fokker (The Hague: Martinus Nijhoff, 1934-1939), 5: 190.

[18] Elie Zahar, "Why Did Einstein's Programme Supersede Lorentz's? (I)," *British Journal for the Philosophy of Science* 24 [1973]: 238. E. Cunningham, an early British expositor of SR, complained,
"If we accept the suggestion [of the deformable electron], it will not be long before we go on to ask the further question, 'Why is this so?' This can only be answered by an analysis of the construction of the electron, in terms of its parts, so that we should at once be carried beyond the stage at which the electron is an ultimate element in our thought" (E. Cunningham, *Relativity, the Electron Theory, and Gravitation*, 2d ed., Monographs on Physics [London: Longmans, Green, & Co., 1921], pp. 25-26).
Lorentz's theory had outstripped the particle physics of his day.

Despite its eventual demise, Lorentz's electron theory, as Russell McCormmach has emphasized, revolutionized classical physics.[19] Lorentz had formulated a new, non-Newtonian mechanics and had discovered the set of equations known as the Lorentz transformations, which allow us to convert spatio-temporal coordinates of events from one inertial frame to another and which constitute the mathematical core of Einstein's Special Theory of Relativity. His electron theory laid the foundation for the investigation of the atomic structure by explaining the Zeeman effect, helped to develop the theory of elementary particles, helped to originate modern solid state physics, and completed the pre-relativistic theory of the electromagnetic field, thus paving the way for relativity theory.[20]

Still it has been a matter of some debate the extent to which Lorentz's theory was *ad hoc*. The electron theory continued to evolve right up to Einstein's publication of his epochal paper in 1905 (and beyond), and Holton lists eleven *ad hoc* assumptions which underlay Lorentz's theory in its 1904 version.[21] On the other hand, the great French mathematician Henri Poincaré, who had earlier complained that "of hypotheses there is never lack,"[22] deemed the assumptions of the 1904 paper to be mutually complementary and productive of a theory with great explanatory power.[23] Zahar has argued that Lorentz's continual modifications of his aether theory are a good example of a progressive research programme.[24] He

---

[19] Russell McCormmach, "H. A. Lorentz and the Electromagnetic View of Nature," *Isis* 61 (1970): 484-485.
[20] See Tetu Hirosige, "Origins of Lorentz' Theory of Electrons and the Concept of the Electromagnetic Field," *Historical Studies in the Physical Sciences* 1 (1969): 151.
[21] Gerald Holton, "On the Origin of the Special Theory of Relativity," in *Thematic Origins of Scientific Thought: Kepler to Einstein* (Cambridge, Mass.: Harvard University Press, 1973), p. 170. Holton lists: restriction to small ratios of velocities v to light velocity c, postulation *a priori* of the transformation equations (rather than their derivation from other postulates), assumption of a stationary ether, assumption that the stationary electron is round, that its charge is uniformly distributed, that all mass is electromagnetic, that the moving electron changes one of its dimensions precisely in the ratio of $(1-v^2/c^2)^{1/2}$ to one, that forces between uncharged particles and between a charged an uncharged particle have the same transformation properties as electrostatic forces in the electrostatic system, that all charges in atoms are in a certain number of separate "electrons," that each of these is acted on only by others in the same atom, and that atoms in motion as a whole deform as electrons themselves do.
[22] Henri Poincaré, *Science and Hypothesis* [1927], in H. Poincaré, *The Foundations of Science*, trans. G. B. Halstead (Science Press: 1913; rep. ed.: Washington, D. C.: University Press of America, 1982), p. 147. What is needed, says Poincaré, is a single explanation that will cover both first and second order effects and that leads us to think that it will suffice for effects of any order. This is precisely what Lorentz claimed to have provided by 1904 (Lorentz, "Electromagnetic Phenomena," p. 174). He later recalled that it was Poincaré's complaint about special hypotheses that prompted him to find a more general theory that would cover all cases less than the speed of light (H. A. Lorentz, "Deux mémoires de Henri Poincaré sur la physique mathématique," *Acta mathematica* 38 [1914]: 293ff, rep. in *Collected Papers*, 7: 259).
[23] Henri Poincaré, "L'état actuel et l'avenir de la physique mathématique," [1904], reprinted as "The Present Crisis of Mathematical Physics," in H. Poincaré, *The Value of Science* [1905], in *Foundations of Science*, chap. 8. Here Poincaré refers to "the principle of relativity," which concerns how we are to interpret the velocity of light. He still complains that Lorentz got the job done "only by accumulating hypotheses" (p. 306). Cf. idem, "The Principles of Mathematical Physics," *Monist* 15 (1905): 1-24, from an address to the International Congress of Arts and Science in St. Louis in September of 1904.
[24] Elie Zahar, "Einstein's Programme (I)," pp. 95-123; idem, "Why Did Einstein's Programme Supersede Lorentz's? (II)," *British Journal for the Philosophy of Science* 24 (1973): 223-262. Zahar's analysis employs the framework of Imre Lakatos's notion of a research programme as an ever-improving sequence of related theories.

delineates three senses in which a change in a theory can be said to be *ad hoc*: (i) the new theory has no novel consequences as compared with its predecessor, (ii) none of the novel consequences predicted by the new theory have been verified, and (iii) the new theory is obtained from its predecessor through a modification of the auxiliary hypotheses which do not accord with the spirit of the heuristic which guides the programme. Noting that Lorentz's theory underwent a three stage development generated by the application of its heuristic (the explanation of phenomena using an aether-based metaphysic) to specific problems, namely,

> T1 = hard core (Maxwell's equations, Newton's laws of motion, etc.) without any assumptions concerning rod contraction or clock retardation
>
> T2 = hard core + Lorentz-FitzGerald contraction
>
> T3 = hard core + Lorentz-FitzGerald contraction + clock retardation,

Zahar argues that none of these shifts was *ad hoc*. Zahar contends that the Lorentz-FitzGerald contraction was deduced from the Molecular Forces Hypothesis, which was itself a plausible auxiliary hypothesis that introduced no alien elements into Lorentz's programme. Moreover, the Molecular Forces Hypothesis "had nothing to do with Michelson's experiment" but arose out of mathematical considerations pertaining to the transformation properties of Maxwell's equations.[25] As a result of the pursuit of his programme, Lorentz was able to make both theoretical and empirical progress, often in advance of Einstein.[26]

Zahar's analysis has drawn sharp criticism from Miller and Schaffner.[27] Miller charges that Zahar's three senses of *ad hoc* are contrived merely to support his contention. The point—which Miller demonstrates pretty convincingly—is that Lorentz's methodology in his works of 1892 and 1895 was to posit the Lorentz contraction specifically to account for the Michelson-Morley experiment and then to posit the Molecular Forces Hypothesis as a plausibility argument in support of the contraction hypothesis. "Thus, the hypothesis of contraction was proposed to explain a single experiment; it was obtained from reasoning unconnected with the electromagnetic theory; its connection with the electro-magnetic theory was at best tenuous—in short the hypothesis of contraction was *ad hoc*."[28] Schaffner agrees that the Molecular Forces Hypothesis was proposed in response to the Michelson-Morley result and charges that Lorentz's extension of the transformation properties to non-electromagnetic forces in 1904 was *ad hoc* either in sense (iii) because it

---

[25] Zahar, "Einstein's Programme (I)," p. 107.
[26] Ibid., p. 122.
[27] Arthur I. Miller, "On Lorentz's Methodology," *British Journal for the Philosophy of Science* 25 (1974): 29-45; Kenneth F. Schaffner, "Einstein versus Lorentz: Research Programmes and the Logic of Comparative Theory Evaluation," *British Journal for the Philosophy of Science* 25 (1974): 45-78; cf. the more sympathetic reaction of P. K. Feyerabend, "Zahar on Einstein," *British Journal for the Philosophy of Science* 25 (1974): 25-28.
[28] Miller, *Einstein's Special Theory*, p. 31.

violated the spirit of the research programme or in sense (ii) because it implied no additional novel facts.

This *Auseinandersetzung* raises the difficult question of what it means to say that a hypothesis is *ad hoc* and, if it is, how serious a detriment that is. I think it could be plausibly maintained that Lorentz's hypothesis was not *ad hoc* in any of Zahar's three senses, since such modifications as Lorentz made seemed certainly in accord with what Zahar characterizes as his heuristic and aimed at making the theory consistent and comprehensive.[29] Granted, in the literal sense of the term his hypotheses were certainly *ad hoc*: directed specifically to handle this or that problem. But nothing seems wrong with that. The Michelson-Morley result could not simply be ignored. Schaffner contends that a hypothesis used to modify a theory is *ad hoc* if it lacks independent support or fails to provide additional interesting or testable results but that if it meets these criteria, then it is accepted as an advance.[30] But, we may ask, interesting to whom? And how long may we wait to see if its results are testable? How strong must the independent support be? It could be argued that the hypothesis of the Lorentz-FitzGerald contraction, though specifically formulated to explain the null result of the Michelson-Morley experiment, was extremely interesting in its results, was testable (the Kennedy-Thorndike experiment of 1932 showed the contraction alone was insufficient to account for the invariance of the velocity of light[31]), and enjoyed the support of the Molecular Forces Hypothesis (which was a plausible conjecture worthy of further investigation). The addition of the clock retardation hypothesis to yield the full electron theory of 1904 cannot be called *ad hoc* because it is a consequence of length contraction, which was already part of the 1895 version.

By way of analogy, it seems that Lorentz's hypothesis was no more *ad hoc* than current adjustments to the standard Big Bang model which posit an early epoch of inflationary expansion. This modification of the standard theory is literally *ad hoc* with regard to the appearance of fine-tuning in the initial conditions of the universe, such as come to expression in the flatness problem, the homogeneity problem, the horizon problem, and so forth.[32] Given the assumption of an era of inflationary expansion, the appearance of fine-tuning can be accounted for. Yet the hypothesis has no observable consequences different from the standard model nor any clearly testable results, since any evidence of that early phase has been pushed out beyond the horizon of the observable universe. Moreover, though the hypothesis can be made somewhat plausible by postulating certain quantum effects on spacetime in the very early history of the universe, the hypothesis has itself been forced to go through a number of revisions, the inflationary model being replaced by the new inflationary

---

[29] Capek remarks that within the context of an absolutist theory of space, the contraction hypothesis makes "perfect sense" (Milic Capek, *The Philosophical Impact of Contemporary Physics* [Princeton: D. Van Nostrand, 1961], p. 145).
[30] Schaffner, "Einstein vs. Lorentz," p. 67.
[31] This point has been emphasized by Adolf Grünbaum, *Philosophical Problems of Space and Time*, 2d ed., Boston Studies in the Philosophy of Science 12 (Dordrecht: D. Reidel, 1973), pp. 720-725; cf. Melbourne G. Evans, "On the Falsity of the Fitzgerald-Lorentz Contraction Hypothesis," *Philosophy of Science* 36 (1969): 254-262.
[32] See Alan Guth, "Inflationary Universe: A Possible Solution to the Horizon and Flatness Problems," *Physical Review* D 23 (1981): 347-356.

model,[33] and that being replaced in turn by a chaotic inflationary model,[34] with no satisfactory hypothesis in sight.[35] Needless to say, all these scenarios are extremely conjectural and uncertain. In short, the hypothesis of an inflationary period of exponential expansion seems just as *ad hoc* as the Lorentz contraction. Yet, despite protests from some quarters that the inflationary hypothesis is metaphysics, not physics, this model has quickly become, in John Leslie's words, "nowadays very much the Standard Cosmos."[36]

It is hard to avoid the impression that a great deal of personal taste comes into play in the judgement that a theory is *ad hoc*. Indeed, Schaffner comes to the conclusion that a hypothesis's being *ad hoc* is really an intersubjective judgement.[37] It is instructive to note Holton's comments in this regard: a hypothesis's being *ad hoc* has to do with a sense of artificiality expressed by all concerned.[38] To make this judgement one must have a certain feel of what he calls the "aesthetics of science." A scientist's sense of what is *ad hoc* can be acceptable or not: if it is, then his hypothesis is deemed "elegant," "natural," "reasonable"; if not, then it is termed "bothersome," "artificial," or "ugly." Thus, what is *ad hoc* is relative to a context and to persons. In all cases, Holton concludes, it is a quite separate question whether the *ad hoc* hypothesis turns out to be right or wrong.

---

[33] A. Albrecht and P. I. Steinhardt, "Cosmology for Grand Unified Theories with Radioactively Induced Symmetry Breaking," *Physical Review Letters* 48 (1982): 1220-1223; A. D. Linde, "A New Inflationary Universe Scenario: A Possible Solution of the Horizon, Flatness, Homogeneity, Isotropy, and Primordial Monopole Problems," *Physics Letters* B 108 (1982): 389-393; idem, "Coleman-Weinberg Theory and the New Inflationary Universe Scenario," *Physics Letters* B 114 (1982): 431-435; S. W. Hawking and I. G. Moss, "Supercooled Phase Transitions in the Very Early Universe," *Physics Letters* B 110 (1982): 35-38.

[34] A. D. Linde, "Chaotic Inflation," *Physics Letters* 8 129 (1983): 177-181; idem, "The Present Status of the Inflationary Universe Scenario," *Comments on Astrophysics* 10/6 (1985): 229-237; see also D. V. Nanopoulos, "The Inflationary Universe," *Comments on Astrophysics* 10/6 (1985): 219-227.

[35] John Earman and Jesus Mosterin have shown that even the newest inflationary scenarios do not "overcome the glaring deficiencies of the original versions of inflationary cosmology" (John Earman and Jesus Mosterin, "A Critical Look at Inflationary Cosmology," *Philosophy of Science* 66 [1999]: 36). Nonetheless, they observe, "Despite the lack of empirical successes, the unkept promises, and the increasingly contrived and speculative character of the models, the inflationary juggernaut has not lost steam" (Ibid., p. 45).

[36] John Leslie, "Risking the World's End," *Canadian Nuclear Society Bulletin* 10 (May/June 1989): 14. Consider the comments of George Gale:
> "At the present moment, it appears that a majority of cosmologists—and their particle physicist allies—prefer one or another of the various versions of the inflationary universe scenario....
> Not all cosmologists accept the inflationary picture. Some of them prefer to stand by the contemporary version of the original hot Big Bang model. These cosmologists accuse the inflationists of proposing non-verifiable, that is to say, metaphysical theories, since existence of the inflationary period could produce no evidence, and, moreover, the hypothesis of the period is motivated solely by desire to avoid violation of the principle of causality, a metaphysical principle if ever there was one" (George Gale, "Cosmos and Conflict: Interactions between Observation, Theory, and Metaphysics in Modern Cosmology," paper presented at the conference "The Origin of the Universe," Colorado State University, 22-25 September, 1988).

[37] Schaffner, "Einstein vs. Lorentz," p. 67.

[38] Gerald Holton, "Einstein, Michelson and the 'Crucial' Experiment," in *Thematic Origins of Scientific Thought*, p. 309.

If Holton is correct, then Lorentz need not have been too disturbed by allegations that his contraction hypothesis was *ad hoc*, for he could simply have replied that his aesthetic intuitions differed from the majority's and that, in any case, his theory, even if *ad hoc*, was right, that it gave a realistic account of what happens to an object in translatory motion.

As noted by Zahar, Lorentz's hypothesis of length contraction needed to be supplemented by the hypothesis of clock retardation as well. In his theory of 1895 Lorentz substituted for the Galilean transformation equations two different sets of equations for dealing with different problems. In reducing electrodynamical to electrostatic problems, he employed a set of transformation equations involving the contraction hypothesis. In reducing problems concerning the optics of moving bodies to optics of bodies at rest, he introduced a new variable t' which he termed the "local time" (*Ortszeit*) in order to differentiate it from the true time t relative to the aether rest frame.

In his 1904 paper Lorentz substituted in the place of the Galilean transformation a set of equations which Poincaré then simplified in 1905 and dubbed the "Lorentz transformations."[39] This transformation provides the means of calculating the space and time coordinates of an event relative to different inertial frames in modern relativistic physics:

Lorentz Transformation Equations

$$x' = \frac{x - vt}{\sqrt{1 - v^2/c^2}} \qquad x = \frac{x' + vt'}{\sqrt{1 - v^2/c^2}}$$

$$y' = y \qquad y = y'$$

$$z' = z \qquad z = z'$$

$$t' = \frac{t - (v/c^2)x}{\sqrt{1 - v^2/c^2}} \qquad t = \frac{t' + (v/c^2)x'}{\sqrt{1 - v^2/c^2}}$$

Although Lorentz thus achieved the mathematical relativization of physical time, apparently he attributed no physical significance to the variable t'. Rindler has

---

[39] Henri Poincaré, "Sur la dynamique de l'électron," *Rendiconti del Circolo matematico di Palermo* 21 (1906): 129-176, in H. Poincaré, *Oeuvres de Henri Poincaré*, 11 vols. (Paris: Gauthiers-Villars, 1934-1953), 9:494-550. Here Poincaré declares,

"It seems that this impossibility of making evident experimentally the absolute motion of the earth is a general law of Nature; we are naturally led to admit this law, which we shall call the *Postulate of Relativity* and to admit it without restriction. Whether this postulate, which until now is in agreement with the evidence, should be later confirmed or disconfirmed by more precise experiments, it is in any case interesting to see what its consequences might be" (Ibid., p. 495).

Poincaré modified Lorentz's hypothesis, which the latter had in 1904 "succeeded in bringing...into accord" with the Relativity Postulate, in "only some detailed points" (Ibid.). "Thus, Lorentz's hypothesis is the only one which is compatible with the impossibility of making absolute motion evident; if one admits this impossibility, one must admit electrons in motion contract in such a way as to become ellipsoids of revolution, two of whose axes remain constant...." (Ibid., p. 535).

insisted that no one prior to Einstein in 1905 had voiced "the slightest suspicion" that moving clocks would run slow.[40] Joseph Larmor, who in 1898 had discovered the transformation independently,[41] made a mathematical discovery of a change in the time scale, but had no conscious idea of clocks' slowing down.[42] As far as Lorentz was concerned, local time was a mathematical artifice; it was not until later that he interpreted it realistically.[43] It must be said, however, that Poincaré, France's greatest living mathematician at that time and collaborator of Lorentz, voiced more than a "slight suspicion" of uniformly moving clocks' slowing down. He seemed to have genuinely anticipated Einstein's relativistic time dilation effect.[44]

---

[40] W. Rindler, "Einstein's Priority in Recognizing Time Dilation Physically," *American Journal of Physics* 38 (1970): 1111-1115; cf. Kenneth F. Schaffner, "Outlines of a Logic of Comparative Theory Evaluation with Special Attention to Pre- and Post-Relativistic Electrodynamics," in *Historical and Philosophical Perspectives of Science*, ed. Roger H. Stuewer, Minnesota Studies in the Philosophy of Science 5 (Minneapolis: University of Minnesota Press, 1970), p. 335.

[41] Joseph Larmor, *Aether and Matter* (Cambridge: Cambridge University Press, 1900).

[42] Eddington pinpoints the difference between Einstein's attitude and Lorentz and Larmor's attitude toward the "local time":
> "Both Larmor and Lorentz had introduced a 'local time' for the moving system. It was clear that for many phenomena this local time would replace the 'real' time; but it was not suggested that the observer in the moving system would be deceived into thinking that it was the real time. Einstein in 1905 founded the modern principle of relativity by postulating that this local time was *the time* for the moving observer, no real or absolute time existed, but only the local times, different for different observers" (Eddington, *Space, Time and Gravitation*, pp. 211-212).

[43] Lorentz later commented on his use of the auxiliary device of local time:
> "...I had the idea that there was an essential difference between the systems x, y, z, t and x', y', z', t'. In one one uses—so I thought—axes of co-ordinates which have a fixed position in the aether and what one could call the 'true' time; in the other system, by contrast, one would have to do with mere auxiliary quantities whose introduction is just a mathematical artifice. In particular, the variable t' could not be called 'time' in the same sense as the variable t" (H. A. Lorentz, "Deux Mémoires ...", 293ff, rep. in *Collected Papers*, 7: 262).

At a Mt. Wilson Observatory conference from February 4-5, 1927, Lorentz reflected:
> "The experimental results could be accounted for by transforming the co-ordinates in a certain manner from one system of co-ordinates to another. A transformation of the time was also necessary. So I introduced the conception of a local time which is different for different systems of reference which are in motion relative to each other. But I never thought that this had anything to do with the real time. The real time for me was still represented by the old classical notion of an absolute time, which is independent of any reference to special frames of co-ordinates. There existed for me only this one true time. I considered my time transformation only as a heuristic working hypothesis" (H. A. Lorentz, "Conference on the Michelson-Morley Experiment," *Astrophysical Journal* 68 [1928]: 380).

Subsequent to Einstein's development of SR, Lorentz came to insist on the reality of time dilation for a clock moving relative to the aether (H. A. Lorentz, "Considérations élémentaires sur le principe de relativité," *Revue générale des Sciences* 25 [1914]: 179ff, rep. in *Collected Papers*, 7: 165).

[44] Consider carefully Poincaré's statement cited on p. 39, for example. It might be rejoined that Poincaré here envisages two clocks at relative rest, whereas Einstein was concerned to show clock retardation for two relatively moving clocks. But Poincaré's logic is easily extended to show that if the prescribed synchronization procedure fails for two clocks moving in tandem, then it certainly will not work for two clocks in relative motion, since then the distances traversed by the light signals are obviously different. Hence, he says that neither clock registers the true time, but merely the local time. Any clock moving relative to the aether frame must run slow. It might be further objected that Poincaré did not conceive of clocks' slowing down as due to intrinsic, dynamical causes resulting from the clocks'

EINSTEIN'S SPECIAL THEORY

Lorentz's research programme was interrupted in 1905 by the appearance of a radically different approach to the problem of reconciling mechanics and electrodynamics, which was laid out in a paper in *Annalen der Physik* entitled "On the Electrodynamics of Moving Bodies," by Albert Einstein, an unknown clerk working in the Patent Office in Berne, Switzerland. One recent commentator has called Einstein's article "the most profoundly revolutionary single paper in the history of physics."[45] The key in Einstein's approach to the complete relativization of classical physics was a daring metaphysical revision of the Newtonian worldview: *he presupposed that absolute space*, both in the sense of a structure which is well-defined independently of any reference frame and in the sense of a universal, privileged reference frame, *does not exist*. This bold, philosophical move allowed him to sweep away both the aether and the rest frame of the aether as "superfluous."[46] This is the fundamental *Voraussetzung* of the Special Theory, for without it, as Grünbaum has emphasized, the two fundamental postulates of theory become "self-contradictory" and "an evident absurdity."[47]

Imagine an inertial system K which is at rest in absolute space and a second inertial system K' which is sliding along the x-axis common to both (Figure 1.5).

---

simply being in motion. That is no doubt correct; but then neither did Einstein. In Einstein's presentation, clock retardation is a deduction based on the postulates of the theory, without any reference to intrinsic causes. The chief difference between Poincaré and Einstein on this score was that Poincaré held to an aether rest frame while Einstein did not. As a result, for Poincaré the clocks moving in tandem do not register the true time, whereas for Einstein there is no true time and, hence, by definition the clocks run normally relative to their common inertial frame. But both would agree, so far as I can see, that a clock moving relatively to another clock will run slow with respect to that clock. Later, of course, the Lorentz-Poincaré interpretation was seen to imply clock retardation as an intrinsic effect due to motion. In this sense Poincaré's shortcoming lay not in his failure to anticipate Einstein but to have clear insight into the implications of Lorentzian theory! Notice that Poincaré does seem to suppose that the moving observer will be deceived by the local time. Miller, *Einstein's Special Theory*, p. 189, appears to misinterpret this passage. Abraham Pais agrees that Poincaré "goes beyond Lorentz in treating local time as a physical concept" (Abraham Pais, *'Subtle is the Lord...': The Science and Life of Albert Einstein* [Oxford: Oxford University Press, 1982], p. 128).

[45] James T. Cushing, *Philosophical Concepts in Physics* (Cambridge: Cambridge University Press, 1998), p. 232.
[46] A. Einstein, "On the Electrodynamics of Moving Bodies," trans. Arthur I. Miller, Appendix to Miller, *Einstein's Special Theory*, p. 392, whose book contains a detailed commentary on the whole paper. This translation is superior to the version published in the Dover edition of *The Principle of Relativity*.
[47] Grünbaum, *Philosophical Problems of Space and Time*, p. 712; cf. P. C. W. Davies, *Space and Time in the Modern Universe* (Cambridge: Cambridge University Press, 1977), p. 33, who calls the invariance of light's velocity "nonsensical" apart from the elimination of absolute space.

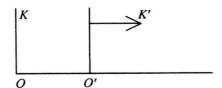

Figure 1.5. An inertial system K' in translatory motion relative to the privileged rest frame K.

When O and O' coincide, let a light ray in the direction of K''s motion be released. If relative to K' the light ray does not travel at $c - v$, then it will be in two places at the same time: it will be at the distance ct from O and also at the distance ct from O'. "This," as Nordenson puts it, "is a clear contradiction which must be eliminated."[48] Einstein will eliminate it by denying that there is any state of absolute rest, K and K' being in relative motion only, thus permitting a radical redefinition of time which will eliminate the frame-independent t. Since an absolute space or privileged reference frame is a sufficient condition of absolute simultaneity, which Einstein means to reject, as we shall see, Capek is thus far justified in viewing the Special Theory as representing at a fundamental level "the dynamization of space." For in Einstein's conception, as he says, "Classical space conceived as a simultaneous juxtaposition of points simply *does not exist*; to admit the contrary would be to admit the absolute simultaneity of distant events, which is precisely what the special theory of relativity denies."[49] By rejecting absolute space, Einstein will now feel free to redefine the notion of simultaneity in such a way as to provide mechanics and electrodynamics the common basis of inertial frame invariance which had so far eluded classical physics.

---

[48] Harald Nordenson, *Relativity, Time, and Reality* (London: George Allen & Unwin, 1969), p. 24.
[49] Capek, *Philosophical Impact of Contemporary Physics*, p. 170; cf. pp. 153-154. He adds, "...there are no instantaneous connections joining distant simultaneous point-events; there are only successive connections, characterizing concrete physical actions, and there is no need to stretch the passive and static space underneath concrete interactions by which physical reality is exhausted" (Ibid., p. 172). Cf. idem, "Time-Space Rather than Spacetime," *Diogenes* (Italy) 123 (1983): 34.

CHAPTER 2

EINSTEIN'S SPECIAL THEORY

INTRODUCTION

Einstein's paper "The Electrodynamics of Moving Bodies" (1905), in which he broached his Special Theory of Relativity, is a model of clarity, concision, and boldness. Einstein opens his paper with a brief statement of the difficulty which prompts his new theory. It is interesting that his point of departure is not the failure to detect the earth's motion relative to the luminiferous aether but rather certain asymmetries in classical physics which he found troublesome. He gives the example, derived from Faraday's experiments in 1831, of a magnet and an electric conductor which are in relative motion to each other. Faraday had shown that whether the magnet was at rest and the conductor moving or the conductor at rest and the magnet moving, their proximate, relative motion to each other induced an electric current in the conductor. But in classical physics, the theoretical explanation for this phenomenon was quite different in the one case than in the other. If the magnet is moving past a conductor at rest, then the current is conceived to be caused by an electric field associated with the moving magnet. But if the magnet is at rest and the conductor moves past it, the current is said to be induced by the force exerted by the magnet on the electric charges in the moving conductor, and no electric field arises in the vicinity of the magnet. Now Einstein found these sorts of asymmetrical explanations for identical phenomena simply "unbearable." As he later explained,

> The thought that one is dealing here with two fundamentally different cases was for me unbearable. The difference between these two cases could be not a real difference but rather, in my conviction, only a difference in the choice of the reference point. Judged from the magnet, there were certainly no electric fields, [whereas] judged from the conducting circuit there certainly was one. The existence of an electric field was therefore a relative one, depending on the state of motion of the coordinate system being used, and a kind of objective reality could be granted only to the *electric and magnetic field together*, quite apart from the state of relative motion of the observer or the coordinate system. The phenomenon of the electromagnetic induction forced me to postulate the (special) relativity principle.[1]

The constancy of the velocity of light, which Einstein mentions in the footnote to the above citation, is the next difficulty in classical physics which he specifies in his 1905 paper. Since Einstein did not give references in his article, it has been disputed

---

[1] Albert Einstein, *Fundamental Ideas and Methods of Relativity Theory, Presented in their Development*, unpublished manuscript cited by Gerald Holton, "On Trying to Understand Scientific Genius," in *Thematic Origins of Scientific Thought: Kepler to Einstein* (Cambridge, Mass.: Harvard University Press, 1973), pp. 363-364.

as to what role the Michelson-Morley experiment played in the development of his thinking. To make matters more complicated, Einstein, when asked about this later, confessed that he could not clearly recall how much he knew of their experiment. But after an interview with Robert Shankland in 1950 in which Einstein said that he had learned of the Michelson-Morley result only after 1905, Einstein wrote Shankland specifically to correct this point, stating that he had known of their negative result through reading Lorentz's 1895 treatise *Toward a Theory of Electrical and Optical Appearances in Moving Bodies*.[2] He said that Lorentz's view led to an interpretation of the Michelson-Morley experiment which seemed to him artificial. But Einstein reiterated that what led him directly to SR was the asymmetry displayed in the magnet-conductor case; he also was guided, he recalled, by the phenomenon of stellar aberration and Fizeau's experiments (which verified Fresnel's dragging coefficient concerning the measured velocity of light). According to Holton, "In reading Lorentz's book of 1895, Einstein will have found that the [Michelson-Morley] experiment was not thought to be the crucial event upon which a new physics must be built: it was only one of several second order experiments that at the time could be explained only by invoking yet another unhappy *ad hoc* hypothesis to all the others on which current theory was built."[3]

However, correspondence from around the turn of the century between Einstein and his fiancée Mileva Maric which has come to light since Holton's essay suggests that Holton underestimated the role which the Michelson-Morley experiment and others like it played in the development of Einstein's thinking. From his correspondence with Maric between 1899 and 1901 we learn that Einstein was enthusiastically preoccupied with experimental attempts to measure the earth's velocity relative to the aether. In his letter of September 1899, he mentions reading "a very interesting paper" by Wilhelm Wien of Aachen.[4] In this paper Wien reviews the Michelson-Morley experiment and also mentions the contraction hypothesis as an attempted explanation of its null result.[5] In September of 1901 Einstein wrote a latter to Marcel Grossmann, in which he refers to the "customary interference experiments" to measure the motion of matter relative to the aether.[6] Writing to Maric in December of that year, Einstein enthusiastically revealed his own plans to design experiments aimed at measuring the Earth's velocity relative to the aether and mentioned his intention to read Lorentz's work. He said that he was on the verge of publishing a "capital paper" on relative motion and electrodynamics[7]—but, as we know, nothing appeared for the next three and a half years. Surveying this

---

[2] R. S. Shankland, "Conversations with Albert Einstein II," *American Journal of Physics* 41 (1973): 896. For discussion see Gerald Holton, "Einstein, Michelson, and the 'Crucial' Experiment," in *Thematic Origins*, pp. 282-285, 315.
[3] Ibid., p. 315. Contrary to Holton, however, even this role strikes one as fairly important. This was also Shankland's impression (R. S. Shankland, "Michelson-Morley Experiment," *American Journal of Physics* 32 [1964]: 16).
[4] Albert Einstein, "Letter 57. To Mileva Maric," in John Stachel, ed. *The Collected Papers of Albert Einstein*, Vol. 1: *The Early Years, 1879-1902* (Princeton, N. J.: Princeton University Press, 1987), p. 233.
[5] Wilhelm Wien, "Ueber die Fragen, welche die translatorische Bewegung des Lichtäthers betreffen." *Annalen der Physik und Chemie* 65, no. 3, Beilage (1898): i-xvii.
[6] Einstein, *Collected Papers*, "Letter 122. To Marcel Grossman," p. 316.
[7] Ibid., "Letter 128. To Mileva Maric," p. 325.

new evidence, John Stachel concludes, "He was very much interested in ether drift experiments and appears to have designed at least two, which he hoped to carry out himself....ideas about ether drift experiments did form an important strand in his thinking about the complex of problems that ultimately led him to develop the special theory of relativity."[8]

Rather than embrace a theory which struck him as *ad hoc* and artificial,[9] Einstein preferred to abandon altogether what he later called a "constructive" approach (such as Lorentz's) to a satisfactory theory and instead to cut the feet out from under the difficulties in classical physics by proposing fundamental revisions of our concepts of space and time so that such problems could not even arise. The first revision was the elimination of absolute space:

> Examples of this sort, together with the unsuccessful attempts to discover any motion of the earth relatively to the 'light medium,' lead to the conjecture that to the concept of absolute rest there correspond no properties of the phenomena, neither in mechanics, nor in electrodynamics...
>
> The introduction of a 'luminiferous ether' will prove to be superfluous because the view here to be developed will introduce neither an 'absolutely resting space' provided with special properties, nor associate a velocity-vector with a point of empty space in which electromagnetic processes occur.[10]

---

[8] John Stachel, "Einstein and Ether Drift Experiments," *Physics Today* (May 1987): 47. This conclusion is consonant with the evidence of Einstein's address of December 14, 1922, in Kyoto, Japan, in which he describes his own planned attempts to measure the velocity of the earth through the ether and says that the Michelson experiment was the first route which led him to relativity (T. Suyoshi Ogawa, "Japanese Evidence for Einstein's Knowledge of the Michelson-Morley Experiment," *Japanese Studies in the History of Science* 18 [1979]: 73-81).

[9] He later wrote, "...that contradiction between theory and experiment was formally resolved by the assumption of H. A. Lorentz and Fitzgerald that moving bodies experience a specific contraction in the direction of their motion. This *ad hoc* assumption, however, appeared to be only an artificial means aimed at saving the theory..." (A. Einstein, "Über das Relativitätsprinzip und die ausdemselben gezogenen Folgerungen," *Jahrbuch der Radioacktivität und Elektronik* 4 [1907]: 413). Again, "This situation was highly unsatisfactory....one was compelled to drafting a thoroughly peculiar hypothesis in order to understand that every relative motion should not make itself discernible" (A. Einstein, "Über die Entwicklung unserer Anschauungen über das Wesen und die Konstitution der Strahlung," *Physikalische Zeitschrift* 10 [1909]: 819).

[10] Albert Einstein, "On the Electrodynamics of Moving Bodies," trans. Arthur I. Miller, Appendix to Arthur I. Miller, *Albert Einstein's Special Theory of Relativity* (Reading, Mass.: Addison-Wesley, 1981), p. 392. Elsewhere Einstein makes it clear that SR not only rendered the aether superfluous, but eliminated it. He asserted that to specify any coordinate system as at rest and therefore as an aether frame is "utterly unnatural" (*überhaupt ganz unnatürlich*) and that a satisfying theory can only be obtained by abstaining from the aether hypothesis. The electromagnetic field is not a state of a hypothetical medium, but an independent reality. When Einstein says that a space free from radiation and matter would appear really empty, he must mean non-existent (Einstein, "Die Entwicklung unserer Anschauungen," p. 819; cf. idem, "Relativitätsprinzip und die Folgerungen," p. 413). Einstein continued to disparage the aether up until 1916, when in response to a letter from Lorentz, in which Lorentz argued that GR is compatible with an aether, Einstein proposed a new relativistic conception of the aether, which is described by the metrical tensor $g_{\mu\nu}$. In an intriguing and carefully documented essay, Kostro traces Einstein's development of this concept from 1916 until his death in 1955. The relativistic ether was essentially identical with spacetime itself and could not therefore serve as a reference frame. The aether of nineteenth century physics, "especially...Lorentz's ether identified with the absolute space conceived as a privileged reference frame," says Kostro, Einstein denied throughout his career after 1905 because it violated his principle of relativity, according to which there is no privileged reference frame (Ludwig

These terse and seemingly modest sentences belie their earth-shattering significance; the foundations of the world have just moved.[11] For by means of the circumlocution of "not introducing" an absolute space, Einstein has in fact excluded it from his ontology. For Einstein is denying, not merely the existence of the material aether, but of absolute space itself, and that not in the sense expressed by the absolute-relational distinction, but more radically in the sense that space neither exists independently of a physical reference frame nor is associated with a privileged reference frame. Spatially, so far as SR is concerned, only local reference spaces exist.[12] Although some have disputed this interpretation, contending that Einstein did not intend in 1905 to take any stance concerning the metaphysical status of space and time, but only concerning their physical status (that is, the behavior of measuring rods and clocks), and that the construal of the theory in metaphysical categories stems rather from Minkowski, who fused space and time into spacetime,[13] and although some support could be given for such a reading of Einstein's paper,[14]

Kostro, "Einstein's New Conception of the Ether," paper presented at the International Conference of the British Society for the Philosophy of Science, "Physical Interpretations of Relativity Theory," Imperial College of Science and Technology, London, 16-19 September, 1988).

[11] In Holton's words,
> "...an agonizing choice had to be made: in order to extend the principle of relativity from mechanics (where it had worked) to all of physics, and at the same time to explain the null results of all optical and electrical ether-drift experiments, one needed 'only' to abandon the notion of the absolute frame of reference and, with it, the ether. But without these the familiar landscape changed suddenly, drastically, and in every detail" (Holton, "The 'Crucial' Experiment," p. 307).

[12] In the words of Abraham Pais, "The one preferred coordinate system in absolute rest is forsaken. Its place is taken by an infinite set of preferred coordinate systems, the inertial frames...There are as many times as there are inertial frames. That is the gist of the June paper's kinematic sections, which rank among the highest achievements of science..." (Abraham Pais, *'Subtle is the Lord... ': The Science and Life of Albert Einstein* [Oxford: Oxford University Press, 1982], pp. 138, 141). On the metaphysic of the 1905 paper, the world has fallen apart, and there is no unity to space or time. There are an infinite number of three-dimensional spaces each enduring through its respective time, the points of which are interrelated via the Lorentz transformations.

[13] So Herbert Dingle, "Time in Philosophy and Physics," *Philosophy* 54 (1979): 99-104. A difficulty for Dingle's dichotomy between "time" and "the time" is that in German the abstract concept "time" is still expressed with the article "*die Zeit.*" See also Stanley Goldberg, *Understanding Relativity: Origin and Impact of a Scientific Revolution* (Boston: Birkhäuser, 1984), p. 68, who asserts that "the theory says nothing about the nature of the world. It only speaks to how measurements are made when we explore questions about that world" (p. 103). What Goldberg fails to appreciate is that in Einstein's theory the measured quantities *are* the reality, as we shall see.

[14] For example, in a reply to Paul Ehrenfest in 1907, Einstein wrote,
> "The Principle of Relativity or—more exactly put—the Principle of Relativity together with the Principle of the Constancy of Light Velocity is not to be conceived as a 'closed system,' indeed, not as a system at all, but merely as a heuristic principle, which considered in and of itself contains only statements about rigid bodies, clocks, and light signals" (A. Einstein, "Bemerkungen zu der Notiz von Hrn. Paul Ehrenfest: 'Die Translation deformierbarer Elektronen und der Flächensatz'," *Annalen der Physik* [4. ser.] 23 [1907]: 206).

But it was because of his conviction that space and time not defined by such operations were senseless and non-existent that Einstein was convinced of the heuristic power of Relativity in guiding further inquiry. Thus, for example, he later spoke of the Special Theory as a "valuable heuristic tool" for the discovery of laws of nature (Albert Einstein, *Über die spezielle und allgemeine Relativitätstheorie*, 4th ed. [1917; rep. ed.: Braunschweig: Friedr. Vieweg & Sohn, 1960], p. 26). Already in 1907 Einstein

Einstein's epistemology, as we shall see later, as well as his explicit statements, strongly imply that he meant to eliminate absolute space and time.

Einstein then proceeds to enunciate two presuppositions (*Voraussetzungen*) or postulates upon which his new theory will be based:[15]

> Postulate 1: "For every reference system in which the laws of mechanics are valid, the laws of electrodynamics and optics are also valid." ("Principle of Relativity")

> Postulate 2: "Light is always propagated in empty space with a definite velocity c which is independent of the state of motion of the emitting body." ("Principle of the Constancy of the Velocity of Light")

Both these postulates or principles are sometimes expressed differently in textbook expositions of SR. A closer examination of each is therefore merited.

Postulate 1, the Principle of Relativity, is simply the statement that all of physics, both mechanics and electrodynamics, is relativized. Not only is it impossible to determine whether one is in uniform motion by means of mechanical experiments; electrodynamical experiments will not reveal it either. But even Lorentz would have agreed with this last statement, given the Lorentz-FitzGerald contraction, which makes the laws of mechanics and electrodynamics *appear* the same in any inertial frame, though in fact it is only in the aether frame that all the physical laws are genuinely descriptive of reality. Einstein goes further: the laws not only appear the same to every inertial observer; they *are* the same, genuinely descriptive of what transpires in that reference frame.

Friedman has tried to give some greater precision to Einstein's Relativity Principle.[16] A popular way of expressing the principle is

> (R) The laws of nature are the same (or take the same form) in all inertial reference frames.

The problem with (R), complains Friedman, is that the notion of "the form" of a physical law and, hence, of a physical law's being "the same" in two reference frames is ambiguous. For the laws of any given spacetime theory can be written in a number of different forms, and even classical electrodynamics can be written in

---

wrote that all that was needed to arrive at a relativistic concept of time was "simply to define...H. A. Lorentz's auxiliary concept, which he called 'local time,' as 'time'" (*"als 'Zeit' schlechthin definieren"*) and that therefore "A time measurement, according to the definition of time given in sec. 1, has meaning only in reference to a reference frame having a specified state of motion" (Einstein, "Über das Relativitätsprinzip und die Folgerungen," pp. 413, 417). In his "Autobiographical Notes," in *Albert-Einstein: Philosopher-Scientist*, 3d, ed. P. A. Schilpp, Library of Living Philosophers 7 (La Salle, Ill.: Open Court, 1970), p. 59, Einstein credits Minkowski merely for introducing a certain formalism into the theory.

[15] Einstein, "Electrodynamics of Moving Bodies," p. 392. He refers to Postulate 2 as the Principle of the Constancy of the Velocity of Light in his "Bemerkungen zu der Notiz von Ehrenfest," p. 206.

[16] Michael Friedman, *Foundations of Spacetime Theories* (Princeton: Princeton University Press, 1983), pp. 149-159.

generally covariant form. Appealing to the idea that inertial frames should be experimentally indistinguishable in a relativistic theory, Friedman suggests replacing (R) with

> (R_1) All inertial reference frames are physically equivalent or indistinguishable.

The problem with (R_1), he admits, is that it fails to capture the full methodological force of the Special Principle of Relativity, since two spacetime theories could be experimentally indistinguishable and yet theoretically distinct in that they posit different theoretical entities. Therefore, he proposes to supplement (R_1) with

> (R_2) If two frames are indistinguishable according to a spacetime theory T, they should be theoretically identical according to T.

The conjunction of (R_1) and (R_2) would disqualify Lorentz's theory as relativistic because although two inertial frames are indistinguishable, they have different absolute velocities.

Postulate 2, the Principle of the Constancy of the Velocity of Light, asserts a wave model for the propagation of light *in vacuo* as opposed to a ballistic model. Light's velocity is independent of the motion of its emitting source: regardless of the direction of motion of the source, the light which it emits will travel at a constant c. Einstein's remark that this postulate is "only apparently irreconcilable" with the first[17] shows that the Light Principle does not assert the invariance of the velocity of light in all inertial frames, for that would be in accord with, rather than apparently incompatible with, the Relativity Principle.[18] The reason the Light Postulate seemed to contradict the Relativity Postulate was that if a signal moved at a finite velocity c regardless of its source's motion, then a receiver moving toward the source at velocity v should measure the signal's velocity at $c + v$ and a receiver moving away from the source should measure the signal's velocity at $c - v$, according to classical mechanics. Only if the signal were travelling with a velocity of zero or infinity would the law of addition (or subtraction) of velocities fail. So the principle that light propagates as a wave motion seemed incompatible with the assertion that all inertial frames were indistinguishable.

Friedman has also drawn attention to the various versions of the Light Principle associated with SR:[19]

> (L) Light is propagated with a constant velocity c independent of the velocity of its source. (Constancy of the Velocity of Light)

---

[17] Einstein, "Electrodynamics of Moving Bodies," p. 392; cf. Einstein, "Entwicklung unserer Anschauungen," p. 819; A. Einstein, "Die Relativitätstheorie," *Vierteljahrsschrift der naturforschenden Gesellschaft in Zürich* 56 (1911): 6, where he calls this a "frightful dilemma." He also makes it clear that the Light Principle (that there is a reference frame relative to which every light beam travels in the vacuum at the universal velocity c) is presupposed by the aether theorists themselves.

[18] See discussion by Kenneth F. Schaffner, "Einstein versus Lorentz: Research Programmes and the Logic of Comparative Theory Evaluation," *British Journal for the Philosophy of Science* 25 (1974): 54-56, and the presentation of the theory by Philipp Frank, *Philosophy of Science* (Englewood Cliffs, N. J.: Prentice-Hall, 1957), pp. 133-135.

[19] Friedman, *Foundations of Spacetime Theories*, pp. 159-165.

(L₁) Light has the same constant velocity c in all inertial reference frames. (Invariance of the Velocity of Light)

(L₂) No "causal" signal can propagate with a velocity greater than that of light. (Limiting character of the Velocity of Light)

We have seen that Einstein himself employed (L), but several expositors of SR use (L_1); in fact, (L_1) alone is sufficient, when combined with Galilean relativity, to generate special relativity, for it asserts, in effect, that electrodynamics is relativized. Interestingly, (L_2) plays no essential role in special relativity; while acceleration to the velocity of light or beyond is forbidden, velocities that are constantly superluminal are not forbidden, and a whole body of literature has grown up concerning the possibility of particles, called tachyons, which travel at superluminal speeds. Einstein's procedure is to regard (L_1) as the consequence of the Relativity Principle plus (L).[20] He asserts, "These two postulates suffice in order to obtain a simple and consistent theory of the electrodynamics of moving bodies taking as a basis Maxwell's theory for bodies at rest."[21]

DEFINITION OF SIMULTANEITY

The next section of Einstein's paper, entitled "Definition of Simultaneity," is the most philosophically interesting section in the article. As Miller notes, "This initial section is, in fact, nothing less than an epistemological analysis of the nature of space and time."[22] Einstein invites us to pick an arbitrary inertial frame and consider it to be at rest. It will be recalled that in classical physics such a supposition was unproblematic because Newtonian mechanics was relativized: any inertial system could be considered to be at rest. Einstein goes on to point out that the designation of a material point at rest in this system presents no difficulty: one has simply to give its Cartesian coordinates x, y, z, representing the three spatial dimensions. But in order to describe the *motion* of a material point we must give its spatial coordinates at different times. "Now," he continues, "we must bear carefully in mind that a mathematical description of this kind has no physical meaning unless

---

[20] In Albert Einstein, *The Meaning of Relativity*, 6th ed. (1922; rep. ed.: London: Chapman & Hall, 1967), p. 26, he explicitly draws the inference that since light is propagated in an inertial system with velocity c, it must be so propagated in all systems.
[21] Einstein, "Electrodynamics of Moving Bodies," p. 392.
[22] Miller, *Einstein's Special Theory*, p. 123. Noting that Einstein has "radically changed the problem from one of physical character into one of epistemology," Nordenson agrees that "when Einstein starts to grapple with a new concept of time he embarks on a philosophical problem in the strictest sense of the word..." (Harald Nordenson, *Relativity, Time, and Reality* [London: George Allen & Unwin, 1969], p. 26); hence, "The validity of his Theory of Relativity depends wholly on the justification of his criticism and rejection of classical time and of the validity of the new concept which he makes it his task to establish." So also Ernst Cassirer, *Substance and Function and Einstein's Theory of Relativity*, trans. W. C. Swabey and M. C. Swabey (1923; rep. ed: New York: Dover, 1953), p. 354, who says that all historical examples of "the real inner connection between epistemological problems and physical problems" are outdone by the Special Theory of Relativity. From "its very beginning" in its analysis of time the theory appeals to an epistemological motive which is of "decisive significance" in its putative advantage over other explanations, like that of Lorentz.

we are quite clear what we will understand by 'time'."[23] Thus, what Einstein is about in this section is giving physical meaning to the concept of time.

All judgements concerning time, he claims, are statements about *simultaneous events*. To use his example: When I say "The train arrives here at 7 o'clock," I mean, "The train's arrival and my watch's pointing to 7 are simultaneous events." So perhaps we can overcome the difficulties in defining time if we just replace "time" with "the reading of my watch." This works fine for defining the time of an event which occurs in the same place as my watch. But when we want to evaluate the times of events at places distant from the watch, things are not so simple. Imagine two widely separated spatial points A and B. If observers at A and B respectively have clocks of identical constitution, then they can each check the time of the events at their location. But how can we know which events at A are simultaneous with given events at B? What we need is a common time for both events at A and events at B.

Einstein then proposes a method to define such a time. We first simply require by definition that the "time" it takes light to travel from A to B is the same as the "time" necessary for light to travel from B and A. Now let the observer at A send a light beam at A-time t_A which is received and reflected back by the observer at B at B-time t_B and which is then received again by the observer at A at a later A-time t_A' (Figure 2.1).

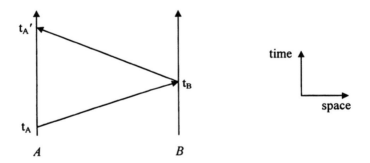

Figure 2.1. A light beam from A is sent to B at t_A and then reflected back from B at t_B and received at A at t_A'.

The clocks at A and B are defined to be synchronized if $t_B - t_A = t_A' - t_B$. Einstein then makes a number of assumptions: (i) that this definition of synchronization is free of any possible contradictions, (ii) that it is applicable to arbitrarily many points, and (iii) that the following relations are universally valid: (a) If the clock at B synchronizes with the clock at A, the clock at A synchronizes with the clock at B; and (b) If the clock at A synchronizes with the clock at B and also with the clock at C, the clocks at B and C also synchronize with each other (or in other words, this method of synchronization is symmetric and transitive).

---

[23] Einstein, "Electrodynamics of Moving Bodies," p. 393.

Einstein concludes this section by stating that we have, with the help of certain (imaginary) physical experiments, defined what we mean by synchronous stationary clocks located at different places and thus have a definition of "simultaneous" (or "synchronous") and "time." "The 'time' of an event is the reading simultaneous with the event of a clock at rest and located at the position of the event, this clock being synchronous, and indeed synchronous for all time determinations, with a specified clock at rest."[24] By this definition he apparently means that the time of an event is the reading simultaneously given by a clock at the same location as the event—"simultaneous" in this case not being defined (in a footnote Einstein begged off discussing "the inexactitude which lurks in the concept of simultaneity of two events at [approximately] the same place"[25])—if and only if that clock is synchronized (and stays synchronized) by the light signal method with a clock that records the time for that inertial frame. In this way, local simultaneity is an undefined given, and distant simultaneity is defined by light signal synchronization of a distant clock with a local clock at rest.

In the second section of his paper Einstein will deduce from his postulates and definitions the relativity of simultaneity among events in different inertial frames; but a number of philosophers of space and time, including Reichenbach, Grünbaum, and Salmon, take him to have already in this first section established something even more fundamental: the conventionality of simultaneity even within a single reference frame.[26] It is important, therefore, before proceeding to section 2 of Einstein's paper, that we consider whether SR implies the conventionality of simultaneity. Defenders of this thesis point out quite correctly that this first section has nothing to do with the simultaneity of events in frames which are in relative motion to each other: the section begins by considering a single coordinate system dubbed "the resting system" and concludes with the words, "...the time now defined being appropriate to the resting system we call 'the time of the resting system'."[27] The burden of the section is to specify what we mean by the simultaneity of distant events within this one inertial system. The answer given to that question comes in terms of a method for synchronizing clocks by means of exchange of light signals. But the light signal method of synchronization rests on an arbitrary convention, namely: the *definition* that the time required for the light beam to go from A to B is equal to the time it takes to go from B to A. That this seemingly innocuous assumption is a convention is evident from the empirical fact that it is impossible to

---

[24] Ibid., p. 394.
[25] Ibid., p. 393. See comment by Miller, *Einstein's Special Theory*, p. 194.
[26] Hans Reichenbach, *The Philosophy of Space and Time* [1928], trans. Maria Reichenbach and Joseph Freund (Rep. ed.: New York: Dover, 1957), pp. 123-134; Adolf Grünbaum, *Philosophical Problems of Space and Time*, 2d ed., Boston Studies in the Philosophy of Science 12 (Dordrecht: D. Reidel, 1973), pp. 342-368, 666-708; Wesley C. Salmon, "The Conventionality of Simultaneity," *Philosophy of Science* 36 (1969): 44-63. For discussion see Friedman, *Foundations of Spacetime Theories*, pp. 165-176, 309-320; cf. idem, "Simultaneity in Newtonian Mechanics and Special Relativity," in *Foundations of Spacetime Theories*, ed. John S. Earman, Clark N. Glymour, and John J. Stachel, Minnesota Studies in the Philosophy of Science 8 (Minneapolis: University of Minnesota Press, 1977), pp. 417-431.
[27] Einstein, "Electrodynamics of Moving Bodies," p. 394.

measure the one-way velocity of light.[28] All we can measure is the round-trip velocity of light, its velocity over an out-and-return journey. For in order to measure its one-way velocity, we should have to know that the clock at the departure point is synchronized with the clock at its terminus, and there is no way to establish that which does not beg the question. Hence, we must rest content with measuring the two-way velocity of light: recording the time it takes for a light beam to travel to a reflector and return to the place of emission. We must simply assume that the beam traveled the same speed in both directions. It is possible that it traveled at a velocity $> c$ on the outward leg and at a velocity $< c$ on the return leg. We must simply *define* the one-way velocity of light to be a constant.

It has been suggested that we could overcome this conventionality by using an alternative method of clock synchronization: slow transport of clocks.[29] This method consists in taking two local, synchronized clocks and slowly transporting (so as to minimize the effects of motion) one of the clocks to the distant point of the events to be timed. In this way, we now have synchronized clocks at a distance and could even use them to measure the one-way velocity of light. If we wanted to be doubly sure that our distant clock remained synchronous with its counterpart, we could slowly transport it back again and compare it with the clock that remained at rest. In this way, non-arbitrary simultaneity relations could be established within a single inertial system.

But the defenders of the conventionality of simultaneity rejoin that even slow transport of a clock may affect that clock, so that by the time it reaches its distant destination it may be no longer synchronized with its counterpart. It does no good to bring it back again to its starting point, for the effects of the return journey could exactly compensate for what occurred on the outward journey. There is thus no non-conventional way to synchronize distant clocks.

Another way of putting this is to substitute for Einstein's definition $t_B - t_A = t_A' - t_B$ the equivalent formula $t_B = t_A + \frac{1}{2}(t_A' - t_A)$. The conventionalists' claim is that any fraction between 0 and 1 is as suitable as ½, so that a variable ε should replace it in the formula: $t_B = t_A + \varepsilon(t_A' - t_A)$. Since ε is only arbitrarily defined to be ½, the simultaneity of distant events in the same reference frame is conventional.

In sorting through the complex debate which has arisen concerning the conventionality of simultaneity, it is critical that two questions be kept distinct: (1) is simultaneity conventional in the context of Einstein's search for a theory which will deal satisfactorily with the problems he has posed, and (2) is simultaneity conventional within the context of SR as a completed theory?[30]

With respect to the second question, a series of authors stretching all the way back to A. A. Robb and including, more recently, philosophers of science like David Malament have shown that given Minkowski's spacetime formulation of SR, $\varepsilon = \frac{1}{2}$

---

[28] See Wesley C. Salmon, "The Philosophical Significance of the One-Way Speed of Light," *Noûs* 11 (1977): 253-292.
[29] B. Ellis and P. Bowman, "Conventionality in Distant Simultaneity," *Philosophy of Science* 34 (1967): 116-136.
[30] These usually conflated questions have been distinguished by Howard Stein, "On Relativity Theory and Openness of the Future," *Philosophy of Science* 58 (1991): 153.

is not arbitrary but is uniquely definable from the metric of Minkowski spacetime.[31] Robb showed that the whole structure of Minkowski spacetime, including simultaneity relations, is definable in terms of causal connectability alone (which is held by conventionalists to be non-conventional), and Malament proved that $\varepsilon = \frac{1}{2}$ is the only simultaneity relation so definable in the context of SR.[32] Hence, as Friedman points out, one cannot dispense with the standard simultaneity relation without dispensing with the entire conformal structure of Minkowski spacetime. "Standard simultaneity...is explicitly definable from the other quantities of relativity theory: it cannot be varied without completely abandoning the basic structure of the theory."[33] Friedman agrees with Reichenbach that the light signal method of clock synchronization cannot yield $\varepsilon = \frac{1}{2}$ in a non-circular way. But since the standard simultaneity relation is intimately connected with the rest of the structure of Minkowski spacetime, we should expect there to be some other non-circular method for establishing simultaneity. He plumps for the slow transport of clocks. We initially synchronize the clocks at A and B for any value of ε. We then transport clocks synchronized with A's clock to B at progressively slower velocities. If the differences between these clocks upon arrival at B and the clock at B converge toward zero, then the two clocks at A and B are synchronized according to the formula $\varepsilon = \frac{1}{2}$. No matter what value of ε is arbitrarily chosen to begin with, one finally obtains $\varepsilon = \frac{1}{2}$ in the end. A clock "transported" with infinitely slow velocity would show no difference between A and B at all. In this way, non-conventional relations of simultaneity within a reference frame can be established in the context of SR.

In the minds of many, this analysis has sufficed to settle question (2). But Lawrence Sklar attempts to push the debate a notch further by considering the question within the wider context of a more radical conventionalism about the geometry of spacetime.[34] This more fundamental conventionalism has its roots in Poincaré and was seconded by Reichenbach and Grünbaum. Poincaré regarded statements about the geometric properties of space and time—about whether space is Euclidian or non-Euclidian, for example, or whether two non-overlapping spatial or temporal intervals are congruent or not—as neither true nor false, but merely as more or less convenient, given the nature of our measuring instruments. Sometimes Poincaré will even say that such statements are meaningless.[35] The basis of

---

[31] A. A. Robb, *A Theory of Time and Space* (Cambridge: Heffer & Sons, 1913); see note 31.
[32] David Malament, "Causal Theories of Time and the Conventionality of Simultaneity," *Noûs* 11 (1977): 293-300.
[33] Friedman, *Foundations of Spacetime Theories*, p. 320; see all of pp. 309-320.
[34] Lawrence Sklar, "The Conventionality of Simultaneity! Again?" Paper presented at the International Conference of the British Society for the Philosophy of Science Conference, "Physical Interpretations of Relativity Theory," Imperial College of Science and Technology, London, 16-19 September 1988.
[35] "...all these affirmations have by themselves no meaning" (Henri Poincaré, "The Measure of Time," *Revue de métaphysique et de morale* 6 [1898]: 1ff., rep. in H. Poincaré, *The Value of Science* [1905], trans. G. B. Halstead, in *The Foundations of Science*, by H. Poincaré [Science Press, 1913; rep. ed.: Washington, D. C.: University Press of America, 1982], p. 228); "...the equality or inequality of two distances are words which have no meaning by themselves; that is precisely why I was obliged to have recourse to a convention in order to give them one" (H. Poincaré, "Sur les principes de la géometrie,"

Poincaré's conventionalism is a verificationist critique of claims that space or time does possess certain metrical properties. If we adjust the laws of nature appropriately, Poincaré observes, we can make all empirical phenomena fit whatever geometry we posit for space and time. "Can we suppose that certain phenomena, possible in Euclidian space, would be impossible in non-Euclidian space," he asks, "such that experience, in establishing these phenomena, would directly contradict the non-Euclidian hypothesis?"[36] No, he answers, "there exists no property which could...be an absolute criterion permitting us to recognize a straight line and to distinguish it from every other line."[37] If all objects, including our measuring instruments, are deformed according to some law, "we will not be able to notice this," says Poincaré, from which he concludes, "In reality space is therefore amorphous, a flaccid form, without rigidity, which is adaptable to everything; it has no properties of its own. To geometrize is to study the properties of our instruments, that is, of solid bodies."[38] The same thing holds for time: if everything went more slowly, we would not notice it. "The properties of time are therefore merely those of our clocks, just as the properties of space are merely those of the measuring instruments."[39] Again, the goal is convenience: "The simultaneity of two events, or the order of their succession, the equality of two durations, are to be so defined that the enunciation of the natural laws be as simple as possible. In other words, all these rules, all these definitions are only the fruit of an unconscious opportunism."[40]

Now Sklar points out that if one adopts this sort of chrono-geometric conventionalism, then one could adopt a model of spacetime which is conformal to Minkowski spacetime (the spacetime of SR[41]), but not isometric to it. In this new, curved spacetime the facts about the causal connectability of events would be preserved, but the Robbian definitions of simultaneity would no longer be correct because the spacetime is not flat. The appearance of flatness is due to posited fields which affect the behavior of clocks and rods. Thus, if it is a matter of mere convention whether spacetime is Minkowskian, it is equally a matter of convention whether two events determined by the metric of Minkowski spacetime to be simultaneous are simultaneous.

In assessing this wider approach to the question of the conventionality of simultaneity within SR, it is important that we keep in mind that the chrono-geometric conventionalist is not saying merely that we cannot know the factual geometry of spacetime, but that factually there is nothing there to be known. So

---

Revue de métaphysique et de morale 8 [1900]: 80); "There is no absolute time; to say two durations are equal is an assertion which has by itself no meaning and which can acquire one only by convention" (H. Poincaré, *Science and Hypothesis* [1902], in *Foundations of Science*, p. 92).

[36] H. Poincaré, "Des fondements de la géométrie," *Revue du métaphysique et de morale* 7 (1899): 266.
[37] Ibid.
[38] Henri Poincaré, "Space and Time," in *Mathematics and Science: Last Essays* [*Dernières pensées*, 1913], trans. John W. Bolduc (New York: Dover, 1963), p. 16.
[39] Ibid., p. 18.
[40] Poincaré, "Measure of Time," p. 234; cf. idem, *Science and Hypothesis*, pp. 64-65.
[41] For more on Hermann Minkowski's reformulation of Einstein's theory in terms of a four-dimensional geometry, see chap. 5.

when a Clark Glymour defends the conventionality of even the topology of spacetime, remarking, "I do not say that there is no truth to the matter, but only that we cannot find it out,"[42] he is not defending genuine conventionalism. Rather as Graham Nerlich emphasizes, "Conventionalism...places limits on what can be in the world and on how the world can be structured."[43] It holds that the world has no objective topological and/or geometric structure. "Maybe we could never know, even in principle, whether the earth or our whole space was Euclidian or spherical or what," Nerlich admits; "But we are not concerned with the problems of finding out about spaces. We are worried about the spaces themselves. Our question was whether the space was determinate..., not whether our knowledge was."[44] Given the radical nature of conventionalism, opines Nerlich, we should expect to find some very powerful arguments as to why we should adopt a conventionalist methodology in our approach to time and space.

> What we will be concluding, after all, is that various widely accepted theories characteristically misdescribe the world by overdetermining it. They give us a false picture of the structure of reality...There is no doubt that we nearly all think that material objects have definite shapes and lengths, that objects are indeed factually, objectively equal or unequal in length, whether they are touching or not. *If these ideas are wrong then it is of the greatest importance to rid ourselves of them, since they are widely believed without question.*[45]

If we are so drastically mistaken in our beliefs about the structure of the world, "then the conventionalist has a momentous surprise to spring on most of us...But any argument for a methodology will have to be very powerful indeed to sustain so massive a conclusion."[46]

So what are these very powerful arguments? As we have seen, Poincaré simply inferred from our inability to discern the geometric structure of space and time that no such structure exists. Already in the last century Bertrand Russell protested that it is one thing to hold that although one geometry is true and the others false, still it is absolutely impossible for us to know which one is true and another thing to hold with Poincaré that the reason one cannot prove Euclidian geometry to be true or false is because it is neither true nor false.[47] To establish the latter thesis, "it in no way suffices to prove that other hypotheses are equally capable of explaining the phenomena insofar as we know them."[48] In his response to Russell, Poincaré insists that apart from a convention of measurement, it is meaningless to speak of the

---

[42] Clark Glymour, "Topology, Cosmology, and Convention," in *Space, Time, and Geometry*, ed. Patrick Suppes, Synthèse Library (Dordrecht: D. Reidel, 1973), p. 194. Glymour points out that different global topologies (*e.g.*, the Euclidian plane, cylinder, torus, Klein bottle) may all admit a geometry which is locally Euclidian. "The significant cases of conventional topologies are those in which the world is one way or the other, not both, but things are so arranged that we cannot discover which way the world is" (Ibid., pp. 213-214). To call such a view conventionalism is a misnomer.
[43] Graham Nerlich, *The Shape of Space* (Cambridge: Cambridge University Press, 1976), p. 127.
[44] Ibid., p. 146.
[45] Ibid., pp. 127-128. Recall that we are talking about objects' shapes and sizes within a single inertial frame.
[46] Ibid., p. 128.
[47] B. Russell, "Sur les axiomes de la géométrie," *Revue de métaphysique et de morale* 7 (1899): 685.
[48] Ibid.

equality or inequality of two distances.[49] But such a response presupposes, not proves, metric conventionalism. Why cannot space and time themselves possess a unique and intrinsic metric? To this Poincaré issues the following challenge: if statements about the congruence of distances are meaningful in themselves, then at least show "how one can represent to oneself the equality of two distances."[50] Since this cannot be unequivocally done, Poincaré boldly affirms that the statement "that the distance from Paris to London is greater than a meter in an *absolute* way and independently of all methods of measurement is neither true nor false; I find this to say nothing; if you esteem it to say something, it seems to me that it is up to you to explain to me what it says."[51] The nonconventionalist will find the meaning of such metrical statements clear; for example, the amount of space separating London from Paris is more than the amount of space separating one end of a meter stick from the other. Poincaré's argument really boils down to the assertion that if one cannot know the unique geometry of space or time, then no such geometry exists. But the problem with such an inference, as J. J. C. Smart notes, is that "It comes from a verificationist theory of meaning which we should reject, or what is perhaps the same thing, a confusion of semantic and epistemological matters. Undetectable differences can still be differences."[52] (Smart's point is especially evident from the theistic point of view: God may be perfectly aware of the topological and geometric properties of the spacetime He has created and sustains, even if we limited creatures have no inkling of these properties.)

So what other powerful arguments for conventionalism are there? In Reichenbach we find the same verificationism.[53] Grünbaum does present independent grounds for his conventionalism, a curious argument based on the continuity of space and time which has, however, received devastating criticism at the hands of his critics.[54] So where are the arguments sufficient to motivate so

---

[49] H. Poincaré, "Sur les principes de la géométrie," *Revue de métaphysique et de morale* 8 (1900): 80.
[50] Ibid.
[51] Ibid., p. 81.
[52] J. J. C. Smart, "Quine on Spacetime," in *The Philosophy of W. V. O. Quine*, Library of Living Philosophers 18, ed. Lewis Edwin Hahn and Paul Arthur Schilpp (LaSalle, Ill.: Open Court, 1986), p. 507.
[53] Reichenbach, *The Philosophy of Space and Time*, pp. 10-37. For example, he writes, "If universal forces are admitted, the measurements may be interpreted in such a way that many different shapes of surfaces are compatible with the same observations...Taken alone, the statement that a certain geometry holds for space is therefore meaningless" (Ibid., pp. 29, 35). One may compare with profit Reichenbach's attempt to avert the standard objection to verification-based conventionalism by appeal to the impossibility of determining whether the standard meter in Paris really is a meter long (Ibid., pp. 28-29) with Kripke's distinction between giving the meaning and fixing the reference (Saul A. Kripke, *Naming and Necessity*, 2d rev. ed. [Oxford: Basil Blackwell, 1980], pp. 54-56). Reichenbach's comments in *Philosophy of Space and Time*, p. 20, seem to accord with Kripke's view that the Parisian meter bar merely fixes the reference of the length of a meter, in which case the impossibility of determining whether the bar is really a meter is purely epistemic and has nothing to do with defining the meter in absolute terms. That Kripke was right seems to have been borne out by the fact that the bar in Paris is no longer used to fix the standard length of the meter.
[54] Adolf Grünbaum, *Philosophical Problems of Space and Time*, 2d ed. Boston Studies in the Philosophy of Science 12 (Dordrecht: D. Reidel, 1973), pp. 8-18; cf. chap. 16. For critiques see Michael Friedman, "Grünbaum on the Conventionality of Geometry," in *Space, Time, and Geometry*, pp. 217-233; Paul Gordon Horwich, "On the Metric and Topology of Time" (Ph. D. dissertation, Cornell University,

radical a doctrine as chrono-geometric conventionalism? They have not been forthcoming. Therefore, the argument for the conventionality of simultaneity within Minkowski spacetime which is based on an appeal to a wider conventionalism about the geometric properties of spacetime fails to persuade.

But perhaps the conventionalist could retrench at this point. Perhaps one could argue that while there are objective facts about the topology and geometry of spacetime, nevertheless, for the reasons mentioned, we cannot know which such structures characterize the universe and that the Minkowski model of spacetime, if not conventional, is at least gratuitous or arbitrary. Therefore, the relations of simultaneity within a reference frame determined by the metric of that spacetime are equally arbitrary.

This objection, however, takes us over to question (1) concerning the context of the search for an adequate spacetime theory. For the sort of conventionalism associated with question (1) does not deny that *given* that spacetime is Minkowskian, simultaneity is non-conventional. It challenges, rather, our justification for holding spacetime to be Minkowskian. In particular, we may ask, how could Einstein, in section 1 of his paper, already assume that spacetime had the structure determined by SR, since he was still at that point attempting to discover the metric of space and time? To justify the definition of distant simultaneity based on clock synchronization via light signals by appealing to the metric of relativistic spacetime would be plainly question-begging.

The raising of question (1) thus helps to bring sharply to the fore the radical view of reality which underlies the Special Theory, namely, its denial of absolute space. For wholly apart from the possible inconstancy of the one-way velocity of light, another way in which it could be the case that $\varepsilon \neq \frac{1}{2}$ if A and B, while at *relative rest*, were nonetheless *moving in tandem* with each other. Let A and B be two rocketships in space both moving at a uniform velocity in the same line of motion (Fig. 2.2).[55]

1975), chap. 3; Philip L. Quinn, "Intrinsic Metrics on Continuous Spatial Manifolds," *Philosophy of Science* 43 (1976): 396-414; Nerlich, *Shape of Space*, chap. 8. These critics argue that continuity of a manifold does not entail the metrical amorphousness of that manifold, that one may not equate a metric's extrinsicality with conventionality, that the metric may be intrinsic even in Grünbaum's sense if it is a counterfactual property, and that Grünbaum's attempt to reformulate his argument using a set-theoretic analysis is misconceived and irrelevant because a continuous interval has a size intrinsically iff each sub-interval does.

[55] Adapted from L. Epstein, *Relativity Visualized* (San Francisco: Insight Press, 1983), p. 57. Professor Epstein has asked that I inform readers that his book is available by mail. To obtain a copy, indicate that you desire a copy of *Relativity Visualized*, provide a mailing address, and send a check for $20.90 to Insight Press, Dept. I, 601 Vermont Street, San Francisco CA, 94107-2636.

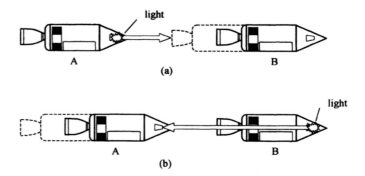

Figure 2.2. Two rocketships A and B moving at a uniform, near light velocity. By the time the observer B receives A's light signal, he will have advanced from the location he occupied when A fired his signal (a). By the time the observer in A receives B's return signal, he will have advanced from the location he occupied when B fired the return signal (b). Thus, the signal takes longer to go from A to B than from B to A, and so $\varepsilon \neq \frac{1}{2}$).

Let the observer in A send a light signal to B. By the time B gets the signal, B will have moved on from where it was when A emitted the light beam. Let B reflect the signal back to A at the instant B receives it. By the time A receives this signal, A will have advanced beyond where it was when B reflected the beam back toward A. Thus, although the beam traveled with a constant one-way velocity from A to B and from B to A, still $\varepsilon \neq \frac{1}{2}$ in this case because the distance the beam traveled from A to B was greater than the distance it traveled from B to A. Hence, it is not true that $t_B - t_A = t_A' - t_B$. Rather because the first leg of the journey was longer than the second leg, $t_B - t_A > t_A' - t_B$.

Given the Relativity Principle, this difference in travel time must be completely indiscernible by A and B. The Relativity Principle conjoined with the Light Principle implies that light travels at a constant velocity c in *every* inertial frame. In classical electrodynamics, the velocity of the light signal from A to B would be $c - v$ relative to B and $c + v$ relative to A. But the Relativity Principle requires that it be simply c relative to both B and A. But this seems crazy (Figure 2.3).[56]

---

[56] See note 54.

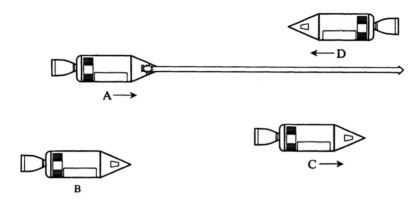

Figure 2.3. Invariance of the velocity of light. Though A is travelling at 90% c, he measures his light signal to have the same velocity as is measured by B, who takes himself to be at rest. C, also traveling at 90% c, observes the signal to overtake him as if he were standing still, as constant c. D, travelling in the opposite direction at 90% c, also measures the signal's velocity as c.

If rocketship A is travelling at, say, .9 c and emits a light beam in its direction of motion, that beam still flies away from A at the velocity of c, the same speed with which it passes by rocketship B which is conventionally taken to be at rest. Moreover, rocketship C, moving in the same direction as A at .9 c also observes the beam flash past it at a full c. Meanwhile, for a fourth rocketship D, travelling in the opposite direction as A and C, the light ray also passes by at the constant c, despite the fact that it is traveling toward the signal at .9 c itself. It needs to be emphasized that the situation is not just that A, B, C, D all *measure* the velocity of light as c in their respective inertial frames (to assert merely that would be a Lorentzian theory), rather the Relativity Principle requires that light really *has* the velocity c in all these frames, despite their relative motion. So, to return to our original illustration, A and B's exchanging signals while moving in tandem is indistinguishable from their doing so when both are at rest; indeed, they *are* at rest relative to each other. The only difference between the two cases lies in the difference in the round trip distance which the light signal traverses; the distance is greater when A and B are moving in tandem than when they are both at rest. But then the claim that $\varepsilon = \frac{1}{2}$ in all cases would seem to be wrong-headed, since if A and B are moving in tandem there will be any number of values for ε depending on the velocity of A and B through space. Clocks moving in tandem within the same frame which are synchronized by light signals would seem not to have identical readings at the same time, though by Einstein's definition they are synchronous. A and B in Figure 2.2, using light signal synchronization, would, unbeknownst to them (since they take $\varepsilon = \frac{1}{2}$), have clocks that do not give identical readings at the same time. Nor could this delusion be overcome through synchronization by slow transport of clocks; for even if the clocks were "transported" infinitely slowly from A to B *relative to the* AB *reference frame*, such clocks would still be in motion and so have different readings at the same time.

The above serves, as I say, to highlight how radical a revision Einstein makes in the traditional ontology of Western thought. I said before that the key to Einstein's theory is his elimination of absolute space or the aether reference frame. Now we see why this is so. Unless he had first eliminated the aether reference frame, then the above considerations would make perfect sense. It was the natural presupposition of the existence of this privileged frame of reference that lay behind the scenarios sketched above. To speak of A and B's being at relative rest but moving in tandem presupposes an absolute space or aether rest frame relative to which they are moving together. It was relative to this frame that the distance from A to B was longer than from B back to A. But Einstein denies the existence of such a frame; there is no such frame as space. All that exists is locally moving frames, and even these do not move relative to space, but only relative to each other. To assert that A and B are moving in tandem or are at rest presupposes that there is such a thing as absolute rest or absolute motion. The reason light signal synchronization can be used to establish simultaneity relations between A and B is that A and B have no absolute motion. They are moving only relative to some other frame of reference, and that sort of motion is inconsequential, since A and B can be considered to be at rest and the other frame moving. (Of course, an observer in the other frame can make just the opposite assumption in order to synchronize his clocks.) It seems difficult to conceive and is perhaps unimaginable that there is no reference frame "space" or preferred frame coextensive with space in which these local frames are moving about. Capek muses, "...we are all unconsciously Newtonians even when we profess to be relativists, and the classical idea of world-wide instants, containing simultaneous spatially separated events, still haunts the subconscious even of relativistic physicists; though verbally rejected, it manifests itself, like a Freudian symbol, in a certain conservatism of language."[57] But if we are to understand Einstein's view of the world for what it is, we must blot out, if we can, from our minds that background space and posit instead only relatively moving frames. When we have done that, we can see why it is no longer contradictory for A, B, C, and D in Figure 2.3 all to observe the velocity of light as c, for light does not travel in the aether frame or background space, but only in the inertial frames. That is why the Light Principle is "only apparently irreconcilable" with the Relativity Principle: A, B, C, and D can each be considered to be at rest in their respective frames, and so light moves relative to them in that frame at c.

By eliminating any absolute reference frame, Einstein was free to conventionally redefine the notion of simultaneity in terms of light signal synchronization. Without that prior move, his attempt to so redefine simultaneity would have appeared ludicrous precisely because the simultaneity relations so defined would have stipulated that clocks having different readings at the same time could nonetheless be synchronous. Only relative to the absolute frame would light synchronization have disclosed true simultaneity relations. Indeed, that was exactly the position

---

[57] Milic Capek, *The Philosophical Impact of Contemporary Physics* (Princeton: D. Van Nostrand, 1961), pp. 190-191.

advocated by Poincaré. In 1904 he argued that the method of light signal synchronization can be used to obtain a quantitative measure of time for two observers at rest in the aether but that for observers in motion, such a method yields only Lorentz's "local time":

> And in fact they mark the same hour at the same physical instant, but on the one condition, that the two stations are fixed. Otherwise the duration of the transmission will not be the same in the two senses, since the station A, for example, moves forward to meet the optical perturbation emanating from B, whereas the station B flees before the perturbation emanating from A. The watches adjusted in that way will not mark, therefore, the true time; they will mark what may be called the *local time*, so that one of them will be slow of the other. It matters little, since we have no means of perceiving it. All the phenomena which happen at A, for example, will be late, but all will be equally so, and the observer will not perceive it, since his watch is slow; so, as the principle of relativity requires, he will have no means of knowing whether he is at rest or in absolute motion.[58]

By using this same method, only in the absence of any absolute frame of reference, Einstein in effect eliminates "true time" and makes Lorentz's local time the only genuine time there is. He writes,

> Surprisingly, it turned out that it was only necessary to grasp the concept of time clearly enough in order to escape the...difficulty. One needed only to recognize that an auxiliary tool introduced by H. A. Lorentz, which he called 'local time,' can simply be defined as 'time.'[59]

The result is that one arrives at a definition of time which is tied to a local reference frame:

> It may be asked what we have thereby accomplished that is especially remarkable, since that all sounds so obvious. What is remarkable lies in the fact that this prescription, in order to arrive at time measurements of a very definite meaning, refers to a system of clocks which is at rest relative to a very definite coordinate system k. We have not obtained simply a time, but a time in relation to the coordinate system k, that is to say, in relation to the coordinate system k along with the regulated clocks at rest relative to k....
> It is not said that the time is absolute, that is, has a meaning which is independent of the state of motion of the reference frame. That is an arbitrary element which had been contained in kinematics.[60]

By defining simultaneity by means of light signal synchronization only *after* eliminating the rest frame of space, Einstein can consistently stipulate $\varepsilon = \frac{1}{2}$ (given the constancy of the one-way velocity of light), so that there is no possibility that clocks so synchronized might not have the same readings at the same time (there being no such time above the "local" time).

Sklar emphasizes that "From the point of view of the aether theory the radar method, slowly transported clocks or the Robbian construction of distant simultaneity are correct only when utilized to determine simultaneity relative to the one and only one true rest frame. All of the other 'relative simultaneities' are not

---

[58] H. Poincaré, *The Value of Science*, in *The Foundations of Science*, p. 307.
[59] Einstein, "Über das Relativitätsprinzip," p. 413.
[60] Einstein, "Die Relativitätstheorie," pp. 8-9.

simultaneities at all."[61] Now the question is: with what justification does Einstein assume that the world is, in effect, Minkowskian? The Lorentzian in particular will regard as gratuitous the assumption that we in fact inhabit a Minkowski spacetime, since phenomenally it is no different than a Lorentzian world. Because in section 1 of his paper Einstein has provided no justification for thinking that spacetime is Minkowskian, his operational definition of distant simultaneity is conventional.

Suppose the Minkowskian tries to turn the tables and claims that a Lorentzian, aether compensatory world can also be re-described as a Minkowski world, thereby removing any special conventionalist onus from relativity theory. The Lorentzian will explain, in Sklar's words, that

> In pre-relativistic spacetime there will be, relative to a given event, one and only one event at each spatial location that is not causally connectable to the given event, the one taken in the standard picture to be simultaneous with that given event. If we allow for the possibility of causal signals propagating from an event to any event in its standard future, any change in which distant events are taken to be simultaneous will demand that in the new description of the world at least some possible causal signals from future to past will have to be tolerated.[62]

In other words, the claim that a Lorentzian world is re-describable as a Minkowski world implies the possibility of backward causation! But whether one will tolerate retro-causation plausibly depends on whether one adopts an A- or a B-Theory of time.[63] Thus, Sklar, recognizing the implication of the possibility of retro-causation by a re-description of a Lorentzian world, muses,

> If we accept the kind of 'timeless' metaphysics familiar to us by now, thinking of all events as timelessly coexisting connected to one another by lawlike regularities,...it is not at all clear that we could not be perfectly happy, in the 'conventionalist' vein, with such an 'alternative representation' of the facts of a pre-relativistic world.[64]

Only if one presupposes a B-Theory of time, then, can the symmetry between Lorentzian and Minkowskian worlds with respect to re-description plausibly be maintained. This unexpected intrusion into the conventionality debate of one of the key metaphysical questions about time is highly significant, for SR is often put forward as the principal proof of the B-Theory of time, when in fact the non-gratuity

---

[61] Sklar, "Conventionality of Simultaneity," p. 531. Cf. R. H. Dicke, "Mach's Principle and Equivalence," in *Evidence for Gravitational Theories*, Proceedings of the International School of Physics "Enrico Fermi," Course xx, ed. C. Møller (New York: Academic Press, 1962), pp. 14-15, who imagines what it would have been like had Lorentz developed an electromagnetic gravitational theory prior to Einstein's geometric theory.

[62] Sklar, "Conventionality of Simultaneity," p. 532. Such retro-causal signaling results from the relativity of simultaneity: two causally connectable events e and e^* will be such that in one reference frame e is earlier than e^* whereas in a different reference frame e^* is earlier than e. SR avoids such retro-causal signals by making c the limit velocity for causal influences.

[63] See William Lane Craig, *Divine Foreknowledge and Human Freedom: The Coherence of Theism I: Omniscience*, Brill's Studies in Intellectual History 19 (Leiden: E. J. Brill, 1991), pp. 150-156. An A-Theory of time is one which posits the objective reality of tensed facts and temporal becoming. The distinction between past, present, and future is observer-independent. By contrast, a B-Theory of time regards all moments of time and events located at such moments as equally real and existent. Tense and temporal becoming are mind-dependent.

[64] Sklar, "Conventionality of Simultaneity," p. 533.

of its stipulated simultaneity relation in opposition to a Lorentzian, aether compensatory theory actually presupposes the B-Theory of time. The B-theorist appealing to SR, then, if he is to avoid a *principio principii* must, it seems, admit that Einstein's operational definitions of simultaneity are arbitrary and are so in a peculiar way not characteristic of Lorentzian theories.

Howard Stein would avert this conclusion by denying that Lorentz or anyone else ever succeeded in finding a viable, aether compensatory theory.[65] So there just is no alternative to Einstein's SR to worry about. But surely this is to misrepresent the situation. Lorentz's own electron theory, as a constructive theory, was scuttled by advances in quantum physics, but neo-Lorentzian theories positing a fundamental reference frame continue to enjoy support among a minority of physicists, as we shall see. Stein brushes these aside on the basis of analogies in an older context which no one would regard as posing "serious alternatives"—for example, one might posit an absolute center of the universe (pseudo-Aristotelian hypothesis) with no other modification of Newtonian physics. "The case of a 'pseudo-ether' hypothesis with a distinguished state of rest (without any other modification of SR)," he says, is in "precisely the same position."[66]

The degree to which Stein's sort of reasoning is persuasive will depend on the closeness of his analogy. The idea that there is no center of the universe is perspicuous within the context of either an infinite, Euclidean space or a curved, boundless space. Moreover, it is even difficult to conceive of how a finite, Euclidean space having an absolute center could exist, a question much discussed by medieval Aristotelian thinkers.[67] Thus, the idea that there is no absolute center of the universe is an intuitive and easily accepted notion. With respect to the idea that there is no privileged reference frame or metaphysical space, however, surely the opposite is the case. It takes a good deal of philosophical effort to conceive of purely relative spaces without a privileged background space, and it is not in any way evident why we ought to think that such a space or fundamental frame does not exist. As Sklar has elsewhere urged,

> The theories presented (Minkowski spacetime, curved spacetime) are novel and radical. Whereas the older theories they are to replace have a certain familiarity and built-in-ness as components of our ordinary way of looking at the world, the new...alternatives present us with a view of the world sufficiently startling that it take [*sic*] us a lot of effort and use of analogy to feel that we really even understand exactly what the picture of the world presented to us by the new theory amounts to.[68]

---

[65] Stein, "On Relativity Theory," p. 154.
[66] Ibid., p. 155.
[67] See Edward Grant, "Medieval and Seventeenth-Century Conceptions of an Infinite Void Space beyond the Cosmos," in *Studies in Medieval and Natural Philosophy* (London: Variorum Reprints, 1981), pp. 39-60; idem, (London: Variorum Reprints, 1981), pp. 39-60; idem, *Physical Science in the Middle Ages*, History of Science (New York: John Wiley & Sons, 1971), pp. 71-82. Newton's conception of an infinite, Euclidean space was anticipated by late medievals like Thomas Bradwardine.
[68] Lawrence Sklar, "Modestly Radical Empiricism," in *Observation, Experiment, and Hypothesis in Modern Physical Science*, ed. Peter Achinstein and Owen Hannaway, Studies from the Johns Hopkins Center for the History and Philosophy of Science (Cambridge, Mass.: MIT Press, 1985), p. 8. Cf. the intriguing remarks by the French physicist Henri Arzeliès: "the theory of relativity leads to profound modifications of certain everyday concepts; an example of this is the notion of absolute time which had

Sklar emphasizes that these absolutist theories are not like "brain in a vat" hypotheses which can be deemed as beyond plausibility. "The alternative accounts...not only are viable scientific hypotheses; they are the hypotheses all reasonable people in the scientific community did hold to..."[69] How, then, can such alternatives be summarily dismissed as not "serious"?

Moreover, Newton thought that there were good metaphysical grounds for holding to absolute space and time, as we shall see, which Stein fails to examine. He confesses, "I am here taking it for granted—perhaps illegitimately—that in the absence of a serious issue within physics, such [metaphysical] arguments against a theory like that of special relativity are futile."[70] But one cannot simply ignore the arguments; one must show them to be unsound. Stein's treatment of the question before us in effect confirms the arbitrariness of Einstein's assumptions regarding distant simultaneity in the face of absolutist theories; but the relativistic outlook has become so deeply ingrained today that some cannot even entertain such positions as genuine possibilities. In sum, while simultaneity relations established on the basis of Einstein's clock synchronization procedure are not conventional within the context of SR, nevertheless they are conventional in the sense that within the context of discovering the structure of space and time the assumption that clock synchronization via light signals discloses relations of simultaneity between relatively stationary observers is gratuitous. Since in section 1 of his paper Einstein cannot yet assume, without begging the question, that spacetime has the structure of Minokowski spacetime, his clock synchronization procedure to define relations of simultaneity within a reference frame is a convention, that is to say, an arbitrarily adopted operational definition.

RELATIVITY OF LENGTHS AND TIMES

Having excluded absolute space from his ontology and defined time in terms of the light signal method of synchronization of clocks, Einstein turns in section 2 of his paper to demonstrating the "Relativity of Lengths and Times." He begins by restating the Relativity Principle and the Light Principle:[71]

1. The laws by which the states of physical systems undergo changes are independent of whether these changes of state are

---

been accepted and employed by every physicist prior to Einstein," whose new ideas still seem "outlandish, unusual, and sometimes unintelligible to many people: even today some eminent physicists figure among these" (*Relativistic Kinematics*, rev. ed. [Oxford: Pergamon Press, 1966], p. 3). He observes that even some physicists who have taught relativity theory for many years (*e.g.*, Jánossy) finally rejected relativistic notions. Why is this? "The unintelligibility of Einstein's theory for certain people is, as I see it, a consequence of their having adopted *a priori* some picture of space and time, more or less unconsciously" (Ibid.). The import of these remarks is that the classical conception of time and space is our intuitive, pre-theoretical conception. It is relativistic *hubris* to assert that this conception is not even a serious alternative to a relativistic conception, which now seems to have taken the former's place as being assumed *a priori*.

[69] Sklar, "Empiricism," p. 9.
[70] Stein, "On Relativity Theory," p. 164.
[71] Einstein, "Electrodynamics of Moving Bodies," p. 395.

referred to one or the other of two coordinate systems moving relatively to each other in uniform translational motion.

2. Any ray of light moves in the 'resting' coordinate system with the definite velocity c, which is independent of whether the ray was emitted by a resting or by a moving body.

On the basis of these principles he will first show the relativity of lengths.

He invites us to consider a rod at rest in the inertial frame referred to in section 1 of his paper, its length being ℓ as measured by a measuring rod also at rest in this frame. We now set the rod in uniform motion along the x-axis of this frame. How then is the length of the moving rod to be determined? Einstein mentions two ways: (1) An observer moves along with the rod and measures it by directly superposing his measuring rod and finds it to have precisely the same length as if it were at rest. (2) Alternatively, the observer may remain at rest and measure the rod by marking in the resting system exactly where the front and back of the rod are at a certain time t, using his synchronized clocks; then at his leisure he may measure the distance between the two marked points in the resting system and so ascertain the length of the rod.

Remarking that classical kinematics tacitly assumes that the length measurements given by these two operations are precisely equal, Einstein declares that, on the contrary, the length of the rod as measured by method (1) will be ℓ, but that its length as measured by method (2) will differ from ℓ. In fact, the length of the moving rod as measured by method (2) will always be shorter by a factor of

$$\sqrt{1-(v^2/c^2)} \qquad (1)$$

exactly the same quantity posited by the Lorentz-FitzGerald contraction hypothesis.

The reason for this surprising result will be more evident if we proceed to explain the relativity of simultaneity. Einstein invites us to imagine that at each end of the rod are clocks which are synchronized with the clocks of the resting system before the rod is set in motion and that they *stay synchronized with the clocks of the resting system*. Moreover, at each end of the rod there is an observer who moves with it, and these observers carry out the light signal method of synchronization. The situation now becomes identical to the one we described above in connection with the conventionality of simultaneity. The observer at A sends out a light beam at t_A, which is reflected back from B at t_B and reaches the observer at A again at t_A'. Since the rod is in motion, the light beam traveled farther from A to B than from B to A (Figure 2.4).

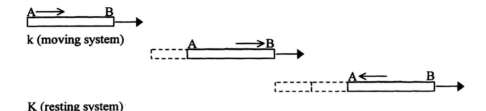

Figure 2.4. Light signal synchonization in a relative moving system.

Einstein puts it this way: letting r_{AB} denote the length of the rod in the resting system, we find that

$$t_B - t_A = \frac{r_{AB}}{c+v} \qquad (2)$$

whereas

$$t_A' - t_B = \frac{r_{AB}}{c-v} \qquad (3)$$

The return leg of the journey thus takes less time than the initial leg. The result of this procedure that if the clocks at A and B stayed in synchrony with the clocks of the resting system, then observers in the moving system associated with the rod would determine that the clocks at A and B are not synchronous after all, though the observer in the rest system would insist that they are. Relativity forbids this, however; otherwise the observer in motion could realize that he is in motion by noticing that his initially synchronized clocks have, after a lapse of some time, gotten out of synchronism. Thus, it must be the case that clocks which are synchronized in a moving system are not in synchronism with clocks in a resting system. The observers in the moving system must therefore regard *different* events as being simultaneous than do the observers in the resting system. Einstein concludes section 2 of his paper,

> Thus we see that we can attribute no *absolute* meaning to the concept of simultaneity, but that two events which examined from a coordinate system, are simultaneous, can no longer be interpreted as simultaneous events when examined from a system which is in motion relatively to that system.[72]

Perhaps the best way to see this is by means of Einstein's later famous illustration. Imagine a train speeding along at a substantial fraction of the speed of light. As an observer in the central compartment of the train passes an observer standing on the embankment, the readings on their watches happen to coincide; it is a case of local simultaneity. At that moment the observer on the ground sees two

---

[72] Ibid., p. 396.

flashes of lightning strike the train in front and in back, scorching the earth at those points. Later measuring the distance between the two points, the ground observer confirms that he was at the mid-point of the train. Using the light signal definition of simultaneity, he reasons that the lightning bolts struck simultaneously, since, in order to travel equal distances at a constant velocity and arrive simultaneously, the light flashes must have been emitted simultaneously. But the passenger in the central compartment was traveling in the direction of the front flash and so sees it first because it has a shorter distance to travel before it reaches him; then he sees the rear flash. Verifying that he is, indeed, at the mid-point of the train, he concludes that the lightning bolts did not strike the train simultaneously. In fact, if we imagine a third bolt to strike the train in the front after the earlier one, the passenger may conclude that it is this later bolt that struck the train simultaneously with the rear bolt because he, being midway between the front and rear, received the light flashes from them at the same time.

Now since we tend to think of the earth as a rest frame, we are apt to think that the ground observer was really correct and that the passenger was fooled due to his state of motion. But once again, we need to be reminded that in Einstein's ontology there is no absolute space or preferred frame of reference. The train could with equal justice be regarded as standing still and the earth spinning past beneath it. Without an absolute frame of reference, all uniform motion becomes relative. If we think of the train as at rest and the earth as in motion, then the passenger is correct that the train was struck first in the front by a single bolt and then simultaneously front and back by a pair of bolts. Neither the passenger nor the ground observer is wrong, since there is no absolute frame of reference from which to pronounce such a judgement. Indeed, they are both correct: the bolts were simultaneous relative to the ground system and non-simultaneous relative to the train system.

This same illustration can be developed to help us understand the relativity of length. Suppose the ground observer decides to measure the length of the moving train. He sets up a line of observers each with a stopwatch along the track and these watches are then synchronized by light signals. As the train roars by, at precisely 3:00, per prior arrangement, whichever observers find themselves directly opposite the front and rear of the train respectively punch their stopwatches. After the train has passed, these two hold their places, while the others measure the distance on the ground between them. This gives the length of the train. Now consider someone on board the train. Stationing himself in the middle of the train, he sends out light signals toward reflectors at the front and rear of the train. When these signals return simultaneously to him, he checks how much time has elapsed since their emission. He then computes the length ℓ of the train as

$$(t_A' - t_A) = \frac{2\ell}{c} \qquad (4)$$

But this length will be longer than the length computed by the ground observer, for from the ground observer's perspective each of the passenger's light signals traveled a longer distance than they would have were the train at rest. The signal toward the front end of the train traveled a longer distance toward the front reflector than it did back to the emitter, and the signal toward the back traveled a longer distance from the reflector back to the emitter than it did from the emitter to the reflector. These out-and-return distances are equal so that the signals return simultaneously, but they are longer than they would be if the train were at rest relative to the ground system. From the passenger's perspective, the ground observer's clocks are out of synchronism: at any given time t, not all of the clocks read t; they all show *different* times. That is because relative to the passenger, the ground observer and his clocks are in motion and so the light signal method of synchronization has, in effect, desynchronized them. No wonder that the ground observer measured the train to be too short! Of course, if the passenger is a well-informed relativist, he will know that the ground observer thinks the *passenger's* clocks are desynchronized and that neither man's opinion is superior. But, the ground observer might protest, if the passenger decided to use method (1) to measure the train, laying a measuring rod along it, would he not discover that the light signal method had given him a distortion of the real length, that the true length was that given by the rod, the same length which the rod measured before the train was set in motion? Not at all, for the length of the rod itself, like the length of the train, is a relative quantity. Therefore, the rod measurement will agree with the light signal method measurement. After all, in the train's inertial system, both these measurements are being carried out in a system whose components are at relative rest, so why would they disagree?

So how long is the train really? The question is malformed, given Einstein's metaphysic. Since there is no absolute reference frame, there is no intrinsic length. All lengths are relative to reference frames. We are prejudiced for the ground observer because we think of the earth as at rest, but as explained before, the train could with equal justification be considered at rest and the earth in motion. Thus, bizarre as it may seem, neither the train nor any physical object has an intrinsic property called "length." How long something is—or, indeed, how wide or how high—is a relational property which depends on which frame of reference one is using in carrying out one's measurements. And again, the relativity of length does not mean merely that objects are *measured* to have different lengths in different frames (that would be a Lorentzian understanding); rather in the absence of any privileged reference frame, objects *have* no absolute length and are therefore really longer or shorter in the various reference frames.

In sum, the key to understanding Einstein's Special Theory of Relativity lies in grasping its metaphysical presuppositions. The theory presupposes that no absolute space or privileged reference frame exists. For if such a space or frame existed, it would constitute a sufficient condition for the existence of absolute time as well and for relations of absolute simultaneity. Only in the absence of absolute space is there no conflict between the Principle of Relativity and the Light Postulate. From these two assumptions the invariance of the velocity of light follows, and by re-defining simultaneity in terms of light signal synchronization of clocks, Einstein was able to found relations of simultaneity which are relative to inertial frames. Finally, in the absence of absolute space or a privileged reference frame, length is defined by operations that imply its relativity as well.

CHAPTER 3

TIME DILATION AND LENGTH CONTRACTION

TIME DILATION

The relativity of simultaneity and the relativity of length lead naturally to the strangest consequences of relativity theory: time dilation and length contraction. Time dilation means that relative to a clock taken to be at rest, a moving clock runs slow, so that relative to the moving clock the amount of time recorded by the clock at rest expands or dilates. Let us suppose that we have two clocks A and B in motion relative to each other (Figure 3.1).

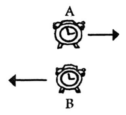

Figure 3.1. Two isolated clocks in relative motion synchronized at the point of coincidence.

When A and B come along side each other, let them have identical readings. Now we may arbitrarily take B to be at rest and A to be in motion, A passing by B from left to right. As we have seen, relativity theory predicts that relative to B, A's clock will gradually get out of synchronism with B, running slow relative to B. *Why* this happens in relativity theory is not clear; it is just a deduction from the postulates. It might be thought to occur as a result of the light signal synchronization method, but that this is not the case is evident from the fact that in our *Gedankenexperiment* nothing but the two clocks exists, so that there is nothing for each to be synchronous with. If each clock were equipped with lights, batteries, and reflectors, signal experiments could be run on each clock, and if this were done, it would be just as if the clock on which each experiment is done were at rest. In our *Gedankenexperiment* each clock simply runs on its own without later re-adjustment or synchronization with the other clock. So A's running slow relative to B is not a result of synchronization procedures to determine distant simultaneity.

What is strange about this phenomenon is that it is *reciprocal*. If we take A to be at rest (which we are free to do), then as B passes it from right to left and moves away into the distance, it will be B that runs slow relative to A. Each clock will, relative to its own inertial frame, record what is called its proper time, and a moving clock will run slow relative to the proper time of another clock by a factor of

$$\sqrt{1-v^2/c^2} \tag{1}$$

again precisely the factor Lorentz used to determine his "local time." This reciprocal time dilation is not self-contradictory because neither A nor B runs slow absolutely; rather A runs slow relative to B and B runs slow relative to A.[1] Such a cryptic sentence as the last is not in itself satisfactory, however. After all, if we were told that in the 100 meters, say, Johnson ran slow relative to Lewis and that Lewis ran slow relative to Johnson, we should certainly find this incomprehensible. In a similar way, we may feel compelled to ask which of the two clocks is really running slow relative to the other, A or B? For example, when each clock has traveled 100 light years from the point of their coincidence, what do their faces read and which one is lagging behind the other?

This is precisely the question so relentlessly pressed by Sir Herbert Dingle, a distinguished expositor of SR who came to reject that theory because he believed it to be self-contradictory.[2] Drawing a distinction between a "clock-reading at an event," which is an unalterable fact of nature, and the "time at which the clock has that reading," which is a convention which we may change freely by changing our coordinate system, Dingle invites us to consider two clocks A and B moving in opposite directions along a straight line at 161,000 miles per second.[3] Call A stationary and B moving. When B passes A, let both clocks read 12:00. When B reads 1:00 and then 2:00, A will read 2:00 and then 4:00. Therefore, B is running slow, since only one hour elapses between the readings, whereas for A two hours elapse. But if we had taken A's readings at 1:00 and then 2:00, B's readings would have been 2:00 and 4:00. But how then can each clock run slower than the other? His answer is blunt: "According to the special theory of relativity, two similar clocks, A and B, which are in uniform relative motion and in which no other differences exist of which the theory takes any account, work at different rates. The

---

[1] Albert Einstein, "On the Electrodynamics of Moving Bodies," trans. Arthur I. Miller, Appendix to Arthur I. Miller, *Albert Einstein's Special Theory of Relativity* (Reading, Mass.: Addison-Wesley, 1981), p. 396.

[2] Of Dingle's extensive writings on SR, see especially "Time in Relativity Theory: Measurement or Coordinate?" in *The Voices of Time*, 2d ed., ed. with a new Introduction by J. T. Fraser (Amherst, Mass.: University of Massachusetts Press, 1981), pp. 455-472 and *Science at the Crossroads* (London: Martin Brian & O'Keefe, 1972). See also his *Auseinandersetzung* with Max Born in Herbert Dingle, "Letters to the Editor: Physics: Special Theory of Relativity," *Nature* 195 (8 September 1962): 985-986; Max Born, "Letters to the Editor: Physics: Special Theory of Relativity," *Nature* 197 (30 March 1963): 1287; Herbert Dingle, "Special Theory of Relativity," *Nature* 197 (30 March 1963): 1248-1249, and with Epstein in Paul S. Epstein, "The Time Concept in Restricted Relativity," *American Journal of Physics* 10 (1942): 1-6; Herbert Dingle, "The Time Concept in Restricted Relativity," *American Journal of Physics* 10 (1942): 203-205; Paul S. Epstein, "The Time Concept in Restricted Relativity—A Rejoinder," *American Journal of Physics* 10 (1942): 205-208; L. Infeld, "Clocks, Rigid Bodies, and Relativity Theory," *American Journal of Physics* 11 (1943): 219-222. For a similar argument see G. Burniston Brown, "What Is Wrong with Relativity?" *Bulletin of the Institute of Physics and the Physical Society* 18 (1967): 71-77, and the reply in "News and Comment," *Physics Bulletin* 19 (1968): 22.

[3] Dingle, "Time in Relativity Theory," p. 465; idem, *Science at the Crossroads*, p. 11.

situation is therefore entirely symmetrical, from which it follows that if A works faster than B, B must work faster than A. Since this is impossible, the theory must be false."[4]

The fallacy of Dingle's objection underscores once more the novelty and subtlety of Einstein's metaphysical world view.[5] The simple word "when" in the clauses, "When each clock has traveled 100 light years..." or even "When B reads 1:00 and then 2:00..." betrays that the inquirer is implicitly assuming an absolute or preferred reference frame relative to which those questions can be asked.[6] To ask, "When each clock has traveled 100 light years, what do their faces read?" assumes a third reference frame in which A and B can be compared and relative to which the question can be answered. It assumes a sort of "God's eye point of view" of the two mutually receding clocks until they reach a separation of 200 light years. But given the problem conditions, no such point of view exists, according to the Special Theory.[7] We could imagine a sort of third reference frame containing the two moving clocks and ask what readings the clocks show when, relative to that frame, they are 200 light years apart. The answer would be that relative to this third frame the two clocks read the same and that both of them are running slow relative to an imaginary clock at rest in the third frame. But this third frame enjoys no privileged status and is purely imaginary in any case.

Similarly, when we say "When B reads 1:00, A reads 2:00; and when A reads 1:00, B reads 2:00," this "when" assumes a common time at which the readings of A

---

[4] Dingle, *Science at the Crossroads*, p. 45.
[5] Often the issue is blurred by reinterpreting Dingle's objection to involve a plurality of A and B clocks (*e.g.*, L. Marder, *Time and the Space-Traveller* [London: George Allen & Unwin, 1971], p. 57; Lewis Carroll Epstein, *Relativity Visualized* [San Francisco: Insight Press, 1981] p. 99), which it does not, thereby concealing the radicalness of the solution.
[6] This is also the fallacy in the defense of absolutism given by J. L. Mackie, "Three Steps toward Absolutism," in *Space, Time, and Causality*, ed. Richard Swinburne, Synthèse Library 157 (Dordrecht: D. Reidel, 1983), pp. 16-20. Mackie invites us to imagine two photons emitted along paths A and B in opposite directions from the point of origin O. Pick an arbitrary point C on path A; there will be a point D on path B corresponding to C. "I do not claim," avers Mackie, "that there is any way of *determining* the point that corresponds to C; I say only that there is one. And nothing but the sort of extreme verificationism which I mentioned, but set aside...would rule out this claim as meaningless" (Ibid., pp. 18-19). At C and D photons sent to the right and left respectively will meet at a point G. By choosing a series of points $C, C_1, C_2,...$ and $D, D_1, D_2,...$, we construct the world line $G, G_1, G_2,...$ of an object which is at absolute rest. Once Jon Dorling, "Reply to Mackie," in *Space, Time, and Causality*, p. 27, concedes that there is a unique point D correlated with C, the game is over. His attempt to elude Mackie's conclusion by saying that "there is a preferred frame of reference associated with that burst of radiation ....But it is not an absolute frame, because it will be different for different bursts of radiation" is incoherent. Rather the problem is that no unique point D exists. Which point on B is correlated with C will vary with reference frames. For further discussion, see J. R. Lucas and P. E. Hodgson, *Spacetime and Electromagnetism* (Oxford: Clarendon Press, 1990), pp. 106-108. It is also worth adding that the same verificationism which Mackie rejects actually underlies the assumption, which he accepts, that the light cone structure is invariant, for this is not the case for aether compensatory theories.
[7] For a clear statement of this point, see Alfred Schild, "The Clock Paradox in Relativity Theory," *American Mathematical Monthly* 66 (1959): 10-11, who substitutes two twins separated at birth for the two clocks: "The two twins can be together once, at birth, and there can compare their ages directly. But they can never do so again in the future. Thus, the whole problem as to which twin is younger and which twin is older disappears completely. When the two twins are flying apart it is a physically meaningless statement to say that one is younger or older than the other."

50   CHAPTER 3

and B exist. It is to assume that there is some point in time at which these contradictory readings take place. Otherwise, there will be no contradiction. If the "when" in "when B reads 1:00" refers only to B time (and analogously for A), then that only repeats that according to B's time A is running slow and that according to A's time B is running slow. But according to SR, there is not any moment when the clocks mutually lag behind each other; indeed, there is no privileged, common moment at which they both exist. This may seem counter-intuitive, but it is the conclusion that follows from Einstein's denial of an absolute reference frame and absolute simultaneity. Unlike Lewis and Johnson, there is no common track on which the two clocks run, so as to be compared at a distance, except in a purely arbitrary way. Once one denies an absolute reference frame, the metaphysical glue holding the world together comes undone, and the world falls into a multitude of only relatively related reference frames.

The peculiarity of the reciprocity of time dilation can be seen in another illustration.[8] Imagine three spaceships traveling in formation across the solar system at

$$\tfrac{1}{2}\sqrt{3}c \qquad (2)$$

(Figure 3.2).

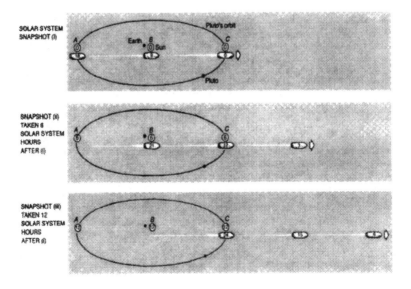

Figure 3.2. Three solar system snapshots of the crossing of a space-ship convoy, showing clock readings.

---

[8]   See Marder, *Time and the Space-Traveller*, pp. 58-60.

TIME DILATION AND LENGTH CONTRACTION

The three spaceships are equally spaced at about 5,600,000,000 km apart and move along the diameter of Pluto's orbit, passing near the sun at intervals of about six hours in solar system time. Each ship takes about 12 hours in solar system time to make the crossing. Figure 3.2 shows the crossing at six hour intervals, with solar system time indicated by three clocks at the center and circumference of the solar system and the formation time indicated by three synchronized clocks each carried by one of the crafts. In frame (1) notice that the clock in the rear ship is 18 hours ahead of the solar system clock at A. But in frame (2) it is only 15 hours ahead of the solar system clock at B. Finally, in frame (3) it is only 12 hours ahead of the solar system clock at C. According to solar system time, then, the spaceship's clock is running slow. But now notice that in frame (1) the solar system clock at C agrees with the clock in the lead spaceship. In frame (2) this same clock at C is six hours behind the spaceship clock nearest it. Finally, in (3) the clock at C is 12 hours slow relative to the nearest spaceship's clock. In this sense, the solar system's clocks are also running slow relatively to the formation's clocks!

Now it might be protested that the tardiness of the solar system's clocks is a sleight of hand, resulting from the fact that the formation's clocks are not synchronized. For they all have different readings! Compared to the clock of just one craft, it is evident that the formation's clocks are going half as fast as the solar system's.

But the problem with such an objection is that, odd as it may seem, the formation's clocks *are* synchronized, according to Einstein's prescription. Light signals from one craft to another will travel different distances relative to the solar system's rest frame, so they appear to the solar system observer to be out of synchronism. If we imagine the formation commander to be situated in the center vessel, then relative to the solar system his radio commands to the point and rear ships will arrive at the rear ship first and the point ship second. If he issues a command to fire their laser weapons, the rear ship will begin firing first. But in the formation's reference frame all begin firing simultaneously. Nor, according to Einstein's theory, is this merely a matter of perspective: in the one reference frame, they *do* all begin firing simultaneously, and in the other the rear ship begins firing *first*. The discrepancy is due to the fact that there is no absolute frame of space in which the relativity is resolved. We tend to favor the solar system frame because it seems to be at rest and its clocks are all set at zero in the illustration. But the solar system is not a preferred frame; and the setting of the clocks reflects the solar system's perspective. As Resnick observes, "Although clocks in a moving frame all appear to go at the same slow rate when observed from a stationary frame with respect to which the clocks move, the moving clocks appear to differ from one another in their readings by a phase constant which depends on their location, that is, they appear to be unsynchronized."[9] The further away a clock in the moving

---

[9] Robert Resnick, *Introduction to Special Relativity* (New York: John Wiley & Sons, 1968), p. 64. See the intriguing illustration of L. Epstein, *Relativity Visualized*, pp. 71-72 (Fig. a).

frame is from a clock in the rest frame, the further behind its reading is. Thus, relative to the formation's inertial frame the solar system's clocks do not, in fact, all read the same, and since that system can be considered to be moving relative to the formation, rather than the other way around, Fig. 3.2 could have been drawn with the formation's clocks all in agreement and the solar system's unsynchronized. That is because different events are simultaneous in each frame, and there is no means in Einstein's theory of resolving that relativity.

Undoubtedly, however, the time dilation effect which is most startling and which has caused the most controversy is the so-called "Twin Paradox."[10] Imagine two twins, one of whom remains on earth, while the other embarks on a voyage into outer space at near light speed. After traveling for a few years, he reverses course and returns to earth. Upon disembarking, he discovers that while he has aged only a few years, his twin who remained on earth is now a tottering old man! Relativity theory predicts that if such a voyage were feasible, this result would actually ensue. It seems scandalous because we have here an absolute effect which emerges as a result of only relative motion.[11] Einstein himself was troubled at this prospect and

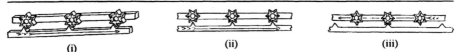

(i)   (ii)   (iii)

Figure a. (i) Clock consisting of gears turned by teeth on a sliding rod. (ii) From the reference frame of gears, the sliding rod is Lorentz-contracted, so that if the gears are to turn synchronously, the teeth will have to be spaced further apart than the gears. (iii) But then, in the reference frame of the rod, it is the gears which are Lorentz-contracted, and so they do not turn synchronously. (See p. 35, n. 54)

Imagine a series of clocks linked together and turned by teeth on a sliding rod. The clock observer will space the teeth further apart than the clocks in order that the Lorentz-contracted rod will turn the clocks in unison; but then to a rod-observer the clocks become desynchronized. If one demands whether the clocks are in unison after six teeth slide by, one has committed Dingle's fallacy, for, odd as it may seem, there is according to SR no absolute moment "after" the six teeth pass by: the clock and rod observers will not agree as to when the teeth have turned the clocks fully around.

[10] See the comprehensive review, to which I am much indebted, by Marder, *Time and the Space Traveller*.
[11] See S. J. Prokhovnik, "The Twin Paradoxes of Special Relativity—Their Resolution and Implications," *Foundations of Physics* (Preprint). As Sherwin explains,

"Supposing, for convenience, that the acceleration takes place in a very small interval and that the clock is unchanged by the acceleration process *per se*, it is clear that essentially all of the phase difference [between the two clocks] is accumulated...during the constant velocity regions of the path. Since this effect is observable without dependence either on the propagation properties of light, or upon any measurement operations using meter sticks, it cannot be dismissed as being an 'apparent effect' having to do somehow with the processes of determining what happens at distant points. One is led therefore to the conclusion that clocks having a velocity in an inertial frame are literally slowed down *by the speed itself.* It is this very deduction which makes the generally accepted prediction regarding the 'clock paradox' unacceptable to Dingle,[22] but which has led both Ives[12] and Builder[21] to consider interpretations of special relativity in which an ether plays an important role at least from the philosophical point of view."

[12] H. E. Ives, J. Opt. Sec. Am. *27*, 305 (1937).
[21] G. Builder, Australian J. Phys. *11*, 279 (1958).

inclined toward the view that the paradox must finally be solved in the General Theory of Relativity by postulating a dynamical cause for the absolute effect.[12] In all other cases we have examined, the effects of relative motion are reciprocal; for example, each of two clocks in relative motion runs slow relative to the reference frame of the other. But now an effect emerges which is not reciprocal: asymmetrical aging. How is this to be accounted for?

It will be helpful to introduce at this point a spacetime diagram on which we may plot each twin's course (Figure 3.3). The vertical axis represents time and the horizontal axis the three dimensions of space.

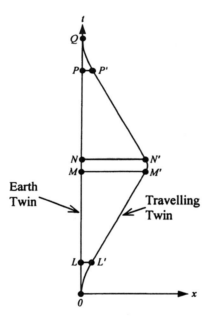

Figure 3.3. The Twin Paradox

Since the earth-bound twin does not leave earth, we shall consider him at rest; therefore his course does not take him away from zero in the direction of the *x* axis, but he simply endures through time. The travelling twin accelerates away from the earth from zero to L', then proceeds at uniform velocity from L' to M', whereupon he decelerates, makes the turn, and reaccelerates back toward earth during the segment M' to N'. From N' to P' he travels at uniform velocity, whereupon he decelerates

---

[22] H. Dingle, *Nature*, *179*, 866 (1957)" (C. W. Sherwin, "Some Recent Experimental Tests of the 'Clock Paradox'," *Physical Review* 120 [1960]: 17-21).

[12] A. Einstein, "Dialog über Einwände gegen die Relativitätstheorie," *Naturwissenschaften* (1918): 697-702. Unfortunately, general relativistic treatments only complicate the problem and often lack physical significance, as shown by Geoffery Builder, "The Resolution of the Clock Paradox," *Australian Journal of Physics* 10 (1957): 246-262.

from P' to Q, to bring him safely back to earth again. In the earth-bound twin's reference frame, the events simultaneous on earth with the above mentioned events on the spaceship are L, M, N, P, Q respectively.

But since time dilation is a reciprocal phenomenon, it seems that we could with equal justice regard the travelling twin as at rest and the earth-bound twin as in motion. In that case, it should be the earth twin who is older. Since during the periods of uniform motion from L to M and N to P the earth twin's clocks run slow relative to the travelling twin's clocks, just as during the periods L' to M' and N' to P' the travelling twin's clocks run slow relative to the earth twin's, should not these effects cancel each other out and the twins upon reunion be the same age?

The periods of acceleration and deceleration cannot be neglected, however. Since the Special Theory covers only uniform motion, the presence of these periods in the travelling twin's world-line reveals that it is, indeed, he who is in motion and not the earth-bound twin. Therefore, his clock is asymmetrically retarded over the whole journey.

It might be said that the periods of acceleration can be made inconsequential by expanding the distances $L'M'$ and $N'P'$ to arbitrarily large lengths, so that the periods of acceleration shrink toward infinitesimal proportions. The reciprocal retardation that occurs during the periods of uniform motion would therefore seem to be the dominant factor. But this response fails to appreciate why the periods of non-uniform motion are so important, regardless of their brevity. Their significance lies, not in their somehow undoing the reciprocal time dilation that occurs during the periods of uniform motion,[13] but rather in their revealing that the space-travelling twin's world-line involves not one, but two inertial frames: one for the outward and one for the return leg of the journey. No matter how comparatively short the bends in his world-line may be, they serve to re-orient the whole direction of his journey, and we shall see how this implies that it will be his clock alone that is retarded.

A useful version of the paradox, due to Lord Halsbury, circumvents the issue of acceleration altogether by turning the paradox of the twins into the paradox of the three brothers (Figure 3.4).

---

[13] As implied by Ray d'Inverno, *Introducing Einstein's Relativity* (Oxford: Clarendon, 1992), p. 38: "The resolution rests on the fact that the accelerations, however brief, have immediate and finite effects on B but not on A who remains inertial throughout."

TIME DILATION AND LENGTH CONTRACTION

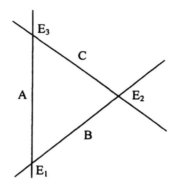

Figure 3.4. Paradox of the three brothers.

Instead of having the space-travelling twin take off from earth, reverse course, and return, we instead suppose, first, that brother A is at rest in an inertial system and that brother B passes him at a uniform velocity at E_1, their clocks agreeing at that moment. After some time, brother C, who is approaching earth at a uniform velocity, passes B at E_2, and at that moment B and C's clocks agree. C continues on his journey and eventually passes A. Do their clocks agree at E_3? Since there are no accelerations at all in this version, must not the time dilations be fully reciprocal and therefore A and C's clocks agree?

Remarkably, relativity theory predicts asymmetrical aging even for this scenario, and the explanation of why this is so will enable us to understand the role of the accelerations in the previous version of the paradox. At the same time, it will reinforce again the queerness of the Einsteinian universe.

The reason the acceleration and deceleration periods, particularly the turn-around period $M'N'$ in Fig. 3.3, are crucial is not because during these times the space traveller's clocks are affected by his non-uniform motion and go whizzing ahead at tremendous speeds to make up for the time lost during the periods of uniform motion; rather it is because during this turn-around time, there is a shift of inertial frames and, consequently, *new relations of simultaneity* established with the earth-bound twin. This element of the story is not eliminated in the version of the three brothers. In the twin version of the paradox, the amount of earth time that elapses during the turn-around period is greater (dilated) in the spaceship's frame of reference than in the earth's own frame of reference (Figure 3.5).

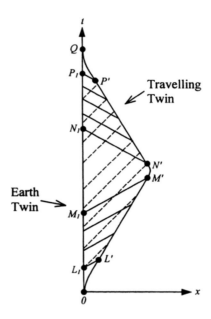

Figure 3.5. Lines of simultaneity between the two twins. Broken lines are calculated by the Earth twin, solid lines by the travelling twin. To the traveller the time of the earth twin dilates between the traveller's M' and N'.

In this figure, the travelling twin draws the lines of simultaneity between earth and spaceship events differently than did his brother. Because his journey involves two inertial frames, there appears a large stretch of earth time that falls between the cracks, so to speak, of the sets of simultaneity lines belonging to his two inertial frames. There will literally be moments of earth time to which no moments of his own time correspond, moments that will be "now" for the earth twin but which are missing from the travelling twin's time. Hence, the earth twin ages more, for he has actually lived through more time than the travelling twin.

The discontinuity in simultaneity relations should not be interpreted as a discontinuity in the travelling twin's observations of events between M_1 and N_1. If the earth-bound twin were sending out periodic light signals, then the travelling twin would receive all of them between O and Q, red-shifted and less frequently during his outbound trip and then blue-shifted and more frequently during his in-bound trip. As Schild emphasizes, there is no physical effect corresponding to the discontinuous jump in time.[14] Rather at the turn-around point, the travelling twin suddenly changes his method of mathematical calculation of simultaneity relations. The only

---

[14] Schild, "Clock Paradox," p. 14; see also Yakov P. Terletskii, *Paradoxes in the Theory of Relativity*, with a Foreword by Banesh Hoffmann (New York: Plenum Press, 1968), p. 41; Lucas and Hodgson, *Spacetime and Electromagnetism*, p. 76.

discontinuity lies in his exchanging one method of calculation for another. The travelling twin observes all the signals sent from earth between M_1 and N_1; but his first method of calculation determined that those events occurred later than his turn-around point, whereas his second method classed them as having occurred earlier than his turn-around. By changing his methods of calculating simultaneity, the traveling twin loses the moments of time simultaneous to the moments between M_1 and N_1, but he does not lose the events which occurred during that time. In effect, those events have moved from being future to being past without ever having been present for the travelling twin.

Indeed, seen in this light, what relativity theory really permits is time travel into the future. The sort of time machine envisioned by H. G. Wells is impossible because it was conceived to move only in time and not also in space and thus involved the contradiction of traveling through, say, 500 years in five minutes' time. But by having the traveller move in space as well as time, relativity theory permits the traveller to live through a proper time which is vastly shorter than the elapsed earth time. This is not due to the fact that he lives through the same time at a slower rate (a Lorentzian view), but because he quite literally lives through a different and shorter time. Thus, the twin paradox goes to illustrate what Bondi has called "the route-dependence of time": the time elapsed will depend on the way one travels from one event to another.[15]

Some attempts have been made to introduce the notion of accelerated reference frames into special relativity, which would enable us to draw in the travelling twin's lines of simultaneity during the turn-around period.[16] One such method, described by Marder, allows one to find for any event on the travelling twin's world-line at least one corresponding simultaneous event on the world-line of the earth-bound twin. Figure 3.6 displays how these lines would be drawn.

---

[15] Hermann Bondi, *Relativity and Common Sense* (New York: Dover Publications, 1964), p. 87; cf. H. Bondi, "The Space Traveller's Youth," *Discovery* 18 (December 1957): 505-510. Davies also interprets the Twin Paradox as a sort of uni-directional time travel (P. C. W. Davies, *Space and Time in the Modern Universe* [Cambridge: Cambridge University Press, 1977], p. 39). Similarly, Fritz Rohrlich, *From Paradox to Reality* (Cambridge: Cambridge University Press, 1987), p. 70: "What is wrong is our concept of time: it is neither absolute (as Newton thought) nor universal (as we as wont to believe). Time *depends on the history of the traveler through space*." Rindler, on the other hand, interprets the situation in a significantly different way. During the twin's initial acceleration from zero to L', he transfers himself into an inertial frame in which the distance to the turn-around point undergoes such tremendous length contraction that more than half his journey is already completed by the time L' is reached. Therefore, he accomplishes his trip in less time (Wolfgang Rindler, *Introduction to Special Relativity* [Oxford: Clarendon Press, 1982], p. 35). See also Capek's struggle to explain how the two times are in some sense contemporaneous (Milic Capek, *The Philosophical Impact of Contemporary Physics* [Princeton: D. Van Nostrand, 1961], pp. 206-212).

[16] An accelerated reference frame for a given observer O is a scheme of labeling of all events with four coordinates (three spatial and one temporal) such that O's space coordinates are (0,0,0) and that all objects with constant space coordinates are (in some sense) at a fixed distance from O.

58   CHAPTER 3

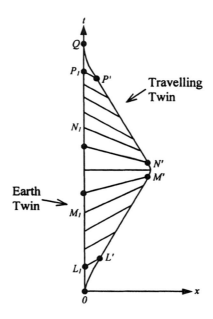

Figure 3.6. Simultaneity lines in accelerated systems for the travelling twin.

But there are still events or moments in the life of the earth twin to which there correspond no simultaneous event or moment in the life of the travelling twin. Therefore, the stay-at-home twin lives through more time than his brother and is thus older than him when they are reunited at Q.

If we take the case of the three brothers, there is no period of deceleration and acceleration, so that accelerated reference frames are superfluous. The shift in lines of simultaneity from B's inertial frame to C's takes place in an instant. All the time in A's world line which lies in between the two sets of simultaneity lines is dilated time which is just not part of B or C's collective time.

Bondi has suggested another means of demonstrating that A's clock will be slower than C's at E_3.[17] Let B emit light flashes every six minutes by his clock toward A, beginning with the moment he passes A and ending when he meets C. If he travels one hour between these meetings, he will thus emit ten flashes. Since B is receding from A at a high velocity, A will receive these signals at intervals greater than six minutes, say, nine minutes, because each successive signal has further to travel. If we suppose the first signal to have been sent—and received, since A and B were locally coincident—at 12:00, then the last signal emitted by B will be received 90 minutes later by A at 1:30. We now suppose C to begin sending light flashes at

---

[17] Bondi, *Relativity and Common Sense*, pp. 80-82. Notice that if bullets were used instead of light signals, relativity would not result because B's bullets would travel slower than C's due to addition of velocities; but because light velocity is invariant, relativity obtains.

six minute intervals, beginning at the instant of his rendezvous with B and ending when he passes A. Since C is approaching A, the latter will receive the signals every four minutes. The first of these will come simultaneously with B's last signal at 1:30 and the last one 40 minutes later at C's meeting A. Therefore, the meeting of A and C will occur at 2:10 by A's clock, but at only 2:00 by C's clock.

What is interesting about the three brothers' case is that since the periods of acceleration have been eliminated, any one of the three can be considered to be at rest and the other two in motion. In each case, clock retardation for the one who is at rest results.[18] Thus, if B is conceived to be at rest and A and C in motion, then A and C pass by B at different times and at identical velocities, so that relative to B both A and C's clocks are running equally slow. On the other hand, if it is C that is at rest, then A and B passed each other at some distance from C at which moment A and B were mutually synchronized but neither synchronized with C, whose clock read an earlier time. B, running slow relative to C, comes to agree with C by the time they meet, and A is lagging behind C by the time they meet. Thus, in each case A and B agree, B and C agree, and A is slow relative to C. Which brother really is at rest is a question which cannot be posed in the context of the Special Theory, since it was precisely to eliminate such asymmetries in cases of uniform motion that Einstein excluded any privileged rest frame.

LENGTH CONTRACTION

The relativity of length leads to an equally interesting phenomenon: *length contraction*. The theory predicts that relative to a frame at rest, a moving object contracts by a factor of

$$\sqrt{1 - v^2/c^2} \qquad (3)$$

Like time dilation, however, this phenomenon is reciprocal. Given two objects in relative motion, each is contracted relative to the other. This reciprocity results because, as we have seen, length is a relative quantity on Einstein's ontology: there is no absolute space in which a thing has a true length. The length of an object in its rest frame is called its proper length and is the maximum length of an object.

Length contraction also leads to some very peculiar situations. Consider once more our formation of three spaceships with the commander's vessel in the center. Suppose the commander gives an order for acceleration. To an observer in the solar system, the command signal reaches the rear ship first and thus it accelerates before the lead ship. During the time lag between the rear ship's acceleration and the lead ship's acceleration, the ships will pull closer to one another. But to the vessels' crews, the signals were received simultaneously, and so the distances separating the three spaceships do not decrease. Now suppose that another acceleration is ordered.

---

[18] Peter Kroes, *Time: Its Structure and Role in Physical Theories*, Synthèse Library 179 (Dordrecht: D. Reidel, 1985), p. 47; Waldyr A. Rodrigues, Jr. and Marcio A. F. Rosa, "The Meaning of Time in the Theory of Relativity and 'Einstein's Later View of the Twin Paradox'," *Foundations of Physics* 19 (1989): 705-724.

Again the rear ship gets the order first, accelerates first, and creeps closer to the lead ship. But to the crews, the vessels all remain equally distant. Suppose yet another acceleration is ordered. How can the crews deny that the rear ship is creeping perilously closer? As Epstein exclaims, "If the ships got close enough to each other, the nose cones would cram into the engine bells, so if the fools didn't see what was going on, they would certainly feel it!"[19]

The solution to the problem lies in the phenomenon of length contraction. As the distance between the ships closes, the ships undergo an exactly proportionate and compensating length contraction. Therefore, the gaps between the vessels remain proportionately the same. Of course, relative to the crews, no such contraction occurs, since the orders are received and executed simultaneously.

Or consider the various scenarios concerning two similar spaceships, A and B, passing each other in opposite directions in space.[20] For a warm-up, consider their respective attempts to measure each other. Since they have the same proper length, B relative to A will be shorter than A; but A relative to B will be shorter than B. Of course, if the ships docked together, then the length of A would equal the length of B. If they want to measure their relative velocity as they pass each other, each could either (1) time how long it takes the nose of his own ship to pass the length of the other ship, or (2) time how long it takes the nose of the other ship to pass from the front to the tail of his own ship. Unfortunately, these two methods do not agree in relativity theory. For method (1) is affected by the length contraction of the other craft, so that the nose of one's own ship passes it more quickly than in the absence of this phenomenon. But length contraction of the other spaceship is irrelevant to method (2), since one is measuring the time it takes a point on the other vessel to pass by one's own uncontracted ship. Suppose then that the commander of A uses method (1) to measure their relative velocity. The corresponding measurement of this quantity would be obtained by the commander of B using method (2). But since method (1) yields a smaller quantity than method (2), this would seem to imply that the commanders do not agree on their ships' relative velocity—which violates relativity. The situation is saved by time dilation—relative to A, B's clock is running slow by exactly the same factor that B is contracted relative to A, so that the commander of B measures the time needed to pass A as the same time which A measured it took to pass B.

Let us embellish the scenario by supposing that B has a tail gun pointing perpendicular to the line of motion (Figure 3.7[a]).

---

[19] Epstein, *Relativity Visualized*, p. 39.
[20] I borrow these examples from Marder, *Time and the Space-Traveller*, pp. 145-163; Edwin F. Taylor and John Archibald Wheeler, *Spacetime Physics* (San Francisco: W. H. Freeman, 1966), p. 18 of "Answers to the Exercises of Chapter I."

Time Dilation and Length Contraction 61

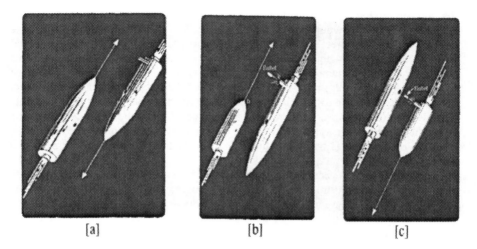

Figure 3.7. Illustration necessitating the conspiration of length contraction and time dilation to avoid contradiction. [a] Two rocket ships A and B passing at high speed. [b] In the frame of B one expects a bullet fired when the nose of B passes the tail of A to miss the other ship. [c] In the frame of A one expects a bullet fired when the nose of B passes the tail of A to hit the other ship.

Suppose the commander fires the gun when the nose of B passes the tail of A. In B's reference frame, A is contracted, so the bullet should miss. But in A's frame of reference, B is contracted (Figure 3.7[b]). Therefore, A expects to be hit. How is this discrepancy to be resolved?

Time dilation again saves the day. It is only in B's frame that the firing of the gun and the coincidence of B's nose with A's tail are simultaneous. In A's frame of reference the gun will have already been fired before B's nose passes A's tail. Therefore, A is not struck by B's bullet.

One interesting sidelight on such scenarios: although discussions of relativity theory often involve expressions like "A observes B to be contracted" or even "A sees B contracted," such expressions must not be taken literally. Early expositors of the theory—perhaps including Einstein—were under the impression that an object moving at relativistic speeds would appear visually to an observer at relative rest as squashed in the direction of motion, a sphere, for example, appearing as an ellipsoid and finally, if moving fast enough, as a pancake.[21] It was not until 50 years later that Terrell demonstrated that the Lorentz contraction is invisible.[22] This does not imply

---

[21] See, for example, Arthur Eddington, *Space, Time and Gravitation*, Cambridge Science Classics (1920; rep. ed.: Cambridge, Cambridge University Press, 1987), pp. 22-23.
[22] James Terrell, "Invisibility of the Lorentz Contraction," *Physical Review* 116 (1959): 1041-1045; cf. V. F. Weisskopf. "The Visual Appearance of Rapidly Moving Objects," *Physics Today* 13 (September 1960), pp. 24-27. The key difference between *seeing* an object and *observing* it via measurement is that

that the contraction is a fiction of measurement, for it is precisely *because* the object is contracted that one would see it as Terrell describes. What one sees visually is the *back* or rear of the object (as though it were rotated toward the observer) conjoined with the contracted length of the object, so that the overall length is the same. This peculiar visual phenomenon results from the fact that the light rays from the back of the object (which would normally be blocked off from the observer by the object itself) are free to come to the observer because the object has moved on so quickly that their path to the observer is no longer blocked. Hence, the Lorentz contraction is invisible. When we speak of *A observing* some event in another inertial frame, what we really mean is that he mathematically calculates the event to be such and such.

Coming down to earth, let us consider some fanciful terrestrial thought experiments which illustrate relativistic length contraction. Take, for example, the case of the pole vaulter and the shed. A vaulter running at near light speed must run through a shed with doors at each end on the way to his vault. Suppose the shed's proper length is ten meters and the pole's proper length is twenty meters. If he runs fast enough, however, the runner's pole will, relative to the reference frame of the shed, be contracted to half its length and so will fit inside the doors of the shed. But by the same token, in the runner's reference frame, the pole will be uncontracted and instead the shed shrunk down to five meters, in which case the runner and his pole cannot be contained inside the doors of the shed. Is he then in the shed or not?

Or, analogously, consider once again Einstein's train. As the train approaches a tunnel at near light speed, ground observers stationed at either end of the tunnel decide to measure whether the train fits into the tunnel or not. Suppose they measure the tunnel to be 100 meters long, and an observer on the train measures the train to be 110 meters in length. When the back end of the train enters the tunnel, the ground observer there flashes a signal to the observer at the tunnel's other end. If he receives this signal before the front of the train emerges from the tunnel, then he knows that the tunnel contained the entire train. Due to the Lorentz contraction, the train shrank to fit in the tunnel. On the other hand, from the passenger's standpoint, the tunnel is contracted to 90 meters, and therefore the uncontracted train will stick out both ends of the tunnel. How can both be true?

Because in Einstein's theory there is no absolute frame of reference, there is no absolute (that is, non-relativistic) truth about whether the pole is in the shed or the train is contained in the tunnel. Those facts are relative. To take the pole vaulter's case, in the frame of reference of the shed, the runner and his pole do fit inside both doors of the shed. But in the runner's frame of reference, the front end of the pole exits the back door of the shed before the rear end of the pole enters the front door (Figure 3.8).

---

the former involves the simultaneous *reception* of light quanta from the object whereas the latter involves simultaneous *emission* of light quanta from it.

Time Dilation and Length Contraction

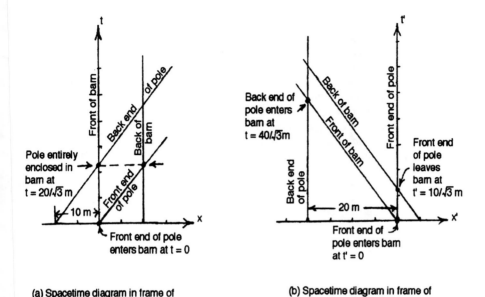

Figure 3.8. The pole vaulter and the shed.

Similarly, in the case of the train, the train does fit into the tunnel in the ground observer's inertial frame, but in the frame of reference of the passenger it does not.

But what, we might ask, would happen if, say, the doors of the shed were to lock if ever the runner were inside or, again, if a landslide would seal the tunnel if the train were contained in it? Would not these experiments reveal the absolute truth of the situation? No, for time dilation comes into play again. In the shed's frame, the doors would close simultaneously and the runner would then burst through the back door with his pole. In the runner's frame, however, he would crash through the back door first and then the front door would close behind him after the rear of the pole had come through. Similarly, from the ground observer's perspective, the landslide would seal off the front and back of the tunnel simultaneously and the train would be smashed within the tunnel against the rocks. But from the passenger's point of view, the landslide at the far end of the tunnel occurs first, the train smashes into it, squashing up like an accordion under the force of its momentum, and then the rear slide seals off the end of the tunnel. The point is that there is no privileged frame of reference according to which one can say that one description constitutes the absolute truth.[23]

---

[23] There does not seem to be any absolute effect due to length contraction as there is due to time dilation in the twin paradox. Although Kroes attempts to find such an effect (Peter Kroes, "The Physical Status of Time Dilation within the Special Theory of Relativity," paper presented at the International

64   CHAPTER 3

There is perhaps a natural tendency to regard time dilation and length contraction, not as facts about time and space themselves, but merely about our *measurements* of time and space,[24] and thus many writers feel constrained to emphasize to their readers that these phenomena are not just appearances, but real. But what such authors often fail to explain is that the reason that they are real and not merely apparent is because in Einstein's theory time and space are defined in terms of the measurements themselves.[25] The measurements *are* the reality; there is no metaphysical or privileged time or space lurking behind the measurements. Hence, an object has no length other than relativistic length; an event occurs at no time other than relativistic time. That is why the phenomena are not merely appearances due to measurements; reality has been re-defined to be what the measurements say.

Conference of the British Society for the Philosophy of Science, "Physical Interpretations of Relativity Theory," Imperial College of Science and Technology, London, 16-19 September, 1988), all he shows is that length contraction leaves permanent traces in *spacetime* distances, not spatial distances.

[24] For example W. G. V. Rosser, *Introductory Relativity* (New York: Plenum Press, 1967), pp. 58-59, speaks of a "contraction" occurring in the *measured* length of an object. Posing the question whether the contraction is "real," Rosser declines to answer, saying only that the measures of particular quantities are different in different co-ordinate systems and that there is no absolute length.

[25] As Arzeliès nicely explains,
  "In pre-relativistic physics, time and space were considered more or less implicitly as the framework within which phenomena occurred; these frameworks were supposed to have an independent existence. The standards of space and time were deemed to *measure* space and time.
  Relativistic physics rejects these inaccessible metaphysical entities, just as it rejects the material aether of the older optics. In relativistic physics *space and time are not measured but instead, standards of length and time are defined*; this definition constitutes the fundamental postulate" (Henri Arzeliès, *Relativistic Kinematics*, rev. ed. [Oxford: Pergamon Press, 1966], p. 45; cf. pp. 120-121).

CHAPTER 4

EMPIRICAL CONFIRMATION OF SR

There are obviously many other ramifications of Special Relativity, but as our interest is in the time concept in relativity, the brief survey in the foregoing chapters will suit our purposes. SR is one of the pillars of modern physics. Bondi has said that "...there is perhaps no other part of physics that has been checked and tested and cross-checked quite as much as the Theory of Relativity."[1] According to J. G. Taylor, "as far as special relativity is concerned all has been worked out and tested;" the theory has enjoyed "remarkable successes, and absolutely no failures."[2] Let us review the principal empirical confirmations of the theory.

1. *The Thomas Precession:* In Newtonian theory an electron will maintain the direction of its spin axis in its orbit about an atomic nucleus with respect to an inertial frame, just like a gyroscope when carried around in a circle. But in 1927 L. H. Thomas showed that the theory of relativity predicts that an electron that revolves once around a nucleus will have its spin axis pointing in a different direction than when it started.[3] This phenomenon—called the Thomas precession—results from the relativity of simultaneity and has observable effects on the emission lines of some atomic spectra.

2. *The Kennedy-Thorndike Experiment.* The Lorentz-FitzGerald contraction hypothesis could explain the result of the Michelson-Morley interferometer experiment. The arm of the interferometer pointing in the direction of motion was contracted to such a degree that the light travel time along it was equal to that along the perpendicular arm. But in 1932 R. J. Kennedy and E. M. Thorndike conducted interferometer experiments over a period of months using a device with arms of unequal length, so that a negative result of their attempt to measure the ether drift could not be explained by the simple contraction hypothesis.[4] Although the experiment ruled out the Lorentz theory without clock retardation, it presented no difficulty to the Lorentz-Larmor theory, which combined length contraction with time dilation. In fact, as H. E. Ives was later to point out, both the Michelson-Morley and Kennedy-Thorndike results could be explained without time dilation if a different sort of length contraction hypothesis were adopted.[5]

3. *Ionization of Charged Particles:* The observed ionization produced by a charged particle moving at near light speeds can only be satisfactorily explained

---

[1] Hermann Bondi, *Relativity and Common Sense* (New York: Dover Publications, 1964), p. 168.
[2] J. G. Taylor, *Special Relativity*, Oxford Physics Series (Oxford: Clarendon Press, 1975), preface.
[3] L. H. Thomas, "The Kinematics of an Electron with an Axis," *Philosophical Magazine* (1927): 1.
[4] Roy J. Kennedy and Edward M. Thorndike, "Experimental Establishment of the Relativity of Time," *Physical Review* 42 (1932): 400.
[5] H. E. Ives, "Historical Note on the Rate of a Moving Clock," *Journal of the Optical Society of America* 37 (1947): 810; cf. L. Jánossy, *Theory of Relativity Based on Physical Reality* (Budapest: Akadémiai Kiadó, 1971), p. 62.

through the Lorentz contraction of that particle's electric lines of force.[6] Without the Lorentz contraction, the particle could not eject electrons from atoms located far from its path, and therefore the ionization would fall below the observed value. The contraction shortens the electric lines of force into a thin, concentrated bundle and so shortens the effective time of action of the force to a time brief enough for the particle to ionize atoms far from its path.

4. *The Ives-Stilwell Experiments.*[7] In 1938 and 1941, Ives and G. R. Stilwell produced what Miller has called "the only positive proof" to this day of time dilation.[8] In classical theory, a Doppler effect occurs when a light source is approaching or receding in the line of sight. If the source is approaching, its light will appear bluer; if receding, redder. But there is no transverse Doppler effect for a source moving across one's line of sight. Relativity theory, however, predicts such a transverse or second order Doppler shift, which, though very small, Einstein predicted might be detectable in the light emitted by fast-moving particles in hydrogen canal rays. Ives and Stilwell's observational evidence for a second order Doppler effect can only be explained by time dilation. Ironically, Ives and Stilwell were themselves neo-Lorentzians and so interpreted their results as a vindication of clock retardation for inertial frames in motion relative to the aether frame. "The conclusion drawn from these experiments is that the change of frequency of a moving light source predicted by the Larmor-Lorentz theory is verified," they concluded.[9] Einsteinian relativists point out, however, that the results can be equally interpreted as a verification of the Einsteinian interpretation of Special Relativity.

5. *Meson Lifetimes:* Particles called μ-mesons or muons have a lifetime at rest of only 2.2×10^{-6} second before they decay into an electron and two neutrinos. Such particles are produced in the upper atmosphere of the earth by cosmic ray interaction with the air. With so short a lifetime, such particles should decay before reaching the surface of the earth. But in fact, they are found in relative abundance at the earth's surface. Relativistic physics explains this via time dilation and length contraction.[10] Relative to the earth's reference frame, an imaginary clock associated with the muon runs slow. Therefore, it records the time taken for the muon to plunge from the upper atmosphere to the earth's surface as less than 2.2×10^{-6}

---

[6] E. J. Williams, "The Loss of Energy by B-Particles and its Distribution between Different Kinds of Collisions," *Proceedings of the Royal Society* A 130 (1931): 328-346.

[7] Hubert E. Ives and G. R. Stilwell, "An Experimental Study of the Rate of a Moving Atomic Clock," *Journal of the Optical Society of America* 28 (1938): 215-226; idem, "An Experimental Study of the Rate of a Moving Atomic Clock. II," *Journal of the Optical Society of America* 31 (1941): 369-374.

[8] Arthur I. Miller, *Albert Einstein's Special Theory of Relativity* (Reading, Mass.: Addison-Wesley, 1981), p. 266. This is because "The various experiments utilizing elementary particles involve a vicious circle because their data analysis depends on special relativity; consequently these experiments test only the consistency of the special theory of relativity." In this judgement, he concurs with Herbert Dingle, "Time in Relativity Theory: Measurement or Coordinate?," in *The Voices of Time*, 2d ed., ed. with a new Introduction by J. T. Fraser (Amherst, Mass.: University of Massachusetts Press, 1981), pp. 469-470; idem, *Science at the Crossroads* (London: Martin Brian and O'Keefe, 1972), pp. 34, 143, 149, 208.

[9] Ives and Stilwell, "Experimental Study. II," p. 374.

[10] Bruno Rossi and J. Barton Hoag, "The Variation of the Hard Component of Cosmic Rays with Height and the Disintegration of Mesotrons," *Physical Review* 57 (1940): 461-469; Bruno Rossi and David B. Hall, "Variation of the Rate of Decay of Mesotrons with Momentum," *Physical Review* 59 (1941): 223-228; David H. Frisch and James H. Smith, "Measurement of the Relativistic Time Dilation Using μ-Mesons," *American Journal of Physics* 31 (1963): 342.

second. On the other hand, relative to the muon's inertial frame, its imaginary clock does not run slow, but the distance it must travel to the earth's surface is Lorentz - contracted to so short a distance that it can cover its course during its momentary existence.

Another type of meson, the π-meson or pion, has a lifetime of 2.54×10^{-8} second while at rest. In cyclotron experiments pions moving near the speed of light traverse distances that would require them to exist many times their proper lifetimes. Without time dilation, most of them would decay after a few meters.[11]

6. *The Mössbauer Effect.* The prediction of asymmetrical aging in the twin paradox has apparently been actually verified in experiments involving the Mössbauer effect. R. L. Mössbauer discovered that an atomic nucleus can emit gamma radiation without any recoil effect upon the nucleus itself and that similarly another nucleus can absorb this radiation without recoil. This effect permits a fantastically precise measurement of the frequency of the radiation, which in turn enables one to demonstrate time dilation of atomic nuclei. The effectiveness of a nucleus to absorb the gamma radiation emitted by the source is highly sensitive to the temperature of both source and absorber, that is to say, to their thermal motions. These thermal motions of the nuclei set up a second order Doppler effect which requires that the frequency of recoilless gamma rays emitted by the source be reduced, that is, its internal nuclear clock runs slow. It has therefore been pointed out that the source and absorber nuclei are like two twins, one of whom ages less rapidly than the other.[12]

The Mössbauer effect can also be used to demonstrate the transverse Doppler effect more directly.[13] A source of gamma radiation is placed around the center of a rotary disk and an absorbing band around the disk's perimeter. In effect, the one twin stays at home while the other makes a round trip journey (literally!). The reduction in frequency is greater in the absorber, since it is more greatly accelerated, which is to say that its imaginary clock runs slow relative to the source. The nuclei on the perimeter thus age less quickly than those in the source and that by the amount predicted by relativity theory.

7. *The Hafele-Keating Experiment:* In 1971 J. C. Hafele and R. E. Keating conducted a remarkable experiment which demonstrated time dilation at even low speeds.[14] They flew four cesium clocks around the world on commercial jetliners, first eastward, then westward. They discovered that on the eastbound trip, the clocks lost about 59 nanoseconds (one nanosecond = 10^{-9} second) and on the westbound trip gained about 273 nanoseconds. The reason for the gain is the

---

[11] R. Durbin, H. H. Loar, and W. W. Havens, "The Lifetime of the π + and π - Mesons," *Physical Review* 88 (1952): 179-183.

[12] Chalmers W. Sherwin, "Some Recent Experimental Tests of the 'Clock Paradox'," *Physical Review* 120 (1960): 17-21.

[13] H. J. Hay, J. P. Schiffer, T. E. Cranshaw, and P. A. Egelstaff, "Measurement of the Red Shift in an Accelerated System Using the Mössbauer Effect in Fe," *Physical Review Letters* 4 (1960): 165.

[14] J. C. Hafele and R. E. Keating, "Around the World Atomic Clocks: Predicted Relativistic Time Gains," *Science* 177 (14 July 1972): 166-168; idem, "Around the World Atomic Clocks: Observed Relativistic Time Gains," *Science* 177 (14 July 1972): 168-170. They conclude: "These results provide an unambiguous empirical resolution of the famous clock 'paradox' with macroscopic clocks" (Ibid., p. 168). Unfortunately the dilation is too small to be detected with non-atomic timepieces.

rotation of the earth. If the earth is rotating with an angular velocity Ω and has a radius R, then a clock at the equator is itself moving with a velocity R Ω relative to an underlying non-rotating inertial frame. This clock on the earth's surface therefore experiences a time dilation effect as well. Once this is taken into account, Hafele and Keating's results accord closely with the predictions: the predicted time difference for the clocks on the eastward journey was –40±23 and for the westward trip 275±21. The verification of these predictions provides evidence that atomic clocks in motion do go slow, even on commercial aircraft.

These experimental data have confirmed that the Lorentz transformations, which constitute the mathematical core of SR, provide the physically correct way of transforming the spatio-temporal co-ordinates of an event from one inertial frame to another.

CHAPTER 5

TWO RELATIVISTIC INTERPRETATIONS

MINKOWSKI SPACETIME

A physical theory is comprised of two components: a mathematical formalism (a set of equations and a set of calculational rules for making predictions that can be compared with experiment) and a physical interpretation (what the theory tells us about the underlying structure of phenomena, that is to say, an ontology).[1] Thus, a single formalism with two different interpretations counts as two theories. Einstein's original formulation of his Special Theory was mathematically algebraic in nature and metaphysically a space and time theory. By this latter characterization, I mean that Einstein presupposed an ontology of spatial objects which endure through time, howbeit that no single, universal time exists. But in 1908 the German mathematician Hermann Minkowski proposed a formulation of SR which was strikingly different mathematically and metaphysically. Minkowski proposed that space and time be united in a four-dimensional mathematical space, three of whose dimensions represent physical space and the fourth time.[2] In this manifold, relativity theory and the Lorentz transformations can be exhibited with great clarity. Events in spacetime are specified by giving their four coordinates, and although the temporal and spatial distances between two specified events will differ from one coordinate system to another (relativity of simultaneity and length), nevertheless the composite spacetime interval between events is absolute. Letting ds represent the spacetime interval, and dt represent the temporal distance, and dx, dy, dz the spatial distance between the events, Minkowski spacetime has a metric of the form $ds^2 = dx^2 + dy^2 + dz^2 - dt^2$, or alternatively expressed, $ds^2 = dt^2 - dx^2 - dy^2 - dz^2$. The notion of spacetime interval between two events can be understood as an extension of spatial interval. In two dimensions x and y, we can calculate the spatial interval between two points by means of the Pythagorean Theorem (Figure 5.1).

---

[1] See James T. Cushing, "The Causal Quantum Theory Program," in *Bohmian Mechanics and Quantum Theory: An Appraisal*, ed. James T. Cushing, Arthur Fine, and Sheldon Goldstein, Boston Studies in the Philosophy of Science 184 (Dordrecht: Kluwer Academic Publishers, 1996), p. 4.
[2] See H. Minkowski, "Space and Time," in *The Principle of Relativity*, by A. Einstein, *et al.*, trans. W. Perrett and G. B. Jeffery (New York: Dover Publications, 1952), pp. 75-91.

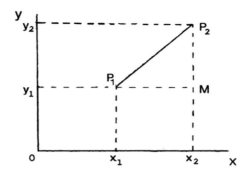

Figure 5.1. The spatial interval between two points P_1 and P_2 calculated by the Pythagorean Theorem. Let the interval $P_1P_2 = S$. So, $S^2 = (P_1M)^2 + (P_2M)^2$. But $P_1M = x_2 - x_1$ and $P_2M = y_2 - y_1$. So, $S^2 = (x_2 - x_1)^2 + (y_2 - y_1)^2$.

We can extend this same method of calculation to three dimensions x, y, and z (Figure 5.2).

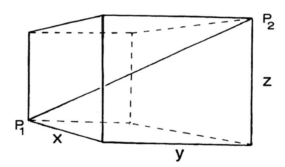

Figure 5.2. Extension of Pythagorean Theorem to three dimensions. Now $S^2 = (x_2 - x_1)^2 + (y_2 - y_1)^2 + (z_2 - z_1)^2$.

Finally, we can extend this to a four-dimensional mathematical space having the coordinates x, y, z, and t. The last of these represents the temporal dimension. Such a four-dimensional geometry is non-visualizable and will be non-Euclidean (unless we employ imaginary numbers for the time co-ordinate), so that the t coordinate has the opposite sign of the spatial coordinates: $s^2 = (x_2 - x_1)^2 + (y_2 - y_1)^2 + (z_2 - z_1)^2 - (t_2 - t_1)^2$. Letting d represent the distance involved, we thus arrive at the metric for Minkowski spacetime: $ds^2 = dx^2 + dy^2 + dz^2 - dt^2$, which fixes the absolute location of any event in spacetime. When ds^2 is a positive number, ds represents the proper time of an inertial observer moving in physical space from one event to another:

$$ds = dt\sqrt{1 - v^2/c^2} \tag{1}$$

Thus, although space and time taken separately are relative, taken together they are absolute, in that given an event's spacetime coordinates its location is fixed for all reference frames.

We cannot build four-dimensional models of spacetime, but we can, by suppressing the three spatial dimensions and substituting a single dimension for them, draw spacetime diagrams, such as we have already encountered in the Twin Paradox. By letting the time dimension be calibrated in terms of units of the time required for light to cover one spatial unit, we find that lines representing the paths of light beams always lie at 45° angles (Figure 5.3).

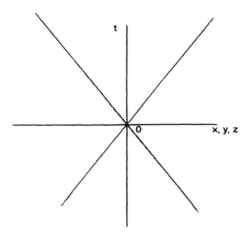

Figure 5.3. Paths of light rays on a co-ordinate system where one axis represents the three spatial dimensions and one axis the temporal dimension and light is conceived to travel one spatial unit in one temporal unit.

If we take the event O lying at the origin of the coordinates, we can see that spacetime falls naturally into three regions with respect to O. Although there is nothing in the mathematical structure to require it, the region above O is usually denominated O's future, that is to say, events *later* than O; whereas the region below O is taken to be O's past, that is to say, events *earlier* than it. So construed, the diagram shows at this spacetime point that one region of spacetime contains all those events to which one can send a signal from O or in other words, which O can causally influence. Since no known causal influence can be propagated faster than light, events outside this region are too far from O to be influenced by O in the time allowed. The briefer the time allowed for the influence to propagate, the narrower is O's range of influence; the more time allowed, the broader its influence. This region of spacetime is called O's absolute future, and the points on the diagram lying on the vector representing light paths is called O's forward light cone. In physical space, this configuration of points would lie on the surface of a sphere of light emanating from the event whose diameter increases with the time elapsed since the event. But on our diagram, with only one spatial dimension, O's absolute future appears as an inverted triangle; if we add another spatial dimension to the model we can represent it as an inverted cone (Figure 5.4). It should be remembered that O is not a substance moving through time; O is the spacetime event itself. For a person S in physical space and time coincident with O, the future is "absolute" only in the sense of "non-relativistic": if S is at rest in the coordinate system x, y, z, t, then at earlier points on the time axis events outside S's future on our diagram were in S's future, and at later points events which are in S's future on the diagram will no longer be so. But S's future at O is absolute in the sense that the set of events constituting S's absolute future are in S's future no matter what reference frame one refers to. Even for a reference frame moving relative to S with nearly the speed of light, the same events lie in S's causal future at the designated spacetime event O. This is because of the invariant velocity of light in all inertial frames. In all inertial frames the light cone for O remains the same; it cannot be tilted one way or the other, since in all frames light propagates from O at the same velocity—one unit of space in one unit of time. The light cone of any event in spacetime is thus a structure existing in spacetime independent of any coordinates that might be used to describe it. Therefore, causal connections between events are invariant or preserved in every reference frame.

The diagram also shows a region below O lying within or on O's back light cone designated O's absolute past. Since O does not send light rays backward in time, the light paths here represent, not signals from O, but signals to O. The past light cone displays the latest time a signal could be sent from any spatial point and be expected to arrive at O. A signal sent from any spatial point on O's past light cone at a later time would not reach event O, but would arrive at some event O' higher up the time axis. Thus, O's absolute past is constituted by all those events which can have had a causal influence on the event O. Again, this region is absolute in that in all reference frames the events lying within or on O's past light cone are the same.

The third region of spacetime exhibited by our diagram is that region which surrounds O and lies outside either its forward or back light cones. Since no event in this region can be considered to occur at the same place as O in any reference frame, Eddington called this region the absolute "elsewhere."[3] These events lie at distances in physical space which are too great for either light signals from them to have the time to reach O or light signals from O to reach them. Of course, eventually signals from these events will reach S at some later point O'; but S will never be able to influence any of the events in this region, for given the limiting character of the velocity of light, it is already is too late by O.

Another way of looking at these regions is in terms of the sort of spacetime separation which exists between O and events in these regions. If $x^2 - t^2 = 0$, then the event having these coordinates lies on either the forward or back light cone (called, hence, the null cone) and its separation from O is called *light-like*. If $x^2 - t^2$

---

[3] Arthur Eddington, *Space, Time and Gravitation*, Cambridge Science Classics (Cambridge: Cambridge University Press, 1920; rep. ed.: 1987), p. 50; idem, *The Nature of the Physical World*, with an Introductory Note by Sir Edmund Whittaker, Everyman's Library (1928; rep. ed.: London: J. M. Dent and Sons, 1964), p. 57.

< 0, then the event lies inside the null cone in either the absolute past or future of O, and its separation from O is termed *timelike*. Finally, if $x^2 - t^2 > 0$, then the relevant event lies in the "elsewhere" region for O, and its separation from O is called spacelike. Figure 5.4 provides a particularly helpful illustration of the various regions into which spacetime may be divided.

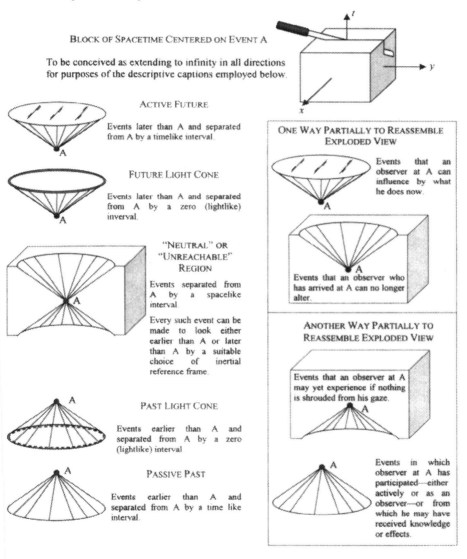

Figure 5.4. Exploded view of the five regions into which the events of spacetime fall apart when classified with respect to a selected event A.

Now our interest focuses on those events having a space-like separation from O. Consider first which events lying in this region are those which S considers to be simultaneous with O. Using the light signal synchronization method, S bounces a signal off an object at relative rest at some event P and determines that $t_p = \frac{1}{2}(t'_0 - t_0)$ (Figure 5.5).

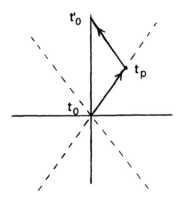

Figure 5.5. Light signal method of synchronization: $t_p = \frac{1}{2}(t'_0 - t_0)$.

Knowing what time his clock read at t_0, S calculates that the event in O-time simultaneous with t_p was $t_0^* = \frac{1}{2}(t_0 + t_0')$. He will therefore draw a line of simultaneity across his spacetime diagram connecting t_0^* and t_p (Figure 5.6).

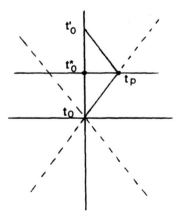

Figure 5.6. S's calculation of a moment which is simultaneous with t_p.

Every time S repeats such an experiment, he finds that the line of simultaneity runs parallel with his x axis. Thus, although S has direct experience only of his proper time τ along his inertial trajectory σ and this time permits him to make temporal comparisons only between events on σ itself (for example, earlier than, simultaneous with, or later than), nevertheless via the light signal method of clock synchronization S can determine indirectly which events are uniquely simultaneous relative to σ for any clock reading in τ and so extend τ to a unique global time relative to σ. He can then slice spacetime into layers of global simultaneity for σ which are orthogonal to σ. In this way S can calculate which events having a space-like separation from him are present (those lying on the line of simultaneity), which future (those above the line), and which past (those below the line).[4]

But now consider the case of a space traveller S' travelling past S at a uniform velocity near to c and whose clock agrees with S's clock at O. Using the same light signal method of synchronization, he bounces a signal off an object at rest relative to him, which event is coincident with P (Figure 5.7).

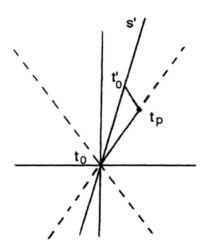

Figure 5.7. Light synchronization procedure as carried out in the relatively moving inertial frame of S'.

But because he is moving relative to S, S' receives his return signal first, according to S's reference frame. Therefore, when S' calculates that the event simultaneous with t_p is $\frac{1}{2}(t_0' - t_0)$, this will from S's perspective appear erroneous. For S' will

---

[4] For discussion, see Michael Friedman, "Simultaneity in Newtonian Mechanics and Special Relativity," in *Foundations of Spacetime Theories*, ed. John S. Earman, Clark N. Glymour, and John J. Stachel, Minnesota Studies in the Philosophy of Science 8 (Minneapolis: University of Minnesota Press, 1977), pp. 409-410; idem, *Foundations of Spacetime Theories* (Princeton: Princeton University Press, 1983), pp. 165, 309.

figure that $t_0^* = \frac{1}{2}(t_0 + t_0')$, and thus locate t_0^* too early by S's reckoning. S' will then draw the line of simultaneity through t_0^* and t_p askew (Figure 5.8). To S it will seem as though S' has forgotten that he is in motion!

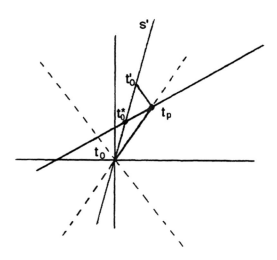

Figure 5.8. S''s calculation of which event on his inertial trajectory is simultaneous with P and his consequent plane of simultaneity.

The result is that S' will regard *different* events as past, present, or future than will S. Some event which S calculates to be present will already lie in the past for S', since it lies below his line of simultaneity.

But S has no basis for thinking S' is mistaken. For S' might with equal right regard himself as being at rest and S in motion. In the absence of any preferred frame of reference, the controversy as to which is correct is meaningless. Events in the "elsewhere" region are not absolutely past or future but only relatively so, depending on the reference frame in question.

In fact, events having a space-like separation can even be measured to occur in inverted temporal order (Figure 5.9).

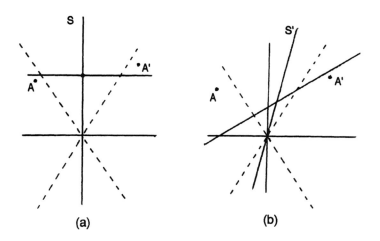

Figure.5.9. Events A and A', having a spacelike separation, are calculated by S to be past and future respectively (a), whereas S' calculates them to be future and past respectively (b).

In S's reference frame, A occurs before A' takes place. But in S''s reference frame, it is event A' which happens prior to A. It perhaps bears repeating that neither S nor S' has any direct experience of simultaneous events other than the relevant event on σ itself, but later when they do receive light rays from events having a space-like separation, they will receive them in different temporal orders.

TWO RELATIVISTIC ONTOLOGIES

Now Minkowski took his spacetime realistically: it was not merely a geometrical representation of the world of space and time as described by Einstein's SR; rather it *was* the world. When he said, "A point of space at a point of time, that is, a system of values x, y, z, t, I will call a *world-point*. The Multiplicity of all thinkable x, y, z, t systems of values we will christen the *world*,"[5] he was making self-consciously a metaphysical statement, proposing a new ontology. Heralding "a metamorphosis of our concept of nature," Minkowski declared, "Henceforth, space by itself, and time by itself, are doomed to fade away into mere shadows, and only a kind of union of the two will preserve an independent reality."[6] Minkowski's spacetime approach to relativity theory, especially with the development of GR, came to be the dominant mode of presentation and discussion of relativity.

Now the question arises whether we should take a realist or merely instrumentalist view of Minkowski spacetime. That is to say, should we agree with Minkowski that the fundamental component of our physical ontology is the

---

[5] Minkowski, "Space and Time," p. 76.
[6] Ibid., pp. 75, 76.

tenselessly existing, four-dimensional spacetime manifold or is Minkowski spacetime merely a geometrical representation of a fundamental ontology involving the temporal becoming of spatial objects which endure through time, such as appeared in Einstein's 1905 paper?

At issue here are two competing theories of time, a dynamic or tensed theory versus a static or tenseless theory, usually called for convenience the A- and B-Theories of time respectively. On an A-Theory of time there is no tenselessly subsisting manifold of events, for such a manifold would be incompatible with the objective reality of temporal becoming. According to the A-Theory, things come into being and pass away, which the constituents and contents of spacetime do not do. Spacetime is therefore a theoretical construct only, a geometrical representation of a theory which is really about physical objects enduring through time. SR is, just as Einstein originally formulated it, a 3+1 dimensional theory, not a 4-dimensional theory. A Minkowski diagram is a helpful tool, but neither depicts reality nor implies an ontology. Graham Nerlich calls such an interpretation of SR the *relativity interpretation* and characterizes it in the following way:

> It has a familiar, classical ontology of continuant spatial objects persisting in time. It uses the idea of a frame of reference, a relative space and a relative time. The frame defines (relatively) the same place at different times and thus the concepts of rest, motion, speed and velocity. It also defines (relatively) the same time (simultaneity) at different places. Light is marked out as having an invariant speed for every frame of the interpretation. Whereas one can do spacetime physics without coordinates, it is meaningless to speak of a frame-free relativity physics. The concepts of the relativity interpretation make sense, in SR, only with respect to some frame.[7]

On this interpretation, Minkowski's advance over Einstein was mathematical, not metaphysical. A good representative of this original Einsteinian perspective is the French physicist Henri Arzeliès. In his *Relativistic Kinematics*, Arzeliès asserts, "The Minkowski continuum is an abstract space of four dimensions, the sole role of which is to interpret in geometrical language statements made in algebraic or tensor form.... The four-dimensional continuum should therefore be regarded as a useful tool, and not as a physical 'reality'."[8] While it is true that relativity theory banishes the notions of absolute spatial and temporal intervals from physics, nonetheless "It is perfectly clear that in relativity, the ordinary three-dimensional space (which is Euclidian in special relativity) and the time of pre-relativistic physics is employed."[9]

There is no gainsaying Arzeliès insofar as Einstein's original formulation of SR is concerned. But it is also indisputable that once having encountered Minkowski's geometrical formulation of the theory, Einstein became an outspoken realist concerning spacetime. Regarding reality as four-dimensional seemed more natural to him than the complicated maneuver of relativizing presentness to reference frames. He wrote,

> Inertial spaces, with their associated times, are only privileged four-dimensional co-ordinate systems, that are linked together by the linear Lorentz transformations. Since

---

[7] Graham Nerlich, *What Spacetime Explains* (Cambridge: Cambridge University Press, 1994), p. 63.
[8] Henri Arzeliès, *Relativistic Kinematics*, rev. ed. (Oxford: Pergamon Press, 1966), p. 258.
[9] Ibid.

there exist in this four-dimensional structure no longer any sections which represent 'now' objectively, the concepts of happening and becoming are indeed not completely suspended, but yet complicated. It appears therefore more natural to think of physical reality as a four-dimensional existence, instead of, as hitherto, the evolution of a three-dimensional existence.[10]

Again, in a work co-authored with Leopold Infeld, Einstein rejects his own original formulation of SR in 3+1 dimensions in favor of the spacetime ontology depicted in two dimensions on a Minkowski diagram: "We must not consider space and time separately in determining the time-space co-ordinates in another CS [coordinate system]. The splitting of the two-dimensional continuum into two one-dimensional ones seems, from the point of view of the relativity theory, to be an arbitrary procedure without objective meaning."[11] True, "Even in the relativity theory we can still use the dynamic picture if we prefer it. But we must remember that this division into time and space has no objective meaning since time is no longer 'absolute'."[12] Thus, relativity theory was "distinctly in favor of the static picture and found in this representation of motion as something existing in time-space a more convenient and more objective picture of reality."[13] As Ludwig Kostro has shown,[14] Einstein came to view the spacetime described in the General Theory of Relativity as a relativistic ether, uncomposed of either particles or moments, in which all things exist and light is propagated, but which is not itself a reference frame and so does not serve to establish absolute simultaneity and length as did the classical aether. It is a breath-taking metaphysical vision of reality, truly justifying the title "Philosopher-Scientist" for Albert Einstein.[15] The seriousness with which Einstein took this conception may be seen in the fact that when his life-long friend Michael Besso died, Einstein sought to comfort his bereaved family by reminding them that for physicists Besso had not ceased to exist but exists tenselessly as a permanent feature of the spacetime reality.[16]

Nerlich calls this second approach to SR the *spacetime interpretation*, and he characterizes it as follows:

---

[10] Albert Einstein, *Relativity: The Special and the General Theory*, 15th ed. (New York: Crown Trade Paperback, 1961), p. 150.
[11] Albert Einstein and Leopold Infeld, *The Evolution of Physics* (New York: Simon & Schuster, 1938), p. 219.
[12] Ibid., p. 220.
[13] Ibid., p. 217.
[14] Ludwig Kostro, "Einstein's New Conception of the Ether," paper presented at the International Conference of the British Society for the Philosophy of Science, "Physical Interpretations of Relativity Theory," Imperial College of Science and Technology, London, 16-19 September, 1988.
[15] When his friend Michael Besso inquired if Einstein never concerned himself with the flow of time, Einstein wrote back, indicting Besso for not taking seriously four-dimensional reality. What Besso calls the "world" is a spacelike section for which relativity theory denies objective reality: the "now" is eliminated in the objective construction of the world. (Albert Einstein and Michele Besso, *Correspondance 1903-1955*, trans. with Notes and an Introduction by Pierre Speziali [Paris: Hermann, 1979], pp. 276-277.)
[16] Of Besso's death Einstein wrote, "This signifies nothing. For us believing physicists the distinction between past, present, and future is only an illusion, even if a stubborn one" (Cited in Banesh Hoffmann with Helen Dukas, *Albert Einstein: Creator and Rebel* [London: Hart-Davis, MacGibbon, 1972], p. 258).

> It has an ontology of events (alternatively of four-dimensional objects) and speaks of a spacetime (but not of a space nor a time). Spacetime in SR has zero curvature and Minkowski metric in which physical quantities appear as 4-vectors or 4-tensors. A very wide range of coordinate systems may be used in the description of spacetime and its physics, but assigning the same time coordinate (or the same space coordinates) to different events is not interpreted as reflecting, even relatively, the concepts of simultaneity (or rest) since these are no part of the scheme of concepts of this interpretation. Further, the use of coordinate systems is not essential to spacetime physics, coordinate-free treatments of it being quite standard. Light is distinguished neither as invariant in speed nor as the fastest signal, since ideas like 'speed' and 'fastest' do not belong in this interpretation. Light is picked out in that light trajectories always lie in cones of null geodesics, one of which is defined at each spacetime point.[17]

This characterization makes evident how remarkable and unusual such an interpretation is, since all the familiar notions of SR as Einstein originally formulated it, such as reference frames, simultaneity, and the speed of light, find no place in it. Rather the central feature of the spacetime interpretation of SR is the lightcone structure, which is independent of reference frames or coordinate systems. In 1911 A. A. Robb was able to recover all the geometric structure of Minkowski spacetime on the basis of the single relation *after* among its points, conjoined with several conditions of that relation.[18] Taking Robb's relation to be extensionally equivalent to some sort of causal relation, recent theorists have defined causally the Lorentz group of transformation equations,[19] orthogonality to a time-line in Minkowski spacetime,[20] and the metrical congruence of intervals in that spacetime.[21] Spacetime realists debate intramurally whether causality is truly constitutive, rather than merely (at best) co-extensive with Robb's fundamental relation,[22] but the point remains that the familiar physical entities of the relativity interpretation make no appearance in the spacetime interpretation.

When it comes to the General Theory of Relativity, these two interpretations present very different pictures. The relativity interpretation sees the key advance over SR as the introduction of "accelerated" frames. Gravitation is construed as a force acting on objects, including clocks, in three-dimensional space. But in the spacetime interpretation it is the introduction of the curvature of spacetime that

---

[17] Nerlich, *What Spacetime Explains*, p. 63; cf. idem, "Time as Spacetime," in *Questions of Time and Tense*, ed. R. Le Poidevin (Oxford: Oxford University Press, 1998), p. 119-134.
[18] A. A. Robb, *A Theory of Time and Space* (Cambridge: Heffer and Sons, 1913).
[19] E. C. Zeeman, "Causality Implies the Lorentz Group," *Journal of Mathematical Physics* 5 (1964): 490-493.
[20] David Malament, "Causal Theories of Time and the Conventionality of Simultaneity," *Noûs* 11 (1977): 293-300.
[21] John Winnie, "The Causal Theory of Spacetime," in *Foundations of Spacetime Theories*, ed. Earman, pp. 134-205.
[22] On this score Nerlich finds himself at odds with theorists such as Reichenbach, Grünbaum, van Fraassen, Salmon, and Winnie. On Nerlich's view the heart of SR is the second-order constraint on all physical laws of Lorentz invariance, which springs from the inherent symmetries of Minkowski spacetime. Temporal relations in that spacetime cannot be reduced to causal relations because the latter presuppose the former.

marks the departure from SR. Gravitation is interpreted, not as a force acting between bodies in space, but as a warping of spacetime itself.[23]

These two interpretations of relativity theory thus present strikingly different metaphysical visions of reality; they are as radically divergent in their ontologies as is relativity theory itself in comparison with the Newtonian physics of absolute time and space.

The A-theorist of time holds to the objective reality of tensed facts and temporal becoming; but how is such a commitment to be integrated with SR? The A-theorist affirms the common sense intuition that only present events are real; but since, according to SR, there is no unique time, how are we to understand the notion of "present events"? The most obvious move for him to make would be to adopt an Einsteinian interpretation of SR and to relativize presentness and, hence, reality to reference frames. Indeed, I suspect that this is the solution preferred by most A-theorists. If one holds that presentness is not absolute, but frame-relative, then which events in a given frame are most plausibly to be taken as present? The answer seems obvious: only and all those events which are simultaneous with any given event when that event becomes present. Since simultaneity is frame-relative in SR, so is presentness, and, given the tight connection in the A-theorist's thinking between presentness and reality, so is the reality of all temporal events.

Because only events which are present are real, and only events which are simultaneous with an event which is present on the world line of some entity are present, and simultaneity is relative to reference frames, presentness and reality are relative to reference frames as well. Presentness cannot, within the context of SR, be possessed by an event absolutely because then that event would be present relative to *every* reference frame, whether or not that event is defined as simultaneous relative to a given frame according to SR's clock synchronization procedure. Given a plurality of absolutely present events, a privileged present or absolute now would exist in contradiction to SR. Given his presentist commitments, the A-theorist's relativizing reality to reference frames is quite natural.

The B-theorist might object at this point, as D. H. Mellor has done, that the simultaneity relation in SR is conventional, so that events determined to be simultaneous relative to a reference frame at any given point on the world line of some observer cannot be regarded as objectively present or real.[24] The objector might seek to force the A-theorist to choose between three equally unacceptable alternatives: either (i) to regard the entire "elsewhere" region as present and real to some observer, or (ii) to shrink the present down to a single spacetime point at which the hypothetical observer is located or (iii) to choose arbitrarily one frame of reference to define simultaneity and, hence, present reality at any time. According to alternative (i), all events having a space-like separation from me-now, that is, lying outside the forward and backward light cones of me-now, are present and real for me-now. The trouble with this construal of simultaneity on the A-Theory,

---

[23] For discussion see Martin Carrier, "Physical Force or Geometrical Curvature?" in *Philosophical Problems of the Internal and External Worlds*, ed. John Earman, Allen I. Janis, Gerald J. Massey, and Nicholas Rescher (Pittsburgh: University of Pittsburgh Press, 1993), pp. 3-21.

[24] D. H. Mellor, "Special Relativity and Present Truth," *Analysis* 34 (1973-1974): 75-76.

Mellor rejoins, is that events co-existent with me-now are not co-existent with each other. For example, if e_1 and e_2 are events outside the light cone structure of me-now and e_2 lies in the absolute future of e_1, then both e_1 and e_2 exist with me-now, since they lie in my "elsewhere" region. Nonetheless, e_1 and e_2 are not co-existent, since e_2 lies in e_1's absolute future. Thus, two events which are equally real for me nonetheless do not co-exist. In one and the same reference frame, e_1 and e_2 both exist (being space-like separated from me-now) and do not both exist (one being absolutely future to the other). Hence, one cannot escape this contradiction by relativizing reality to reference frames. Similarly, Robert Weingard argues that if all events outside my light cone structure are equally real, then this result can be easily generalized to yield the conclusion that all spacetime events are real.[25] For e_1's "elsewhere" region will include events in my absolute future and past, which must therefore be real, too. Again it does no good to relativize reality to reference frames because in this case all the events are *ex hypothesi* being regarded from the standpoint of the same frame. Moreover, Mellor points out, on this alternative tensed sentences can turn out to be both true and false. Suppose that e_1 is rain in London and e_2 is clear weather. Since both events are equally real, the tensed sentence "It is raining in London" is both true and false for me-now. Therefore, alternative (i) is untenable.

According to alternative (ii), we are to take as present only events that are present in every acceptable frame of reference. On this view, only events are present which every observer at some point in space agrees are present. Thus, I-now and you-now, though associated with different reference frames at the same location, will agree that our "now's" are coincident and, hence, equally present. But then, Mellor objects, the present becomes shrunk down to what is here at this point-instant. As Lawrence Sklar explains,

> We first reduce 'reality' to the lived experience of the observer; that is, we first fall...into solipsism. Then seeing that our own future experiences and past experiences are as remote *now* from us as the spatially distant, the non-immediately sensed, etc., we fall from solipsism into solipsism of the present moment. Reality has now been reduced to a point![26]

If this is not self-evidently absurd, Mellor also points out that on this view if I-now am never located in London, then the bizarre conclusion follows that the tensed sentence "It is raining in London" is never true. Sklar notes the further strange consequence of this alternative that we must say that there will be events which are now such that they will be in my real past at some future time but which will never have a present reality to me at all! That is because they are now in my future light cone and will be in my past light cone, but have no reality in between. Since neither past nor future events exist, how is it that such events become past without ever having been actualized in the present? Indeed, is it not self-contradictory to assert

---

[25] Robert Weingard, "Relativity and the Reality of Past and Future Events," *British Journal for the Philosophy of Science* 23 (1972): 119-121; similarly, Veselin Petkov, "Simultaneity, Conventionality, and Existence," *British Journal for the Philosophy of Science* 40 (1989): 69-76.

[26] Lawrence Sklar, "Time, Reality, and Relativity," in *Reduction, Time and Reality*, ed. Richard Healey (Cambridge: Cambridge University Press, 1981), p. 140.

that the sentence "The Battle of Waterloo occurred" is true and yet that it was never true that "The Battle of Waterloo is occurring"? Thus, this second alternative is also untenable.

According to alternative (iii), we choose arbitrarily one frame of reference to define simultaneity and, hence, present existence and truth at any time. In other words, we deny that there are no privileged observers but single out one reference frame as privileged and, hence, definitive for the "now." But Mellor objects that such a move is *ad hoc*, picking out a present merely to preserve the A-theorist's account of existence and truth. Special Relativity allows no more than a conventional distinction between one such present and any other; but the distinction between what exists and what does not is surely more than conventional. Mellor concludes that for A-theorists, therefore, "it becomes conventional what exists at any time, and hence, for tense-logic, it becomes conventional what is presently true."[27]

The A-theorist may agree with Mellor that the three proffered alternatives to the standard simultaneity relation are unacceptable. But he may plausibly dispute Mellor's presupposition that the standard simultaneity relation is conventional. For, as we have seen, the standard simultaneity relation is *not* conventional *within the context of SR*.[28] The simultaneity relation as calculated by the clock synchronization procedure is an intrinsic feature of Minkowski spacetime. The question of conventionalism arises only within the context of discovery of whether we inhabit a spacetime which is Minkowskian. Since the objection to the A-Theory from SR is based on the assumption that spacetime is (locally) Minkowskian, the B-theorist can hardly try to force the A-theorist to embrace alternatives incompatible with that theory. Accordingly, if the A-theorist wishes to relativize presentness and temporal becoming to reference frames, then, given that events simultaneous with a present event are also present, he will regard the world-wide "now" and the edge of becoming as coincident with the hyperplane of simultaneous events inherent in the structure of Minkowski spacetime, relative to a reference frame, on which a locally present event is located.

Now the A-theoretical attempt to relativize tense and temporal becoming to reference frames presupposes the relativity interpretation of SR with its instrumentalist understanding of Minkowski spacetime. But while such a relativization seems quite natural within the context of an A-Theory of time, given the close connection between presentness and existence in that theory, nevertheless what I now want to argue is that the resultant interpretation of SR is implausible and deficient. That is to say, Einstein's SR is more plausibly construed along the lines of the spacetime interpretation.

First, the pluralistic fragmentation of reality into distinct spaces and times associated with reference frames is an ontology which is fantastic.[29] The idea that

---

[27] Mellor, "Special Relativity," p. 75.
[28] See pp. 30-38.
[29] This is a complaint which has recently been voiced by many spacetime realists, *e.g.*, Steven F. Savitt, "There's No Time like the Present (in Minkowski Spacetime)," paper presented at the symposium "The Prospects for Presentism in Spacetime Theories," Philosophy of Science Association biennial meeting,

there is a single world, an objective reality independent of observers, that if we both exist then what co-exists with me co-exists with you, is a powerful intuition. It is fantastic to think that you and I, occupying the same location in space and time, but in relative motion, should in virtue of that motion literally dwell in two different worlds, which intersect only at a point. Yet SR requires that even if we are merely passing each other in automobiles, our hyperplanes of simultaneity do not coincide, and at sufficient distances empirically distinguishable events and things are occurring and exist for me which are future and therefore unreal for you. Other events which are in my future and therefore unreal are already actual for you. But if I decelerate and we come to relative rest, then we share the same reality; events which were once present and real in relation to me are now non-existent and future. They remain real relative to my former reference frame, but I have switched frames, and in my new frame they do not yet exist. According to SR, after all, at any point there are an infinite number of reference frames with which hypothetical observers can be imagined to exist, and there is a different reality, a different space and time, relative to each one of these reference frames. One can change frames and, hence, realities just by changing one's relative motion.

By contrast, on the spacetime interpretation, which does not link simultaneity to reality, events do not pop in and out of existence as I switch reference frames. All that changes is which class of events is orthogonal to my worldline in spacetime at a designated point and, hence, which events I reckon to be simultaneous with my present. All the events subsist tenselessly, and different hyperplanes in spacetime serve merely to mark out which events count as simultaneous relative to my inertial frame. There is a shared, objective reality which exists independently of observers or reference frames, and we all inhabit the same spacetime world; we just reckon different events in that one world to stand in the relation of simultaneity with one another.

The spacetime interpretation has been nicely illustrated by Taylor and Wheeler in their parable of the surveyors:

> Once upon a time there was a Daytime surveyor who measured off the king's lands. He took his directions of north and east from a magnetic compass needle. Eastward directions from the center of the town square he measured in meters (x in meters). Northward directions were sacred and were measured in a different unit, in miles (y in miles). His records were complete and accurate and were often consulted by the Daytimers.
>
> Nighttimers used the services of another surveyor. His north and east directions were based on the North Star. He too measured distances eastward from the center of the town square in meters (x' in meters) and sacred distances north in miles (y' in miles). His records were complete and accurate. Every corner of a plot appeared in his book with its two coordinates, x' and y'.
>
> One fall a student of surveying turned up with novel openmindedness.... In defiance of tradition, the student took the daring and heretical step to convert northward measurements, previously expressed always in miles, into meters by multiplication with a constant conversion factor, k. He then discovered that the quantity $[(x_A)^2+(ky_A)^2]^{1/2}$

Kansas City, October, 1998; Craig Callender, "Is Presentism Worth its Price?" Paper presented at the symposium "The Prospects for Presentism in Spacetime Theories;" and especially Yuri Balashov, "Enduring and Perduring Objects in Minkowki Spacetime," *Philosophical Studies* (forthcoming).

based on Daytime measurements of the position of gate A had exactly the same numerical value as the quantity $[(x_A')^2+(ky_A')^2]^{1/2}$ computed from the readings of the Nighttime surveyor for gate A. He tried the same comparison on the readings computed from the recorded positions of gate B, and found agreement here too. The student's excitement grew as he checked his scheme of comparison for all the other town gates and found everywhere agreement. He decided to give his discovery a name. He called the quantity

(1) $\quad [(x)^2+(ky)^2]^{1/2}$

the *distance* of the point (x, y) from the center of town. He said that he had discovered the *principle of the invariance of distance*; that one gets exactly the same distances from the Daytime coordinates as from the Nighttime coordinates, despite the fact that the two sets of surveyors' numbers are quite different.[30]

Taylor and Wheeler point out that the student's discovery of the concept of distance is paralleled by the Einstein-Poincaré discovery of the idea of spacetime interval. The interval as calculated from one observer's measurements agrees with the interval as calculated from another observer's measurements even though the separate coordinates employed in their calculations do not agree. The presupposition of this analogy is that all points of spacetime are as real and existent as the points on a spatial surface. Thus, Taylor and Wheeler comment,

> *The invariance of the interval*—its independence from the choice of the reference frame—forces one to recognize that time cannot be separated from space. Space and time are part of the single entity, *spacetime*. The geometry of spacetime is truly four-dimensional. In one way of speaking, the 'direction of the time axis' depends upon the state of motion of the observer, just as the directions of the y axes employed by the surveyors depend upon their different standards of 'north.'[31]

On a spacetime ontology, there is thus a unified, independent reality which is merely *measured* differently by observers using different coordinate systems. But on the relativity interpretation reality literally falls apart, and there is no one way the world is.

What all this amounts to saying is that under the relativity interpretation of SR the ontology implied by Einstein's denial of the existence of Newton's absolute space or the aether frame is unbelievable. On the spacetime interpretation the change from Newtonian spacetime to Minkowski spacetime is not really so radical ontologically, chiefly involving the denial of any preferred foliation of spacetime into hyperplanes of simultaneity.[32] There is an attendant change in the metric of spacetime, but the reality of spacetime points or events remains unaffected. But in the absence of any all-embracing spacetime, the "dynamization of space," to recall Capek's phrase,[33] which is involved in the change from Newtonian to relativistic physics is so radical as to be fantastic. Space is replaced by an infinity of disjoint spaces, each associated with a reference frame and each constituting a different

---

[30] Edwin F. Taylor and John Archibald Wheeler, *Spacetime Physics* (San Francisco: W. H. Freeman, 1966), pp. 1-2.
[31] Ibid., p. 3.
[32] See Friedman, *Spacetime Theories*, p. 125.
[33] Milic Capek, *The Philosophical Impact of Contemporary Physics* (Princeton: D. Van Nostrand, 1961), p. 170.

reality. There is no encompassing space in which all these separate spaces exist and of which they are but three-dimensional slices. Time is similarly replaced by an infinity of separate times, each associated with a reference frame and each containing different sets of events at every respective moment. Certain events will actually be instantiated in reality in different sequences depending on the relevant reference frames, and this is not a matter of mere bookkeeping: reality is literally actualized in a reverse order in some frames in comparison to others. It is little wonder that Swinburne can complain that a universe to which the concept of absolute simultaneity has no application is nearly inconceivable;[34] coming from an anti-realist interpretation of spacetime he is surely correct. As Christensen observes, "Simultaneity is not seen in ordinary thought as being relative for the simple reason that existence is not. To the question, 'What events in different regions of the universe are simultaneous with one that is now happening here?' the ordinary view answers, 'All and only those that exist!'"[35] But on the relativity interpretation there is no single answer as to what exists now because existence, like simultaneity, is relative. At the distance of the planet Neptune from Earth, for example, any event occurring there within a span of about eight hours could be regarded by us as happening now. On the relativity interpretation, this is not merely a matter of persons situated at the Earth's location using different coordinate systems to calculate what currently exists or is happening on Neptune; rather for different Earthlings in relative motion different states of affairs are actual on Neptune; there are now literally different realities on Neptune for these various Earth observers. Nor does this relativity of reality concern only greatly distant events. At the distance of the Earth's diameter, anything happening within about 1/10th of a second could be happening now for us; there is no unique instant corresponding to our instant here. On the relativity interpretation someone on the other side of the Earth, the Chinese premier Zhu Rongji, for example, may literally be dead for some of us and alive for others who are in relative motion at our shared location. I suspect that it is due to the counter-intuitive ontology of the relativity interpretation—coupled, perhaps, with a lack of serious appreciation of the spacetime interpretation—, that most of us are, as Capek said, unconsciously Newtonians, even when we profess to be relativists.[36] Indeed, I venture to speculate that much of the initial scepticism toward SR—Lorentz and Poincaré were never convinced of it—lay in the implicit assumption that SR presupposed the pluralistic ontology of the relativity interpretation, Minkowski's spacetime being construed instrumentally as a mere formalism. An anti-realist construal of spacetime, conjoined with a justifiable incredulity concerning the ontology of the relativity interpretation, left one with no choice but to reject Einsteinian relativity. Thus, for example, although he argued

---

[34] Richard Swinburne, *Space and Time*, 2d ed. (London: Macmillan, 1981), p. 201. Cf. Charles Hartshorne's bewilderment: "This relativistic conception [of contemporaneity] is quite different from the older concepts not only of simultaneity but also of contemporaneity. It is one of the strangest ideas ever introduced by science. No wonder some physicists almost went out of their minds trying to assimilate it" (Charles Hartshorne, "Bell's Theorem and Stapp's Revised View of Space-Time," *Process Studies* 7 [1977]: 184).
[35] F. M. Christensen, *Space-like Time* (Toronto: University of Toronto Press, 1993), p. 270.
[36] Capek, *Philosophical Impact of Contemporary Physics*, pp. 190-191.

(notoriously) against the consistency of SR, Herbert Dingle's conversion from being an expositor to being a determined opponent of relativity theory sprang fundamentally from his anti-realism concerning spacetime and his realization that the relativity interpretation required more than the "pure relativity" he formerly expounded. He found it incredible that absolute effects should arise from purely relative motion, and his abortive attempts to prove an inconsistency within SR were based on an inability to assimilate the strange, fragmented world assumed by the relativity interpretation. As for Minkowski spacetime, he dismissed this as "essentially metaphorical,"[37] "a purely gratuitous interpretation of an arbitrarily adopted mathematical formula."[38] "To symbolize that association [of time and space] by the image of co-ordinates in an imaginary continuum and then to endow the symbol with reality and call what is symbolized a shadow, is a proceeding which has only to be understood to be at once condemned."[39]

The difficulty posed by such an anti-realist reading of spacetime is that without real spacetime SR's denial of absolute space and time (in the sense of a privileged reference frame and attendant proper time) yields an ontology which seems truly outrageous.

Second, the relativity interpretation is explanatorily deficient. Nerlich complains,

> Without the realism of a physical Euclidean space, the ontic foundation of the relativity of uniform motion was anything but transparent. Nothing sustained it.... Special relativity before Minkowski deepened this obscurity since, in an ontology of enduring three-dimensional objects and time-extended processes there is this anomaly: three-dimensional objects have none of their three-dimensional properties (shape, mass or duration—though charge is an exception) well-defined intrinsically, despite the fact that each takes its existence, so to speak, as a shaped or massive or enduring thing. The fundamental material entities were continuants, but had no intrinsic continuant (metrical) properties. Each property had to be related to one of an infinite set of privileged frames of reference, yet the basis of this privilege remained mysterious.[40]

On the relativity interpretation physical objects have properties of shape, mass, and duration only extrinsically, relative to inertial frames, yet why this is so is not explained. Moreover, it is unclear why three-dimensional objects enduring through

---

[37] Herbert Dingle, *The Special Theory of Relativity* (London: Methuen, 1940), p. 86.
[38] Herbert Dingle, *Science at the Crossroads* (London: Martin Brian & O'Keefe, 1972), p. 172; cf. idem, "Time in Philosophy and Physics," *Philosophy* 54 (1979): 103-104, where he inveighs against Minkowski's conclusion as a "famous piece of mysticism" which has wrought inestimable harm in general thought, in philosophy, and in the understanding of physics.
[39] Dingle, *Special Theory of Relativity*, p. 86.
[40] Nerlich, *What Spacetime Explains*, p. 5. Cf. Christensen, *Space-like Time*, p. 260:
"Without the Minkowskian 'absolutes', we may be forced to hold that there are no intrinsic features in the world at all. For once again, the only non-invariant feature in original SR is electric charge, and its very uniqueness suggests that even it is really observer-dependent, being merely the same relative to all observers.... An entity must, under SR, have an *identity* that is not observer-relative, since we have to say that different observers get the same or different results for numerically the same particulars. But all of the *qualitative features* of an entity would be extrinsic, in the sense involved here. The Minkowski version of Relativity, with its world that is ultimately observer-independent, might seem highly preferable to any who find this prospect troubling."

time suffer relativistic effects such as length contraction and time dilation in virtue of their being in relative motion. It is important to realize that under the 3+1 dimensional ontology of the relativity interpretation these relativistic effects are *just as much real, physical effects* as under aether compensatory theories, such as Lorentz's. On the relativity interpretation length contraction and clock retardation cannot be dismissed as merely apparent phenomena on the analogy of the mutual observation of diminishing size when two observers retreat from each other. Admittedly, since length contraction and time dilation are reciprocal and the result of merely relative motion, it does seem incredible that they could be anything more than mere appearances, just as the so-called "pure relativists" like Bergson and Dingle insisted.[41] But Einstein realized right from the start that these effects described in his theory were real, not apparent, and could be shown to be real by various *Gedankenexperimente*. When V. Varicak asserted in 1911 that length contraction in Einstein's theory is, in contrast to Lorentz's theory, "only an apparent, subjective appearance," "psychological, not physical,"[42] Einstein responded with a *Gedankenexperiment* designed to show that the contraction does not have its roots in the arbitrary establishment of clock synchronization and length measurement procedures.[43] Consider the following situation (Figure 5.10):

Figure 5.10. Twin rod experiment. The length A*B*, marked by the coincidence of A', A" and B', B" is shorter than either A'B' or A"B".

Measuring rods A'B' and A"B" are two rods having the same length at rest and moving uniformly along the same axis of a stationary system in opposite directions. Now as they pass each other, let A* be marked in the stationary system as the point at which A' and A" coincide and B* be marked as the point in the stationary system at which B' and B" coincide. In contradiction to Newtonian physics, relativity theory predicts that the length A*B* will be shorter than either A'B' or A"B", which can be verified by placing either of the rods at rest along the length A*B*. Hence,

---

[41] For Bergson's view, see *Duration and Simultaneity*, trans. Leon Jacobson, Library of Liberal Arts (Indianapolis: Bobbs-Merrill, 1965). For Dingle, see notes 37 and 38. According to Chang, Dingle came to see that he was wrong in his claims that SR did not entail asymmetrical aging; but he demanded some causal-physical explanation for the absolute effects predicted by SR. All he received in response were explanations in terms of the geometry of Minkowski spacetime—which was precisely what he did not want! Chang himself muses, "In my opinion, Dingle's question remains unanswered to this day" (Hasok Chang, "A Misunderstood Rebellion: The Twin Paradox Controversy and Herbert Dingle's Vision of Science," *Studies in the History and Philosophy of Science* 24 [1993]: 782).
[42] V. Varicak, "Zum Ehrenfestschen Paradoxen," *Physikalische Zeitschrift* 12 (1911): 169.
[43] A. Einstein, "Zum Ehrenfestschen Paradoxen," *Physikalische Zeitschrift* 12 (1911): 509-510.

the rods in motion really do contract, just as the Lorentzian theory also predicts, and that contraction has nothing to do with synchronization procedures.

In a discussion of Einstein's twin-rod experiment, John Winnie points out the subtlety of this *Gedankenexperiment*. To require the rods to move at equal speeds would have presupposed a definition of simultaneity, but Einstein does not require this. Nor does Einstein say that the length A*B* will be the length of one of the rods multiplied by a factor of $(1 - v^2/c^2)$, but only that A*B* < A'B' (or A"B"). Regardless of the speed of the rods, classical kinematics and SR predict conflicting physical results. "According to classical kinematics the marked off length A*B* will always be equal to A'B' (or A"B"), regardless of the speeds of the moving rods, while according to the special theory, A*B* will be less than A'B' (or A"B"), again regardless of the (nonzero) speeds of the moving rods."[44]

Even more engaging thought experiments have been devised to illustrate the point. Consider, for example, the case of a rod which is too long to fall through a hole when both are at rest.[45] Now let the rod be placed in motion with the velocity v along the x axis of an inertial frame which is considered to be at rest, and a plate with a hole in it centered on the y axis be moved upward along that axis with the velocity u. The center of the rod arrives at the origin of the inertial system at the same time in that system as the rising hole reaches the plane y=0 (Figure 5.11).

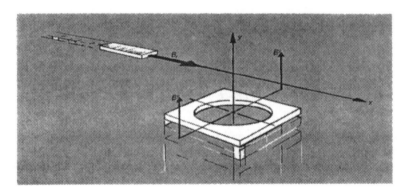

Figure 5.11. The camel and the needle's eye. Relativistic length contraction avoids a collision between the measuring rod and the plate when they both arrive at the origin of the co-ordinate system.

Classically, there should be a collision at that point between the rod and the plate. But when relativistic phenomena are taken into account, the result is strangely different. To an observer at rest in the x, y, z inertial frame, the length of the moving

---

[44] John A. Winnie, "The Twin-Rod Thought Experiment," *American Journal of Physics* 40 (1972): 1091-1094.
[45] Adapted from R. Shaw, "The Length Contraction Paradox," *American Journal of Physics* 30 (1962): 72, this example is explained in Dieter Lorenz, "Über die Realität der FitzGerald-Lorentz Kontraktion," *Zeitschrift für allgemeine Wissenschaftstheorie* 13/2 (1982): 308-312.

rod is Lorentz-contracted, while the diameter of the hole is not, and so the rod neatly passes through the hole in the resting plate and continues on its path. Similarly, an observer associated with the plate will observe the descending rod to be contracted so that it will pass through the hole in the plate. From the reference frame of the rod, however, an observer calculates that it is the diameter of the hole that is contracted rather than himself, so that the hole which was already too small for the rod to fit through when both were at rest has now become even smaller, thus apparently ensuring a collision. But in fact no collision occurs because the observer associated with the rod will determine that the rising plate is not in fact parallel to the rod, but is approaching it at an angle, so that despite the contraction of the hole, the rod can still pass through it (Figure 5.12).

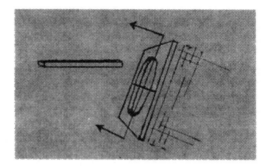

Figure 5.12. The scenario of Figure 5.11 as calculated from the inertial frame of the rod. Despite the contraction of the hole's diameter to less than the length of the rod, the collision is averted because the plate approaches the rod at an angle.

In this way, relativity theory enables the proverbial camel to pass through the eye of a needle!

Though examples of this sort are widely known, it seems that not all physicists have assimilated their significance. J. S. Bell recounts an interesting anecdote in this connection. He invites us to imagine three spaceships, A, B, and C, drifting freely in a region of space remote from other matter, without rotation and without relative motion, with B and C equidistant from A. On reception of a signal from A, the motors of B and C are ignited and they gently accelerate. Let ships B and C be similar and have similar acceleration programs. Then, as reckoned by an observer in A, they will have at every moment the same velocity and so remain separated from one another by the same distance. Suppose that a fragile thread is initially tied between B and C, just long enough to span the distance between them. Then, as the rockets speed up, the thread will be Lorentz-contracted and become too short and so must finally break. Having explained the situation, Bell recounts,

> This old problem came up for discussion once in the CERN canteen. A distinguished experimental physicist refused to accept that the thread would break, and regarded my assertion, that indeed it would, as a personal misinterpretation of special relativity. We decided to appeal to the CERN Theory Division for arbitration, and made a (not very

systematic) canvas of opinion in it. There emerged a clear consensus that the thread would *not* break!

Of course many people who give this wrong answer at first get the right answer on further reflection. They usually feel obliged to work out how things look to observers B or C. They find that B, for example, sees C drifting further and further behind, so that a given piece of thread can no longer span the distance. It is only after working this out, and perhaps only with a residual feeling of unease, that such people finally accept a conclusion which is perfectly trivial in terms of A's account of things, including the Fitzgerald contraction. It is my impression that those with a more classical education, knowing something of the reasoning of Larmor, Lorentz, and Poincaré, as well as that of Einstein, have stronger and sounder instincts.[46]

It is interesting that this very *Gedankenexperiment* is employed by the neo-Lorentzian physicist L. Jánossy as illustrative of the real Lorentz contraction in connected systems.[47] He points out that if the thread is strong enough, then the entire connected system will suffer a Lorentz-contraction, whereas if the thread is too fragile, then it will break, leaving the entire system uncontracted. He also furnishes an illustration involving two wheels on an axle (Figure 5.13).

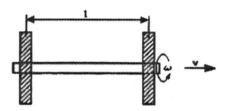

Figure 5.13. The Lorentz deformation in connected and non-connected systems.

The wheels are mounted on the axle at a distance l from each other and made to rotate with an angular velocity ω. If the system is accelerated in the direction of the axis so as to move finally with a velocity v in this direction, then the wheels slow down and their angular velocity becomes

$$\omega^* = \omega \sqrt{1 - v^2/c^2} \qquad (2)$$

Provided the wheels are rotating freely on the axle, they react similarly to the accelerating forces and no shift arises between the phases of the rotating wheels. The angular velocities which the wheels take up at time t are thus always

$$\omega^*_1(t) = \omega^*_2(t) = \omega \sqrt{1 - v(t)^2/c^2} \qquad (3)$$

---

[46] J. S. Bell, "How to Teach Special Relativity," in *Speakable and Unspeakable in Quantum Mechanics* (Cambridge: Cambridge University Press, 1987), p. 68.
[47] L. Jánossy, *Theory of Relativity Based on Physical Reality* (Budapest: Akadémiae Kiadó, 1971), pp. 128-131.

But now suppose the wheels to be fixed on the common axle. The rotating axle when accelerated shows a tendency to twist and thus causes a shift in the phases of the rotating wheels. If the forces exceed the cohesive forces of the system, then the wheels may break off the axle and rotate freely, or the axle breaks in two between the wheels or the axle deforms permanently under the stress. In any of these cases, the reality of the Lorentz-contraction will be evident.

Examples such as these could be multiplied to prove that, perhaps contrary to expectation, the relativity interpretation of relativity theory involves real, physical length contraction and clock retardation, just as much as does the aether-compensatory Lorentzian theory. Podlaha concludes,

> In the relativity theory, the length contraction and time dilatation in all frames is often viewed as a consequence of a 'perspective of observation,' similarly as a rod seems to change its length as observed under different angles...however, it is seen that the results of relativistic experiments have their origin in the length contraction and time dilatation effects which are so real as a change of the length of a rod caused by the change of temperature.[48]

Podlaha's comments are quite correct, at least as far as the relativity interpretation of SR is concerned.

The reality of these relativistic effects under the relativity interpretation reinforces our first point about the fantastic ontology associated with this interpretation, for it requires us to say, for example, that a plate really does shrink up physically in one frame but experiences no contraction at all in another and that the reality of such physical effects depends merely on which reference frame one chooses to occupy. But let that pass; the point I wish to make here is that the relativity interpretation is explanatorily deficient in that it provides, indeed, permits, no causal account of these real, physical distortions of three-dimensional continuants. In contrast to the Lorentzian theory, which sought for a causal explanation for these relativistic effects, in the primitive Einsteinian relativity interpretation they are just deduced from the theory's postulates. Lorentz himself remarked on this difference between the theories:

> His [Einstein's] results concerning electromagnetic and optical phenomena...agree in the main with those which we have obtained in the preceding pages. The chief difference being that Einstein simply postulates what we have deduced, with some difficulty and not altogether satisfactorily, from the fundamental equations of the electrodynamic field. By doing so, he may certainly take credit for making us see in the negative results of experiments like those of Michelson,...not a fortuitous compensation of opposing effects, but the manifestation of a general and fundamental principle. Yet, I think something may also be claimed in favor of the form in which I have presented the theory. I cannot but regard the ether which can be the seat of an electromagnetic field with energy and its vibrations, as endowed with a certain degree of substantiality, however different it may be from ordinary matter.[49]

According to Arthur Miller, "The principle of relativity of Bucherer, Lorentz, and Poincaré resulted from careful study of a large number of experiments, and it was on

---

[48] M. F. Podlaha, "Length Contraction and Time Dilatation in the Special Theory of Relativity—Real or Apparent Phenomena?" *Indian Journal of Theoretical Physics* 25 (1975): 74-75.

[49] H. A. Lorentz, *The Theory of Electrons* [1909] (rep. ed.: New York: Dover, 1952), pp. 229-230.

the basis of a theory in which empirical data could be explained to have been *caused* by electrons interacting with an ether. Einstein's principle of relativity excluded the ether of electromagnetic theory and did not explain anything."[50]

Arzeliès emphasizes that on the relativity interpretation "it is not possible to regard the phenomenon [of length contraction] as a deformation due to forces;" rather it is entirely a result of "the way in which the dimensions of a body in motion are measured."[51] He points out that in some cases the body undergoing contraction may not be subject to any forces and, moreover, even when it is, the pressure due to the force F and the resulting elastic deformation of the rod are the *same* in all Galilean systems. Similarly, with respect to clock retardation, "We no longer think of it as an elastic deformation due to certain forces. The body is subjected to no stress, no force, and thus undergoes no structural modification; only the way in which measurements are made introduces changes of shape as we transfer from one system of measurement to the other."[52] This might lead one to think these relativistic effects are merely apparent; but such an inference is erroneous. Arzeliès remarks,

> Some authors have stated that the Lorentz contraction only seems to occur, and is not real. This idea is false. So far as relativistic theory is concerned, this contraction is just as real as any other phenomenon. Admittedly...it is not absolute, but depends upon the system employed for the measurements; it seems that we might call it an apparent contraction which varies with the system. This is merely playing with the words, however. We must not confuse the reality of a phenomenon with the independence of this phenomenon of a change of system (absolute nature).[53]

In fact, we have seen in our above *Gedankenexperimente* that absolute structural changes do result from these relativistic effects, and these changes cannot be explained as the result of the accelerations involved (recall the Twin Paradox). These relativistic phenomena are, on the relativity interpretation, real deformations of three-dimensional objects enduring through time which have no causal explanation but are merely correlated with different reference frames. They come simply as deductions from the two postulates of the theory.

By contrast, on the spacetime interpretation, three-dimensional objects do not suffer length contraction or time dilation for the simple reason that three-dimensional objects do not exist. Reality is four-dimensional, [54] and the account of

---

[50] Arthur I. Miller, "On Some Other Approaches to Electrodynamics in 1905," in *Some Strangeness in Proportion*, ed. Harry Woolf (Reading, Mass.: Addison-Wesley, 1980), p. 85.
[51] Arzeliès, *Relativistic Kinematics*, p. 121.
[52] Ibid., p. 128.
[53] Ibid., p. 120.
[54] It is important to note that accepting the superiority of the spacetime interpretation does not without further ado commit one to spacetime substantivalism. Spacetime realists disagree among themselves whether spacetime should be construed substantivally or relationally, substantivalists maintaining that spacetime is a substance existing independently of objects and events located in it, relationalists holding that spacetime is not a substance, but depends for its existence on the existence of objects and events. Both are united in opposition to the relativity interpretation in contending that all spacetime points are equally real. For discussion see Paul Horwich, "On the Existence of Time, Space, and Spacetime," *Noûs* 12 (1978): 397-419. Embracing spacetime realism does not even commit one automatically to so-called four-dimensionalism or perdurantism in contrast to three-dimensionalism or endurantism with respect to familiar physical objects, for Mellor, a spacetime realist, espouses three-dimensionalism and endurantism.

relativistic phenomena is, on this interpretation, much more akin to the change of perspective imagined by the pure relativists. Given a spacetime ontology, the different length and time measurements given for specific objects and events by various observers is, as Taylor and Wheeler put it, just a matter of looking at them "from several angles."[55] A four-dimensional object viewed from a certain angle will appear fore-shortened in comparison with the object as viewed from a different angle. Less metaphorically, because co-ordinate time and length intervals, as opposed to proper time and length intervals, are not invariant under the Lorentz transformation equations, they vary from co-ordinate system to co-ordinate system. A Lorentz transformation may be regarded merely as a rotation of the co-ordinate axes in four-dimensional spacetime, with the result that different co-ordinate lengths and temporal intervals will be assigned to the same four-dimensional object. Although this rotation of the axes is obscured in a two-dimensional Minkowski diagram, it may be more perspicuously exhibited by means of a Loedel diagram.[56] Thus, length contraction is not the mysterious (reciprocal!) shrinking of enduring, three-dimensional objects in relative, uniform motion, but merely the result of our applying different coordinate systems to the unchanging, four-dimensional, spatio-temporal object and subtracting the values of the spatial co-ordinates of its endpoints. Similarly, time dilation does not involve a literal slowing down of relatively moving clocks as they endure through time but rather results from the application of different co-ordinate systems to the changeless four-dimensional object and calculating the difference between the temporal co-ordinates of two events. Moreover, in the pseudo-Euclidian Minkowski spacetime a curved worldline between two points is the shortest distance, and hence, the clock of an observer tracing out such a path through spacetime will record less time than a clock following a straight path. Thus in the Twin Paradox, although the path of the travelling twin appears in the diagram to be longer than that of his stationary counterpart, that impression is due to the two-dimensionality of the diagram. In four-dimensional spacetime, the path of the travelling sibling is actually shorter than that of the earth bound twin.[57] Therefore, it is not surprising that his clock records

---

He presumably takes spacetime realism to be the doctrine that all points in spacetime are equally real, which does not entail that objects are spatio-temporal wholes. I have elsewhere argued that Mellor's union of spacetime realism with endurantism is, however, ultimately untenable (William Lane Craig, *The Tenseless Theory of Time: a Critical Examination*, Synthèse Library [Dordrecht: Kluwer Academic Publishers, forthcoming], chap. 8). A consistent spacetime realist will thus view objects as spatio-temporal entities which perdure.
[55] Taylor and Wheeler, *Spacetime Physics*, p. 4.
[56] See Roger Angel, *Relativity: The Theory and its Philosophy*, Foundations and Philosophy of Science and Technology (Oxford: Pergamon Press, 1980), p. 89.
[57] See Taylor and Wheeler, *Spacetime Physics*, p. 34; cf. Hermann Bondi, *Relativity and Common Sense* (New York: Dover, 1964), p. 67; L. Marder, *Time and the Space-Traveller* (London: George Allen & Unwin, 1971), p. 78, who paradoxically also endorses Rindler's interpretation that the traveller's path is Lorentz-contracted. With respect to the situation envisioned by Bell, while the worldline of A is a straight line in Minkowski spacetime, the worldlines of B and C are parallel curves. Similarly the worldline of every point of the connecting thread describes a curve. For that reason, the proper length of the thread cannot be preserved, and thus the thread breaks. By contrast a thread connecting A to another spaceship D at rest relative to A would not break, despite the fact that relative to B and C it is Lorentz-

less time and that he is younger than his brother at their reunion. Since length measurements are inherently dependent on time measurements (length's being a matter not simply of computing $x_2 - x_1$, but $x_2(t_2) - x_1(t_1)$ on the assumption that $t_2 = t_1$), it is not surprising that a relatively moving rod, as viewed by a particular observer, should be calculated to suffer contraction. But, in a sense, the contraction of rods and the retardation of clocks is on the spacetime interpretation ultimately a chimera, since three-dimensional, enduring objects like rods and clocks do not exist.

Advocates of the relativity interpretation (and neo-Lorentzians) are apt to regard the spacetime account as a pseudo-explanation masquerading as a genuine explanation. But that is to fail to take seriously the four-dimensional ontology of the spacetime realist. Given that such an ontology of physical objects is correct, as opposed to the ontology of continuants presupposed by proponents of the relativity interpretation and neo-Lorentzians, one can appreciate why Nerlich, an enthusiastic spacetime realist, finds "the elegant spacetime interpretation of the Fitzgerald contraction and the 'slowing of moving clocks' extraordinarily insightful and intuitive."[58] By contrast the relativity interpretation is markedly explanatorily deficient.

But are we justified in assuming that the relativity interpretation in fact precludes causal explanations of the sort envisioned by neo-Lorentzians? Dennis Dieks, while admitting that it is not natural to seek for causal explanations of relativistic phenomena in Einstein's SR, nevertheless protests that there is no bar to such an investigation:

> Some authors have suggested that it is illegitimate to ask...for an explanation of the length contraction. There seems to be no justification for such a point of view. The questions as to the factors determining the contraction can meaningfully be asked within the framework of the special theory of relativity; and the theory is perfectly able to provide an answer. In fact, it would be a serious shortcoming of the theory if no such answer were forthcoming. We would then be in the situation that the theory would predict a contraction but would not be able to account for it on the basis of physical laws; in that case a relativistic molecular theory of matter would be impossible.[59]

Dieks maintains that the Lorentz contraction in the Einsteinian interpretation can be understood as the consequence of physical processes, the result of changes in the intermolecular forces which occur when a body is set in motion, in exactly the same way as it is in the neo-Lorentzian interpretation. Similarly many authors, in explaining time dilation in Einstein's SR, appeal to the illustration of the light clock to show how clocks in motion do actually run slower than relatively stationary clocks. Hence, the Einsteinian version is not deficient in causal explanatory power.

Kroes, however, disputes Dieks's claim in this regard. He charges that Dieks has overlooked the fact that in Einstein's theory these relativistic effects are *reciprocal*. Lorentz's dynamical interpretations of the shortening of rods and of time dilation only make sense because they are not reciprocal.

---

contracted. As in the Twin Paradox, only the genuinely accelerating system suffers asymmetrical effects. My thanks to Michael Friedman and Graham Nerlich for illuminating discussions of this case.

[58] Graham Nerlich, personal communication, June 20, 1991.

[59] Dennis Dieks, "The 'Reality' of the Lorentz Contraction," *Zeitschrift für allgemeine Wissenschaftstheorie* 15/2 (1984): 340.

> The shortening is an objective feature of a rod which is due to a change in the forces acting between the constituents of the rod and this change in the forces is caused by the motion relative to the ether. But since this motion with respect to the ether is absolute, the shortening of the rod is not reciprocal. And because of this, the Lorentz contraction and its dynamical interpretation cannot be transplanted just like that from Lorentz's to Einstein's theory.
> ...the reciprocity of length contraction and time dilation rules out dynamical interpretations of the Lorentzian type.[60]

With regard to time dilation, Kroes calls the well-known illustration of the light clock (Figure 9.1) a "pseudo-explanation." For a vital element in the "explanation" of the time dilation of a light clock is the constancy of the velocity of light for all inertial observers, which is precisely the crucial point involved in the synchronization of distant clocks. "So it is more or less a coincidence that in the case of the light clock time dilation can be 'explained': the operation of the light clock and the synchronization of distant clocks are based upon the same kind of physical process (and its properties), namely the propagation of light in vacuum."[61] If Kroes is correct in this analysis, then the relativity interpretation of SR not only does not seek but actually excludes causal explanations for the relativistic effects in question.

But how does the reciprocity of length contraction and time dilation evacuate dynamical explanations of meaning? In the absence of metaphysical space, any inertial frame can be conventionally considered to be at rest. The dynamical explanations of contraction and dilation phenomena relative to this frame seem to make as much sense as they do in Lorentz's theory relative to the aether. All that they lack is the same absoluteness. For the moving frame could be considered to be at rest and the stationary frame as moving, and then the dynamical explanation would be reversed. To think that dynamical explanations cannot be reciprocal seems to revert surreptitiously to the conception of Newtonian metaphysical space with respect to which alone dynamical accounts are valid. But if no such fundamental frame exists, then why should, odd as it seems, reciprocal dynamical explanations be ruled out?

Kroes admits that it is possible to achieve reciprocity within Lorentz's theory by introducing Einstein's definition of simultaneity. This is because according to Einstein's method of clock synchronization, it is true for *all* clocks that $t_B - t_A = t'_A - t'_B$, whereas for Lorentz this formula and, hence, the synchronization procedure based on it, holds good only for clocks at rest relative to the aether. By introducing Einstein's method into Lorentz's theory one arrives at judgements of simultaneity relations which the Lorentzian will regard as skewed, and this will in turn affect judgements concerning length measurements, since in order to measure the length of a moving rod one must determine the positions of the endpoints at the same time. A rod at rest in the aether would, if a Lorentzian synchronization procedure were used,

---

[60] Peter Kroes, "The Physical Status of Time Dilation within the Special Theory of Relativity," paper delivered at the International Conference of the British Society for the Philosophy of Science, "Physical Interpreatations of Relativity Theory," Imperial College of Science and Technology, London, September 16-19, 1988.
[61] Ibid.

appear to be expanded to a moving observer, but would, given an Einsteinian procedure, appear contracted (just as the moving rod will appear contracted from the reference frame of the aether-based observer).[62] Hence, the Einsteinian synchronization procedure will result in reciprocal relativistic phenomena even though dynamical causes are at work. Of course, for the Lorentzian this reciprocity will be in appearance only, and so Kroes protests that Einstein's synchronization procedure would be unacceptable to a Lorentzian. No doubt; but that is because an absolute frame exists. But what would happen if the aether frame were regarded as no longer absolute? Here we enter into the realm of counterfactual discourse where intuitions can become very unclear. For example, is it true that

1. If the aether frame were not absolute, then the reciprocal relativistic effects which would obtain between it and other frames would still have dynamical causes,

as Dieks's view allows, or is it true that

2. If the aether frame were not absolute, then the reciprocal relativistic effects which would obtain between it and other frames would no longer have dynamical causes,

as Kroes seems to hold? In favor of (1) it might be said that the dynamical causal factors present in the aether-based theory would not vanish just because the aether frame would no longer be absolute, for their existence does not depend on the absoluteness of that frame, but on properties of the rods and clocks themselves. On the other hand, in favor of (2) it might be said that the dynamical causes do depend in some way on the existence of an absolute frame. For if the aether frame were not absolute, then it would be in motion relative to other frames and relativistic effects would occur in it which would be as real as those obtaining in other frames. But then on the above reasoning the same dynamical causes would be present in it as in other frames. But where did they come from? The defender of (1) must believe that the dynamical causes sprang into being even though no intrinsic property of things in the frame was changed, but only its absoluteness. Since the relativistic effects in this frame are not due to intrinsic, dynamical causes, neither are similar effects with respect to all other frames.

What seems evident from this dispute is that both parties must agree that the existence of dynamical causes is counterfactually (*N.B.*, not causally) dependent upon the absoluteness of a certain frame. For given the truth of the Lorentzian theory, the defender of (1) must hold that

3. If the aether frame were not absolute, then dynamical causes would be as present in it as they are in all other frames,

and the defender of (2) must accept that

---

[62] See discussion in Karl R. Popper, "A Note on the Difference between the Lorentz-Fitzgerald Contraction and the Einstein Contraction," *British Journal for the Philosophy of Science* 16 (1966): 332-333; Jan Dorling, "Length Contraction and Clock Synchronization: The Empirical Equivalence of the Einsteinian and Lorentzian Theories," *British Journal for the Philosophy of Science* 19 (1968): 67-69.

4. If the aether frame were not absolute, then dynamical causes would be as absent from all other frames as they are from it.

But which is true? Under the condition of the relativization of the aether frame, would the dynamical causes be universally present or absent? Conversely, given the truth of the Einsteinian theory, the defender of (1) must maintain that

5. If an arbitrary frame were to be absolute, then dynamical causes would no longer be present in it,

and the defender of (2) must hold that

6. If an arbitrary frame were to be absolute, then dynamical causes would no longer be absent from all other frames.

But which of these is true? Would the absolutization of an arbitrary frame counterfactually imply that dynamical causes are no longer present in it or still absent from it?

Perhaps some headway could be made by recalling Einstein's illustration of the magnet and the conductor. There, it will be remembered, an electric current is generated by the relative motion of the conductor and the magnet. The Lorentzian and Einsteinian are in complete agreement about this. But, according to the Lorentzian, if the magnet is at rest, then the current is induced by the magnet's force on the conductor's electric charges, whereas if the conductor is at rest the current is caused by an electric field associated with the magnet. The causal explanations for the observed effect are very different. The Lorentzian and Einsteinian would both affirm counterfactuals such as

7. If the magnet were at rest in the aether, the current would be caused by its force on the conductor's charges

and

8. If the conductor were at rest in the aether, the current would be caused by the magnet's electric field.

But suppose there were no absolute state of rest, as the Einsteinian claims. Then the question as to whether the field or the force causes the current becomes malformed. The cause of the current is simply the magnet and the conductor's relative motion. To ask for intrinsic causes is simply not to understand the situation.

Now something very much like this seems to be going on in the search for causal explanations of length contraction and time dilation in the context of the relativity interpretation. The Einsteinian cannot affirm that such phenomena are the products of intrinsic, dynamical causes because such a statement is malformed. These effects, like the electric current, are the result of relative motion. If there were absolute rest, then intrinsic causes would be identifiable. But in the absence of metaphysical space or a fundamental frame, the question disappears because it makes no sense. Therefore, the relativity interpretation does, indeed, preclude causal explanations of relativistic phenomena.[63]

Now some SR theorists will respond that within the context of SR no causal explanation of relativistic phenomena is necessary, and, therefore, failure to provide such explanations is no shortcoming. Grünbaum, for instance, argues that it is inept to demand of a theory explanations for phenomena which are simply considered natural in the context of that theory. Questions which may be well-posed in the context of one theory may be entirely out of place in the context of another theory. He gives the example of the demand for a cause of uniform motion in Aristotelian as opposed to Newtonian physics:

> According to Aristotle's physics, an external force is needed as the cause of a body's uniform motion in a straight line. In his physics the demand for such a perturbational (disturbing) external cause to explain uniform motion arises from the following assumption: when a body is not acted on by an external force, its *natural*, spontaneous unperturbed behavior is to be at rest. Yet, as we know, Galileo's analysis of the motions of spheres on inclined planes led him to conclude that the empirical evidence speaks against just this Aristotelian assumption. As Newton's First Law of Motion tells us, uniform motion does not require any external force as its cause; only accelerated motion does.... But, if so, then the Aristotelian demand for an *explanation* of uniform motion *by reference to an EXTERNAL, perturbing force* begs the explanatory question by means of *a false underlying assumption*, rather than asks a well-posed legitimate question as to the 'why' of uniform motion. By the same token, Galileo and Newton could only shrug their shoulders or throw up their hands in despair, if an Aristotelian told them that he has a solution to the 'problem' of the external cause of uniform motion, whereas they do not.[64]

Applying this lesson to the case at hand, Grünbaum asserts that Lorentz made the mistake of thinking that Newtonian physics defines what is natural and so postulated *causes* to explain the result of the Michelson-Morley experiment, whereas Einstein saw that these results do not require causes after all because they are integral to the normal behavior of things.[65] Arthur Miller comments, "Einstein's two axioms of relativity theory do not explain the failure of the ether-drift experiments or, equivalently, why the measured velocity of light always turns out to be c, or why one cannot catch up with a light wave. Rather it is axiomatic that the space of every inertial reference system is homogeneous and isotropic for the propagation of light."[66] From this relativistic effects follow naturally.

---

[63] See further Alfons Grieder, "Relativity, Causality and the 'Substratum'," *British Journal for the Philosophy of Science* 28 (1977): 35-48, who concludes: "...Einstein's theory gives rise to a causal anomaly: not only is it unable to give rise to a causal explanation, but it also blocks the way to a substratum theory of relativity and hence to what appears to be the most attractive causal approach to relativistic phenomena" (Ibid., p. 46).

[64] Adolf Grünbaum, "The Pseudo-Problem of Creation in Physical Cosmology," *Philosophy of Science* 56 (1989): 386.

[65] Adolf Grünbaum, "Science and Ideology," *Scientific Monthly* 79 (July 1954): 16-17.

[66] Miller, *Einstein's Special Theory*, p. 167. At the same time, Miller concedes that Einstein, like Langevin, was troubled by the question of the cause of relativistic effects and preferred a dynamical cause for time dilation, to be sought in GR, but that Builder demonstrated that general relativistic treatments of the clock paradox succeeded in only complicating the problem and often lacked physical significance (Ibid., pp. 272-273). See A. Einstein, "Dialog über Einwände gegen die Relativitätstheorie,"

Writing specifically on the reality of the Lorentz contraction, Dieks argues that while the Lorentzian and Einsteinian theories are empirically equivalent, there is a crucial philosophical difference between them in their ontology that renders questions about length contraction which are well-posed for the Lorentzian ill-posed for the Einsteinian:

> The philosophical difference between Lorentz's electron theory and the special theory of relativity is exactly the different ontology of the two theories. Whereas an adherent of the electron theory works with the classical spacetime relations, and therefore feels obliged to *explain* deviations from these relations which occur if moving rods are used, the relativist will, following Einstein's postulates, take it for granted that this world is relativistic, and not classical.[67]

According to Dieks the Principle of Relativity assumes a central explanatory role within Einstein's SR comparable to the role of the classical ontology in the electron theory. The question why moving rods do not behave classically therefore does not arise in a natural manner within Einstein's theory. Since there is no role in the theory for the classical ontology, there is no need to ask for explanations of deviations from it. The transformation properties are simply taken for granted and require no explanation.

In a fine discussion of the reality of time dilation in SR, Kroes similarly objects to the language of clock *retardation* (and length *contraction*). For although time dilation is real, it is not in any way the result of dynamical causes, of moving clocks' running more slowly than stationary ones, but arises out of Einstein's synchronization procedure for distant clocks. In the Twin Paradox, the travelling clock (clock 2) does not run at a slower rate than the stationary clock (clock 1):

> The idea that the rate of clock 2...must somehow have undergone a slowing down during its motion in order to be able to explain the time difference at [the point of their reunion] C, can only be saved by making a tacit appeal to the notion of absolute time. This goes as follows. Suppose that a clock which behaves according to classical kinematics moves along with clock 2. This classical clock will show the same reading at C as clock 1. With regard to this classical clock, which measures absolute time, it does make sense to claim that the rate of clock 2 is slowed down.
>
> But as soon as the notion of absolute time is abandoned, the claim that the rate of clock 2 has slowed down, becomes meaningless. In my opinion, therefore, it is best to forget within the context of relativity theory about 'changing rates of moving clocks' altogether. Abandoning the notion of absolute time precisely implies that ideal, i.e. undisturbed, clocks can measure different amounts of time between events A and C. It also implies giving up the search for a cause for the time difference between the two clocks at C.... The difference in time between the two clocks at C is a 'brute' fact about the behavior of clocks, just as the triangle inequality is a brute fact about the behaviour of rods. In both cases, the search for an explanation of these facts is superfluous and pointless.[68]

Naturwissenschaften (1918): 697-702; Geoffery Builder, "The Resolution of the Clock Paradox," *Australian Journal of Physics* 10 (1957): 246-262.
[67] Dieks, "'Reality' of the Lorentz Contraction," p. 341.
[68] Kroes, "Physical Status of Time Dilation."

The claim that the relativity interpretation of SR is explanatorily deficient because it fails to give causal explanations for length contraction and time dilation is therefore misconceived.

This counter-rebuttal on behalf of the relativity interpretation has to be very carefully weighed. It is true that length contraction and time dilation follow automatically from the theory's denials of absolute time and space and its redefinition of key concepts. Given the theory's presuppositions and postulates, these phenomena are deducible. But the very fact that there is an interpretation of SR, namely, the spacetime interpretation, which does provide an explanation for these phenomena demonstrates that the quest for an explanation of them is neither inappropriate nor misguided. Now it is true that *within* the relativity interpretation, it is inappropriate (as we have seen) to look for causes of the relativistic phenomena. Similarly, *within* the Lorentzian interpretation, causal explanations should and can be sought. But presumably the issue here is not whether theoreticians are acting consistently *within* the framework of their respective theories or interpretations; rather the question is a meta- or extra-theoretical query posed by someone *without*, or outside, the theories. It is a question of theory-assessment, a question of which theory or interpretation is to be adopted.

Here the claim on behalf of the spacetime interpretation is that it possesses in some sense greater explanatory power than the relativity interpretation. Not that the relativity interpretation does not account adequately for all the phenomena, for it does. But a theory can account for all the phenomena without having much explanatory power. For example, the phenomena could be simply postulated in the axioms of the theory itself, in which case no phenomena would go unaccounted for, but the explanatory power of the theory *vis à vis* those phenomena would be nil.[69] Similarly, by gratuitously postulating as *natural* some phenomenon that would be regarded extra-theoretically as puzzling, one can avoid having to give an explanation of virtually anything. The question is why we should adopt the theory *with* its assumptions. One reason would be that the theory under consideration increases our explanatory power, it involves an increase of intelligibility over a rival theory which has to appeal to more brute facts or givens. If what is simply a brute fact in one theory can be given an explanation in another theory, then we have an increase in intelligibility that counts in favor of the second theory. Now something of this sort can be plausibly claimed on behalf of the spacetime interpretation versus the relativity interpretation. What is simply deduced from basic postulates and definitions in the one is given an explanation in the other, facts that are simply "brute" and "taken for granted" are accounted for in terms of spacetime structures in the other.

---

[69] As John Worrall points out,
"...*any* theory can be made to have correct empirical consequences by 'writing those consequences into it'; the cases that have traditionally induced realist-inclinations in even the most hard-headed are cases of theories, designed with one set of data in mind, that have turned out to predict entirely unexpectedly some further general phenomena" (John Worrall, "How to Remain (Reasonably) Optimistic: Scientific Realism and the 'Luminiferous Ether,'" in *PSA 1994*, ed. David Hull, Micky Forbes, and Richard M. Burian [East Lansing, Mich.: PSA, 1994], p. 335).

Of course, if there are good empirical grounds for accepting the postulates of a theory, then we may be justified or even forced to regard certain phenomena deduced therefrom as natural and not in need of any explanation. Grünbaum's example of Newtonian versus Aristotelian accounts of inertial motion would be a case in point. But it hardly needs to be said that there is nothing in Einstein's postulates which provides empirical grounds for preferring the relativity interpretation, which takes length contraction and time dilation as brute facts, over the spacetime interpretation, which explains such phenomena in terms of spacetime structure.[70]

Thus, it seems to me that the spacetime interpretation of SR has greater explanatory power than the original relativity interpretation. Phenomena which are simply brute facts according to the relativity interpretation can be explained within the spacetime interpretation, thus increasing our understanding of the physical world.

CONCLUDING REMARKS

In summary, if what I have argued is correct, the relativity interpretation of SR, with its pluralistic ontology and contracting and retarded three-dimensional continuants, is fantastic and explanatorily impoverished. In that case, the move to relativize tense and temporal becoming and, hence, reality to reference frames is a desperate expedient. On the other hand, the great advantage of the relativity interpretation is that by retaining a classical 3+1 dimensional ontology of space and time, it makes room for the objective reality of tense and temporal becoming. By contrast, the spacetime interpretation through its commitment to spacetime realism precludes temporal becoming and seems to permit no room for tensed properties or modes of existence like presentness. The spacetime interpretation thus flies in the face of our experience of time. Psychologist William Friedman, who has made a career of the study of time consciousness, reports, "Like [temporal] order and the causal priority principle, the division between past, present, and future so deeply permeates our experience that it is hard to imagine its absence."[71] We have, he writes, "an irresistible tendency to believe in a present. Most of us find quite startling the claim of some physicists and philosophers that the present has no special status in the physical world, that there is only a sequence of times, that the past, present, and future are only distinguishable in human consciousness."[72] As a result, virtually all philosophers of space and time, including proponents of B-Theories of tenseless time, admit that the view of the common man is that time involves a past, present, and future which are objectively real and that things or events really do come to be and pass away in time.[73] Powerful considerations in

---

[70] This is not to say that the explanation must be causal; see Robert Di Salle, "Spacetime Theory as Physical Geometry," *Erkenntnis* 42 (1995): 317-337.
[71] William Friedman, *About Time* (Cambridge, Mass.: MIT Press, 1990), p. 92.
[72] Ibid., p. 2.
[73] For example, Oaklander grants that "non-philosophers have never doubted that there is such a phenomenon as temporal flow or passage...." (L. Nathan Oaklander, *Temporal Relations and Temporal*

favor of the objectivity of tensed facts and temporal becoming include (i) the indispensability and irreducibility of linguistic tense, which mirrors the tensed facts which are characteristic of reality,[74] and (ii) our experience of presentness and temporal becoming, which like our experience of the external world, may be plausibly taken to ground properly basic beliefs in the reality of tensed time.[75] The premier defender of the tenseless theory of time D. H. Mellor frankly admits that "Tense is so striking an aspect of reality that only the most compelling argument justifies denying it: namely, that the tensed view of time is self-contradictory and so cannot be true."[76] Powerful objections can, in turn, be lodged against theories of tenseless time, including (i) in the absence of objective distinctions between past, present and future, the relations ordering events are only gratuitously regarded as genuinely temporal relations of *earlier/later than*,[77] (ii) even the subjective illusion of temporal becoming involves itself an objective temporal becoming of contents of consciousness,[78] and (iii) spacetime realism implies perdurantism, the view that objects have spatio-temporal parts, a doctrine which is metaphysically counter-intuitive, incompatible with moral accountability, and entails the bizarre counterpart theory of transworld identity.[79] Discussion of these several points lies far beyond the scope of this book, but I think it is evident that any interpretation of reality which finds itself obliged to deny the reality of tensed facts and temporal becoming faces *prima facie* a disadvantage.

Certain spacetime theorists have therefore argued that their interpretation of SR can be combined with an A-Theory of tensed time. Thus, Howard Stein,[80] in response to Putnam's claim that SR has eliminated the notion of becoming, proposes to relativize becoming to spacetime points:

Becoming [Lanham, Maryland: University Press of America, 1984], p. 1). Horwich muses that "The quintessential property of time, it may seem, is the difference between past and future" (Paul Horwich, *Asymmetries in Time* [Cambridge, Mass.: MIT Press, 1987], p. 15). Coburn confesses, "If the existence of A-facts [*i.e.*, tensed facts] is an illusion, it is one of our most stubborn ones" (Robert C. Coburn, *The Strangeness of the Ordinary: Problems and Issues in Contemporary Metaphysics* [Savage, Maryland: Rowman & Littlefield, 1990], p. 118). Nerlich is forced to conclude, "Something in experience deludes us about time, something obvious and pervasive—there obtrusively in every thought and perception, yet elusive to analysis" (Nerlich, "Time as Spacetime," p. 130). Mellor acknowledges, "the experienced presence of experience, is the crux of the tensed view of time...and the tenseless camp must somehow explain it away" (D. H. Mellor, *Real Time* [Cambridge: Cambridge University Press, 1981], p. 6).
[74] See my *The Tensed Theory of Time: a Critical Examination*, Synthèse Library (Dordrecht: Kluwer Academic Publishers, forthcoming), Pt. I, sec. 1. The outstanding spokesman for this point of view is Quentin Smith, *Language and Time* (New York: Oxford University Press, 1993).
[75] See my *Tensed Theory of Time*, Pt. I, sec, 2. One of the most eloquent spokesmen for this point of view has been George Schlesinger, *Aspects of Time* (Indianapolis: Hackett, 1980), pp. 34-39, 138-139.
[76] Mellor, *Real Time*, pp. 4-5.
[77] See my *The Tenseless Theory of Time: a Critical Examination*, Synthèse Library (Dordrecht: Kluwer Academic Publishers, forthcoming), Pt. II, sec. 1, chap. 7.
[78] See my *Tenseless Theory of Time*, Pt. II, sec. 1, chap. 8.
[79] See my *Tenseless Theory of Time*, Pt. II, sec. 1, chap. 9. See also the study by Trenton Merricks, "Endurance and Indiscernibility," *Journal of Philosophy* 91 (1994): 165-184; see also Delmas Lewis, "Persons, Morality, and Tenselessness," *Philosophy and Phenomenological Research* 47 (1986): 305-309 and the incisive piece by Peter Van Inwagen, "Four Dimensional Objects," *Noûs*, 24 (1990): 245-255.
[80] Howard Stein, "On Einstein-Minkowski Space-Time," *Journal of Philosophy* 65 (1968): 14; idem, "On Relativity Theory and the Openness of the Future," *Philosophy of Science* 58 (1991): 146-167.

> For an event at spacetime point a, those events, and only those, *have already become* (real or determinate), which occur at points in the topological closure of the past of a.

According to this explication, reality is relative to spacetime points and consists in all those events lying inside or on the past-directed light cone of a, including a itself. According to Stein, in the context of SR we can only think of the temporal evolution of the world from the chronological perspective of each spacetime point. There exists no knife-edge of becoming on this view, even for individual reference frames, but mere pinpoints of becoming constituted by the vertex of any hypothetical observer's past light cone, none of which is privileged.[81] Putnam's error lay in thinking that the relation R "is real to" is transitive, such that an event at spacetime point c, which is real for an observer at spacetime point b, is also real to an observer at spacetime point a, if events at b are real for a. Just as "is simultaneous with" is not a transitive relation, neither is R.

One of the major difficulties with such an attempt to combine temporal becoming with spacetime realism is that this proposal relapses into the same fragmentation of reality characteristic of the Einsteinian interpretation. It thus succumbs to the same criticisms lodged by Mellor against attempts to relativize tense and temporal becoming to what is here and now.[82] Worse, since all events in the topological closure of any spacetime point's past are real, the proposal runs smack into McTaggart's famous paradox.[83] Spacetime events would have somehow to remain identical over time even though they possess at different times different tense determinations like pastness and presentness, and apparently impossible feat. Thus, the attempt to combine a theory of tensed time with a spacetime interpretation of SR seems abortive.

The spacetime interpretation of SR thus faces the dilemma of either denying the objective reality of tensed facts and temporal becoming or else succumbing to the same disintegration of reality that plagues the relativity interpretation and, if McTaggart is right, ultimately to incoherence. What is desirable is a physical interpretation of the SR formalism predicated upon a coherent metaphysic which permits us to affirm both a theory of tensed time and a unified view of reality. Therefore, it behooves us to probe more deeply into the philosophical foundations of SR and the concept of time to see if in fact Einstein has so transformed that conception as to make a presentist metaphysic untenable.

---

[81] See Rob Clifton and Mark Hogarth, "The Definability of Objective Becoming in Minkowski Spacetime," *Synthèse* 103 (1995): 355-387.

[82] See the withering critique of Stein's view by Craig Callender, "Shedding Light on Time," (preprint). Arguing that Stein's defined relation R "is not remotely close to the traditional becoming relation," Callender concludes that "R cannot possibly be a relation of serious philosophical interest to the philosophy of time"; hence, "any notion of becoming *remotely similar to that found among advocates of the tensed view of time* is not compatible with Minkowski spacetime." The attempt by Mauro Dorato, *Time and Reality: Spacetime Physics and the Objectivity of Temporal Becoming*, Collana di Studi Epistemologici 11 (Bologna: CLUEB, 1995), to interpret Stein's theory indexically evacuates temporal becoming of any objective status and thus replaces it with what Le Poidevin rightly calls "a tenseless *ersatz* becoming" (Robin Le Poidevin, critical notice of *Time and Reality*, by Mauro Dorato, *Studies in History and Philosophy of Modern Physics* 28 B [1997]: 545).

[83] See my "McTaggart's Paradox and the Problem of Temporary Intrinsics," *Analysis* 58 (1998): 122-127.

CHAPTER 6

THE CLASSICAL CONCEPT OF TIME

P robably not too many physicists and philosophers of science would disagree with Wolfgang Rindler's judgement that with the development of SR Einstein took the step that would "destroy the classical concept of time."[1] But what is or was the classical concept of time, and how did Einstein's critique render it untenable? In order to answer those questions, we need to recur to the fountainhead of the classical

---

[1] Wolfgang Rindler, "Einstein's Priority in Recognizing Time Dilation Physically," *American Journal of Physics* 38 (1970): 1112. Cf. the verdicts of W. G. V. Rosser: "The theory of special relativity necessitated the abandonment of the concept of absolute time...." (W. G. V. Rosser, *An Introduction to the Theory of Relativity* [London: Butterworths, 1964], p. 397); Stephen Hawking: "...the theory of relativity put an end to the idea of absolute time! ...The theory of relativity does...force us to change fundamentally our ideas of space and time. We must accept that time is not completely separate from and independent of space, but is combined with it to form an object called spacetime" (Stephen W. Hawking, *A Brief History of Time: From the Big Bang to Black Holes*, with an Introduction by Carl Sagan [New York: Bantam Books, 1988], pp. 21, 23); E. F. Taylor and J. A. Wheeler, who note that the invariance of the spacetime interval "forces one to recognize that time cannot be separated from space" (Edwin F. Taylor and John Archibald Wheeler, *Spacetime Physics* [San Francisco: W. H. Freeman, 1966], p. 3); Fritjof Capra, who opines that relativity theory "shattered...the notion of absolute space and time," for space and time are not separate entities, but are both "intimately connected and form a four-dimensional continuum, 'spacetime'" (Fritjof Capra, *The Tao of Physics*, 2d ed. [London: Fontana, 1983], pp. 69-71); W. V. O. Quine, who concludes that relativity theory "leaves no reasonable alternative to treating time as space-like" (Willard Van Orman Quine, *Word and Object* [Cambridge, Mass.: MIT Press, 1960], p. 172); A. d'Abro: "...relativity compels us to abandon our traditional understanding of space and time.... this discovery of the relativity of simultaneity marks a date of the same momentous importance as did the discovery of the Copernican system in astronomy" (A. d'Abro, *The Evolution of Scientific Thought*, 2d rev. ed [n. p.: Dover Publications, 1950], pp. ix, 171); Howard Stein: "...the results of physics leave the *ordinary* notion of 'absolute simultaneity' as empty of real content...as is, the earth being round, the old ordinary notion of the 'absolutely' vertical" (Howard Stein, Critical notice of *The Language of Time* by Richard M. Gale, *Journal of Philosophy* 66 [1969]: 355); John Sinks: "...the special theory forces one to recognize that what is in the future is not in the future *simpliciter*, but in the future relative to some inertial reference frame" (John D. Sinks, "On Some Accounts about the Future," *Journal of Critical Analysis* 2 [1971]: 15); Paul Fitzgerald, who believes that relativity theory "forces us to revise the old views of the 'future'": "Crudely speaking, events that are 'now' in your future may be 'now' in my past..." (Paul Fitzgerald, "The Truth about Tomorrow's Sea Fight," *Journal of Philosophy* 66 [1969]: 310-311); Hermann Bondi, who concludes that SR "forces revisions in the common sense concept of time"—we have to get used to the idea that time is a private matter; "That is to say, that MY time is what MY watch tells me" (Hermann Bondi, *Relativity and Common Sense* [New York: Dover, 1964], p. 65); J. G. Taylor: "We have obtained a time distortion formula which destroys Newton's idea of a universal time" (J. G. Taylor, *Special Relativity*, Oxford Physical Series [Oxford: Clarendon Press, 1975], p. 15); P. C. W. Davies, who concurs that Newton's universal and absolute time was wrong because relativity theory reveals that clock rates depend on the motion and gravitational situation of the observer (P. C. W. Davies, "Spacetime Singularities in Cosmology and Black Hole Evaporations," in *The Study of Time III*, ed. J. T. Fraser, N. Lawrence, and D. Park [Berlin: Springer, 1978], pp. 74-75); Julian Barbour: "Newton's absolute time was completely overthrown by the revolution of the Special Theory of Relativity" (Julian B. Barbour, *Absolute or Relative Motion?*, vol. 1: *The Discovery of Dynamics* [Cambridge: University Press, 1989], p. 633).

concept of time: Isaac Newton and his *Philosophiae naturalis principia mathematica*.

NEWTON'S DISTINCTION BETWEEN ABSOLUTE AND RELATIVE TIME

The *Scholium* to his Definitions in the *Principia* is the *locus classicus* of Newton's exposition of his concepts of time and space.[2] Newton observes that such quantities as time, space, place, and motion are "popularly conceived solely with reference to the objects of sense perception" in terms of "the relation they bear to sensible objects," and thence arise certain prejudices.[3] In order to overcome these, Newton draws a dichotomy with respect to these quantities between "absolute and relative, true and apparent, mathematical and common." With regard to time and space he asserts:

> 1. Absolute, true, and mathematical time, in and of itself and of its own nature, without reference to anything external, flows uniformly and by another name is called duration. Relative, apparent, and common time is any sensible and external measure (precise or imprecise) of duration by means of motion; such a measure--for example, an hour, a day, a month, a year--is commonly used instead of true time.
>
> 2. Absolute space, of its own nature without reference to anything external, always remains homogeneous and immovable. Relative space is any movable measure or dimension of this absolute space; such a measure of dimension is determined by our senses from the situation of the space with respect to bodies and is popularly used for immovable space, as in the case of space under the earth or in the air or in the heaven, where the dimension is determined from the situation of the space with respect to the earth. Absolute and relative space are the same in species and in magnitude, but they do not always remain the same numerically. For example, if the earth moves, the space of our air, which in a relative sense and with respect to the earth always remains the same, will now be one part of the absolute space into which the air passes, now another part of it, and thus will be changing continually in an absolute sense.[4]

---

[2] Isaac Newton, *The Principia*, trans. I. Bernard Cohen and Anne Whitman, with a Guide by I. Bernard Cohen (Berkeley: University of California Press, 1999), pp. 408-415.

[3] Ibid., p. 408. "vulgus quantitates hasce nonaliter quam ex relatione ad sensibilia concipiat." (The critical edition of the *Principia* is Isaac Newton, *Philosophiae Naturalis Principia Mathematica*, 3d ed. [1726], ed. Alexandre Koyré and I. Bernard Cohen, 2 vols. [Cambridge, Mass.: Harvard University Press, 1972]; for the *Scholium* on time and space see vol. 1, p. 46.)

[4] Ibid.

> "I. Tempus absolutum, verum, & mathematicum, in se & natura sua, sine relatione ad externum quodvis, aequabiliter fluit, alioque nomine dicitur Duratio: Relativum, apparens, & vulgare est sensibilis & externa quaevis durationis per motum mensura (seu accurata seu inaequabilis) qua vulgus vice veri temporis utitur, ut hora, dies, mensis, annus.
>
> II. Spatium Absolutum, natura sua sine relatione ad externum quodvis, semper manet similare & immobile: Relativum est Spatii hujus mensura, seu dimensio quaelibet mobilis, quae a sensibus nostris per situm suum ad corpora definitur, & à vulgo pro spatio immobili usurpatur: uti dimensio spatii subterranei, aerii vel coelestis definita per situm suum ad terram. Idem sunt spatium absolutum & relativum, specie & magnitudine; sed non permanent idem semper numero. Nam si terra, verbi gratia, moveatur, spatium aeris nostri, quod relative & respectu terrae semper manet idem, nunc erit una pars spatii absoluti in quam aer transit, nunc alia pars ejus; & sic absolute mutabitur perpetuo."

Though much misunderstood and greatly vilified, Newton's distinction deserves our thoughtful consideration.[5] The most evident feature of this distinction is the independence of absolute time and space from the relative measures thereof. Absolute time or simple duration exists regardless of the sensible and external measurements which we try, more or less successfully, to make of it. In other words, clock time may or may not register the true time. J. R. Lucas calls this the "rational theory of clocks."[6] We cannot measure time directly because we cannot take an interval of, say, an hour and lay it out against another interval, and compare the two in the way that we compare lengths. Rather we must measure time indirectly by having rules which enable us to pick out pairs of instants and to say that the interval between one pair is equal to, or greater than, or twice as great as, the interval between another pair. If two intervals are isochronous, or equal in duration, it is not because of our fiat, but because they really are. The witness to this fact is that our clocks are corrigible. Thus, "...time is not what the clocks say, but what they are trying to tell, are there to tell."[7] It may seem rather puzzling that Newton calls only absolute time mathematical, since clock time certainly seems mathematical in character. But the contrast which Newton has in mind here is, I think, the same as the difference between a geometrical circle and a physical circle: the latter is only a more or less accurate approximation to the former, true circle. Like mathematical objects, absolute time is not constituted by or dependent upon sensible approximations thereto. In stating that absolute time "flows uniformly" without relation to anything external, Newton implies that time itself is not metrically amorphous. As Kroes explains, Newtonian absolute time possesses its own intrinsic metric with which our clocks seek to stay in harmony:

---

[5] With respect to the opprobrium heaped upon Newton at this point, John Earman remarks,
"What I find especially disturbing about such condemnations of Newton is not the injustice they do to Newton but rather the fact that they are possible only after an abdication of philosophical responsibility. In all the philosophical literature with which I am acquainted, there is precious little attempt to give reasonably clear and precise answers to the questions which are central to the cluster of philosophical issues which revolve around Newton's conception of space and time....
...it seems to me that Newton demonstrated a much deeper understanding of the nature of space and time than Berkeley, Leibniz, and Mach. And so far as I can see, neither modern philosophers of science...nor the people identified by modern philosophers as major philosophical figures of the 17th, 18th, and 19th centuries, have succeeded in raising any compelling philosophical objections to absolute space, absolute time, or absolute spacetime..." (John Earman, "Who's Afraid of Absolute Space?" *Australasian Journal of Philosophy* 48 [1970]: 288, 317).

[6] J. R. Lucas, *A Treatise on Time and Space* (London: Methuen, 1973), p. 62-64, 69.

[7] Ibid., p. 64. He concludes,
"The fact that we have a rational theory of clocks vindicates Newton's doctrine of absolute time. If we really regarded time simply as the measure of process, we should have no warrant for regarding some processes as regular and others as irregular.... Even our best clocks are subject to correction. So long as we are prepared to assess the time-keeping qualities of a clock, and are prepared in principle to replace it by a more regular one, if it could be obtained, we are committed to an idea of absolute time which is not simply what the clocks actually say" (Ibid., p. 91).
Cf. the judgement of Richard Swinburne, *Space and Time*, 2d ed. (London: Macmillan, 1981), p. 202, who agrees that Newton was correct: "there is a true time which might or might not be recorded by actual measuring instruments."

> To Newton it was self-evident that there exists just one fundamental metric for time: the intrinsic metric of absolute time. Physical processes of whatever kind could provide a more or less accurate 'sensible measure' thereof. A 'true' sensible measure of absolute time could only be reached in the case of a perfectly isolated, completely undisturbed periodical system which would constitute an ideal clock. In a certain sense, all physical processes had to obey, according to Newton, the rhythm of absolute time; an ideal clock, of whatever nature (mechanical, gravitational, etc.) could provide an exact measure of the unique, fundamental metric of absolute time.[8]

The clock retardation that occurs in the context of SR and GR would not have disturbed Newton, since, as we shall see, he freely concedes that we may not have any accurate measure of time. Newton, of course, did not anticipate the relativistic phenomenon of clock retardation, but it would have troubled him little to learn that clocks in motion or in gravitational fields run slowly.

Kroes concurs that if we reject Newton's distinction between time and its measures, then the statement that two time intervals are isochronous can only be conventionally true.[9] Without absolute time, our temporal metrics as determined by, say, mechanical, gravitational, or electromagnetic clocks, may not stay in synchronism, and the unity of time becomes a mere assumption.[10]

Similarly, Newtonian space is absolute in the sense that it is distinct from the relatively moving spaces associated with inertial frames. Max Jammer has emphasized that the absolute statement of Newton's first law of motion,

> I. *Every body perserveres in its state of being at rest or of moving uniformly straight forward, except in so far as it is compelled to change its state by forces impressed*[11]

requires the assumption of absolute space as a prerequisite of its validity.[12] The classical principle of inertia becomes meaningless apart from a state of absolute rest

---

[8] Peter Kroes, *Time: Its Structure and Role in Physical Theories*, Synthèse Library 179 (Dordrecht: D. Reidel, 1985), p. 49. Cf. John Earman, *World Enough and Spacetime* (Cambridge, Mass.: MIT Press, 1989), p. 8.

[9] Thus Sklar observes that Newton is on to something "of vital importance" in drawing his distinction (Lawrence Sklar, "Real Quantities and their Sensible Measures," in *Philosophical Perspectives on Newtonian Science*, ed. Phillip Bricker and R. I. G. Hughes [Cambridge, Mass.: MIT Press, 1990], p. 61). Sklar notes that there are natural measures of time which yield simple, elegant laws of nature and that a wide variety of clocks will not only agree with each other in their metric of time but will measure time in a way that approximates the natural measure. If, on the other hand, time itself is not distinguished from its measures, then *any* process has equal right to the status of the standard measure, regardless of how sporadic it might be relative to the concordant "natural" measures. See further J. R. Lucas and P. E. Hodgson, *Spactime and Electromagnetism* (Oxford: Clarendon Press, 1990), p. 239.

[10] David Park seems oblivious to the vicious circularity of his reasoning when he says,
   "*Time is what is measured by a clock.* What is a clock?
   A clock is a device whose law of motion is known.... How is it that we can define time in terms of clocks and clocks in terms of time without running into trouble? There is an assumption of regularity in the world that underlies the definition. We assume, in fact, that there is a universal time that governs all motion.... And the basis for the assumption is our knowledge of the world. It might have been otherwise—at least one can easily imagine it otherwise—but it is not" (David Park, *The Image of Eternity* [Amherst, Mass.: University of Massachusetts Press, 1980], p. 40).

[11] Newton, *Principia*, p. 416. "Corpus omne perseverare in statu suo quiesciendi vel movendi uniformiter in directum, nisi quatenus illud a viribus impressis cogitur statum suum mutare."

relative to which bodies may be said to be in a state of motion. Since in Corollary V Newton affirms Galilean relativity, however, the first law of motion, while necessitating the existence of absolute space, provides no means by which that space can be experimentally distinguished from relative spaces. Since the laws of motion hold in all inertial frames, "it turns out that the condition that absolute space be that space in which Newton's laws hold fails to specify a unique space as absolute space. The condition picks out an infinite set of spaces, the inertial spaces, which are the relative spaces of a set of observers moving uniformly with respect to one another in inertial motion."[13] Jammer muses that "If Newton had been a confirmed positivist he would have acknowledged all uniformly moving inertial systems as equivalent to each other. As it was, only one absolute space existed for him."[14]

Newton proposed to pick out experimentally the privileged frame of absolute space, not by his laws of motion, but by means of rotation, which is experientially distinguished from merely relative motion with respect to its effects.[15] Modern commentators have observed that Newton's *Gedankenexperimente* of the rotating bucket and the revolving globes only succeed in demonstrating the existence of absolute motion, not absolute space. A spinning bicycle tire, for example, will be rotating in every inertial frame, whether, for example, the bicycle is taken to be moving and the earth at rest, or the earth moving and the bicycle at rest; but such absolute motion does not serve to pick out any reference frame as preferred.

Absolute space, in contrast to the plurality of relative spaces, is one and immovable. In stating that absolute and relative space are the same in species and magnitude, differing only in that a volume of relative space moves through different (equal) volumes of absolute space, Newton shows that he has no suspicion of what came to be known as the Lorentz-FitzGerald contraction. Sklar observes the oddity that Newton does not mention measuring rods as a more or less accurate measure of the absolute metric of space.[16] But it is implied that absolute space, like absolute time, is endowed with an intrinsic metric which determines whether the distances we measure are equal. Because "these parts of space cannot be seen," Newton explains, we use in their stead "sensible measures," so that "we define all places on the basis of the positions and distances of things from some body that we regard as immovable,."[17] On the other hand, "in philosophy abstraction from the senses is required."[18] Again, if we reject Newton's distinction, then it becomes impossible to

---

[12] Max Jammer, *Concepts of Space* (Cambridge, Mass.: Harvard University Press, 1954), pp. 99-103. Cf. Einstein's verdict that this realization was "one of Newton's greatest achievements" (Albert Einstein, "Foreword" to *Concepts of Space*, by Max Jammer, p. xiv). See also Fritz Rohrlich, *From Paradox to Reality* [Cambridge: Cambridge University Press, 1987], chap. 5.
[13] John D. Norton, "Philosophy of Space and Time," in *Introduction to the Philosophy of Science*, ed. Merrilee Salmon (Englewood Cliffs, N. J.: Prentice-Hall, 1992), p. 181.
[14] Janner, *Concepts of Space*, pp. 100-101.
[15] Newton, *Principia*, p. 412-414.
[16] Sklar, "Real Quantities," p. 61.
[17] Newton, *Principia*, p. 410.
   "Verum quoniam hae Spatii partes videri nequeunt, & ab invicem per sensus nostros distingui; earum vice adhibemus mensuras sensibilis. Ex positionibus enim & distantiis rerum a corpore aliquo, quod spectamus ut immobile, definimus loca universa."
[18] Ibid. "in Philosophicis autem abstrahendum est a sensibus."

hold that two spatial distances are objectively congruent, and we must swallow metric conventionalism with respect to space.[19]

Up to this point our discussion of Newton's distinction between absolute and relative quantities has concerned the difference between time and space as independent of their sensible measures and time and space as measured. To recall our original discussion of the variety of senses in which time and space may be said to be "absolute," Newton's distinction thus far discussed concerns only the "absolute-measured" distinction, according to which true time and space are distinct from measured time and space. But, of course, Newton also conceived of time and space as absolute in a more profound sense, which is expressed in the "absolute-relational" distinction; namely, he held that time and space are absolute in the sense that they exist independently of any physical objects whatsoever. He claims, as we have seen, that absolute time "in and of itself and of its own nature, without reference to anything external, flows uniformly" and that absolute space "of its own nature without reference to anything external, always remains homogeneous and immovable."

Usually, this is interpreted to mean that time and space would exist even if nothing else existed, that there exists a possible world which is completely empty except for the container of absolute space and the flow of absolute time. But here we must be very careful. Modern secular scholars have tended frequently to forget how ardent a theist Newton was and how central a role this theism played in his metaphysical outlook. Noting that Newton considered God to be temporal and therefore time to be everlasting, David Griffin observes that "Most commentators have ignored Newton's heterodox theology, and his talk of 'absolute time' has been generally misunderstood to mean that time is not in any sense a relation and hence can exist apart from actual events."[20] In fact, Newton made quite clear in the General Scholium to the *Principia*, which he added in 1713, that absolute time and space are constituted by the divine attributes of eternity and omnipresence:

> He is eternal and infinite..., that is, he endures from eternity to eternity and he is presently from infinity to infinity....he is not eternity and infinity, but eternal and infinite; he is not duration and space, but he endures and is present. He endures always and is present everywhere, and by existing always and everywhere he constitutes duration and space. Since each and every particle of space is *always*, and each and every indivisible moment of duration is *everywhere*, certainly the maker and lord of all things will not be *never* or *nowhere*.[21]

---

[19] For a compelling demonstration of just how indigestible metric conventionalism is see Graham Nerlich, *The Shape of Space*, 2d ed. (Cambridge: Cambridge University Press, 1994).
[20] David Ray Griffin, "Introduction: Time and the Fallacy of Misplaced Concreteness," in *Physics and the Ultimate Significance of Time*, ed. David R. Griffin (Albany, N. Y.: State University of New York Press, 1986), pp. 6-7. See also W.-H. Newton-Smith, "Space, Time, and Spacetime: A Philosopher's View," in *The Nature of Time*, ed. Raymond Flood and Michael Lockwood (Oxford: Basil Blackwell, 1986), p. 27, who errs, however, in saying that space and time are aspects of God, rather than concomitants of God.
[21] Newton, *Principia*, p. 941.
> "Aeternus est & Infinitus,...id est, durat ab aeterno in aeternum, & adest ab infinito in infinitum.... Non est aeternitas & infinitas, sed aeternus & infinitus; non est duratio & spatium, sed durat & adest. Durat semper, & adest ubique; & existendo semper & ubique, durationem & spatium constituit. Cum unaquaeque spatii particula sit *semper*,

Because God is eternal, there exists an everlasting duration, and because He is omnipresent, there exists an infinite space. Absolute time and space are therefore relational in that they are contingent upon the existence of God.

This may at first seem puzzling, since we have seen that Newton held that absolute time and space have their respective properties of themselves and of their own natures, without relation to anything external. Two things may be said. (i) Newton does not say that absolute time and space *exist* by their own natures and without relation to anything external, but merely that absolute time *flows uniformly* of its own nature and non-relationally and that absolute space *remains* of its own nature and non-relationally *isotropic and immovable*. Absolute time and space could possess these properties essentially and yet not exist in every possible world. (ii) The key to Newton's thinking lies in another of God's attributes, namely, divine necessity. Since God is a necessary being, He exists in all possible worlds, and therefore (since He is also essentially eternal and omnipresent) time and space also exist in every possible world, though their existence is contingent upon God's existence. Newton writes,

> God is one and the same God always and everywhere. He is omnipresent not only *virtually* but also *substantially*; for action requires substance. In him all things are contained and move,* but he does not act on them nor they on him. God experiences nothing from the motions of bodies; the bodies feel no resistance from God's omnipresence.
>
> It is agreed that the supreme God necessarily exists, and by the same necessity he is *always* and *everywhere*.

---

*This opinion was held by the Ancients....also by the sacred writers....[22]

Since God exists in all possible worlds and is essentially eternal and omnipresent, absolute time and space also exist in all possible worlds, whether in those worlds there exists any creation at all. Thus, time and space are absolute in the sense that they exist independently of any physical objects, but that is not to say that they exist wholly non-relationally.[23]

& unumquodque durationis indivisibile momentum *ubique,* certe rerum omnium Fabricator ac Dominus non erit *nunquam, nusquam.*"

[22] Ibid., pp. 941-942.

"Deus est unus & idem Deus semper & ubique. Omnipraesens est non per *virtutem* solam, sed etiam per *substantiam:* nam virtus sine substantia subsistere non potest. In ipso continentur & moventur universa, sed sine mutua passione. Deus nihil patitur ex corporum motibus: illa nullam sentiunt resistentiam ex omnipraesentia Dei. Deum summum necessario existere in confesso est: et eadem necessitate *semper* est & *ubique*.

*Ita sentiebant veteres.... Ita etiam scriptores sacri...."

[23] According to Earman, by "without relation to anything external," Newton meant without relation to material bodies (Earman, "Absolute Space," p. 289). It is worth drawing attention to the fact that while the absolute-relational distinction implies the absolute-measured distinction, the converse is not the case. That is to say, if time is absolute in the non-relational sense, then it is also absolute in the sense that its metric is the standard for isochronous intervals; but it is not obviously the case that if time transcends our attempts to measure it then time exists non-relationally, wholly independently of all events. One could quite consistently maintain, as Leibniz did, that time would not exist in the utter absence of events but is in some sense constituted by the fact of change (perhaps in God's actions or thoughts), and also maintain that there is a true time (God's time) which our physical clocks approximate. Similarly, one could maintain that space is absolute in the sense that it is approximated but not constituted by our

In Newton's thinking, metaphysical time, God's time, is A-theoretic time and the foundation of becoming. We have seen that metaphysical time "of its own nature, without reference to anything external flows uniformly." To Newton therefore it makes perfectly good sense to imagine that all ideal clocks and physical processes in the universe should start to run twice as fast as they previously had.[24] God could simply speed up everything so that events occur as in a film being run on fast-forward. His time, metaphysical time, would remain unaffected by this alteration in physical time. In the absence of metaphysical time, such a speed-up becomes literally meaningless. Since clocks would define what a temporal interval is, it would be nonsense to say that clocks could all run twice as fast as they had, for the minute hand's circling the face of a clock one time *is* the temporal interval we call an hour. The hand would not circle the clock's face twice in an hour; rather two hours would elapse. That is not to say that two hours would elapse in an interval previously designated as one hour, for that is to presuppose the standard of metaphysical time to which physical time is relative. Rather in the absence of metaphysical time, it is just two hours elapsing in two hours' time, and any talk of a speed-up is meaningless. For Newton, however, it would seem obvious that it lies within God's power to make all physical processes occur more quickly, which implies the existence of metaphysical time.

NEWTON'S THEISM AND THE CLASSICAL CONCEPT OF TIME

Scholars writing during the positivist era tended to treat Newton's theological disquisitions dismissively as incidental intrusions into otherwise sober scientific thinking. E. W. Strong, for example, writing in 1952, contended that the absence of any references to God in the original editions of the *Principia* and *Opticks* showed that God was irrelevant to works written by a scientist for a scientific purpose:

> Newton nowhere asserts that space and time are postulated as absolute because they are the sensorium of God and hence cannot be other than absolute in natural philosophy without impropriety.... When it is said that he constitutes duration and space, it may, with equal logic be said that he constitutes bodies also in their composition and in their motions relative to one another.[25]

Similarly Jammer seems to see Newton's references to God as indications of the onset of senility: "So a comparison of the first and later editions of the *Principia* shows that the identification of absolute space with God, or with one of his attributes, came into the foreground of Newton's thought only toward the end of his

---

measurements of it without holding that it is absolute in the sense of either "non-relational" or "non-dynamical." Newton never provides any argument, so far as I can determine, as to why God's existence must be spatial. *Contra* Newton, one could consistently hold that in the absence of physical objects space would not exist and also that the metaphysical space which does exist is non-Euclidean globally and locally and yet maintain that it is absolute in the sense that it constitutes a privileged fundamental frame which provides the background for local reference frames (relative spaces), against which absolute motion or rest is determined.

[24] For interesting discussion see George N. Schlesinger, "What does the Denial of Absolute Space Mean?" *Australasian Journal of Philosophy* 45 (1967): 44-60; Adolf Grünbaum, "The Denial of Absolute Space and the Hypothesis of a Universal Nocturnal Expansion: A Rejoinder to George Schlesinger," *Australasian Journal of Philosophy* 45 (1967): 61-91.

[25] E. W. Strong, "Newton and God," *Journal of the History of Ideas* 13 (1952): 154.

life, that is, at the beginning of the eighteenth century."[26] Even a contemporary philosopher like Howard Stein insists that Newton's ideas are not based in theology; rather a certain theology was acceptable because its conceptions agreed with those required by mechanics.[27]

With the demise of positivism and the awakening of interest in the history of science, contemporary historians and philosophers of science have come to realize that Newton's statements concerning God need to be taken seriously, and this has led to a new appreciation of the role of Newton's theology in his scientific thinking. Observing that "Metaphysics is both a starting point and the final aim of Newton's physics" and that his declared commitments on the subject of God's place in his system must therefore be taken at face value, Zev Bechler reflects the consensus of contemporary Newtonian historiography when he writes,

> The effort to present Newton as a sober positivistic, no-nonsense scientist fit for incorporation into the venerable origins of the twentieth century tradition leads scholars to attribute to him ideas about the autonomy of science which are strangely out of tune with his declared commitments.[28]

In editing the *Principia*, Bernard Cohen discovered that in Book III, Proposition 8, Corollary 5 of the first edition, Newton explicitly states that God placed the planets in their orbits at appropriate distances, but that this reference to the deity was dropped in subsequent editions.[29] "The result of this alteration has been that almost all commentators of Newton have erroneously assumed that Newton mentioned God in the *Principia* only in the later editions," specifically in the General Scholium, Cohen notes; "We may now reject wholly the view that Newton's introduction of God into the *Principia* was a result of senility, intellectual decline, or even a later development" after the first edition.[30] In fact, on 10 December 1692, Newton confided to Richard Bentley, "When I wrote my Treatise about our System, I had an Eye upon such Principles as might work with considering Men, for the Belief of a Deity, and nothing can rejoice me more than to find it useful for that purpose."[31] Similarly, in the Latin edition of the *Opticks* (1706), Newton declares space to be "the Sensorium of a Being incorporeal, living and intelligent, who sees the things themselves intimately, and thoroughly perceives them, and comprehends them

---

[26] Jammer, *Concepts of Space*, p. 108.
[27] Howard Stein, "Newtonian Spacetime," *Texas Quarterly* 10 (1967): 198. Cf. the emotional reactions in L. A. Whitt, "Absolute Space: Did Newton Take Leave of His (Classical) Empirical Senses?" *Canadian Journal of Philosophy* 12 (1982): 709-710, to Newton's giving theological reasons for absolute space (which the author mislocates as being in the Scholium to the Definitions!). Whitt huffs: "Newton has given way (or perhaps vent) to his theological scruples and has, for the brief space of this scholium, gone off the (empirical) rails" (Ibid., p. 711).
[28] Zev Bechler, "Introduction: Some Issues of Newtonian Historiography," in *Contemporary Newtonian Research*, ed. Zev Bechler, Studies in the History of Modern Science 9 (Dordrecht: D. Reidel, 1982), p. 13. Cf. Alan Gabbey, "Newton and Natural Philosophy," in *Companion to the History of Modern Science*, ed. R. C. Olby, G. N. Cantor, J. R. R. Christie, and M. J. S. Lodge (London: Routledge, 1990).
[29] Newton, *Principia*, $E_1\{405.35\}$, p. 383.
[30] I. Bernard Cohen, *Introduction to Newton's 'Principia'* (Cambridge: Cambridge University Press, 1971), p. 156.
[31] Isaac Newton to Richard Bentley, December 10, 1692, in *Isaac Newton's Papers and Letters on Natural Philosophy*, 2d ed., ed. with a General Introduction by I. Bernard Cohen (Cambridge, Mass.: Harvard University Press, 1978), p. 280.

wholly by their immediate presence to himself...."[32] But the most important factor in the reassessment of the role of Newton's theology in his science was undoubtedly the scholarly excavation of Newton's unpublished manuscripts, roughly a third of which are theological in character, which disclose that the themes of the General Scholium were lifelong concerns of Newton. The Yahuda manuscripts 15.3, f. 59 and 15.5, f. 98, housed at the Jewish National and University Library in Israel, are in part virtually identical to the text of the General Scholium.[33] Newton's "On the Gravity and Equilibrium of Fluids" (1666-70) is an especially rich resource for his ideas on space, time, and divinity, which are echoed in the General Scholium.[34] McGuire has also identified an untitled manuscript on time, place, and God in the Cambridge University Portsmouth Collection which constitutes an intermediate stage between these two.[35]

In *De gravitatione* Newton argues that space (and by implication time) is neither substance, nor accident, nor nothing at all. It cannot be nothing because it has properties, such as infinity and isotropy. It cannot be an accident because it can exist without bodies. Neither is it a substance:

> It is not substance; on the one hand, because it is not absolute in itself, but is as it were an emanent effect of God, or a disposition of all being; on the other hand, because it is not among the proper dispositions that denote substance, namely actions, such as thoughts in the mind or motions in the body.[36]

Contrary to the conventional understanding, Newton here declares explicitly that space is *not* in itself absolute (*non absoluta per se*) and therefore not a substance. Rather it is an emanent—or emanative—effect of God (*Dei effectus emanativus*). The notion of emanative causality had played a key role in the metaphysic of the Cambridge Platonist Henry More. In his Axiom XVI More explained, "By an *Emanative Cause* is understood such a Cause as merely by Being, no other activity or causality interposed, produces an Effect."[37] Correlatively, More explained in his Axiom XVII:

> An Emanative Effect is coexistent with the very Substance of that which is said to be the Cause thereof.

---

[32] See Richard S. Westfall, *Never at Rest: A Biography of Isaac Newton* (Cambridge: Cambridge University Press, 1980), p. 647.
[33] See Westfall, *Never at Rest*, p. 749.
[34] Isaac Newton, "On the Gravity and Equilibrium of Fluids," [*De gravitatione et aequipondio fluidorum*] in *Unpublished Scientific Papers of Isaac Newton*, ed. A. Rupert Hall and Marie Boas Hall (Cambridge: Cambridge University Press, 1962), pp. 89-156.
[35] Manuscript Add. 3965, section 13, folios 541^r-542^r; 545^r-546^r. See J. E. McGuire, "Newton on Place, Time, and God: An Unpublished Source," *British Journal for the History of Science* 11 (1978): 114-129.
[36] Newton, *De gravitatione*, p. 132.
> "Non est substantia tum quia non absolute per se, sed tanquam Dei effectus emanativus, et omnis entis affectio quaedam subsistit; tum quia non substat ejusmodi proprijs affectionibus quae substantiam denominant, hoc est actionibus, quales sunt cogitationes in mente et motus in corpore."

[37] Henry More, "The Immortality of the Soul," in *Philosophical Writings of Henry More*, ed. with an Introduction and Notes by Flora Isabel MacKinnon, Wellesley Semi-Centennial Series (New York: Oxford University Press, 1925), p. 74.

> This must needs be true, because that very substance which is said to be the Cause, is the adequate and immediate Cause, and wants nothing to be adjoyned to its bare essence for the production of the Effect; and therefore by the same reason the Effect is at any time, it must be at all times, or so long as that substance does exist.[38]

Three times in *De gravitatione* Newton calls space an emanative effect of God. It is uncreated and co-existent with God and yet ontologically dependent upon Him for its being. Newton calls space a disposition or affection of being (*entis affectio*). To be is to be spatially and, as Newton goes on to explain, temporally.

> Space is a disposition of being *qua* being. No being exists or can exist which is not related to space in some way. God is everywhere, created minds are somewhere, and body is in the space that it occupies: and whatever is neither everywhere nor anywhere does not exist. And hence it follows that space is an effect arising from the first existence of being [*entis primario existentis effectus emanativus*], because when any being is postulated, space is postulated. And the same may be asserted of duration: for certainly both are dispositions of being or attributes according to which we denominate quantitatively the presence and duration of any existing individual thing. So the quantity of the existence of God was eternal, in relation to duration, and infinite in relation to the space in which he is present; and the quantity of the existence of a created being was as great, in relation to duration, as the duration since the beginning of its existence, and in relation to the size of its presence as great as the space belonging to it.[39]

Both space and time are thus inherent to being. God's infinite being thus has as its consequence infinite time and space, which represent the quantity of His duration and presence.

In the Neo-Platonic tradition represented by More the doctrine of emanation is associated with pantheism or panentheism. But, as Newton makes clear in the *General Scholium*, he does not conceive of space or time as in any way aspects of God Himself. Rather absolute space and time are concomitant consequences of God's existence.

Although on Newton's view space (and by implication time) is not a substance in the sense that it is an independently existing entity, nevertheless it is very much like a substance. McGuire explains,

> Newton refers to the nature of space as an existing thing, and has in mind a decided list of properties which he considers belong properly to its nature: it has inherent parts that constitute its structure; in all directions it possesses unlimited extension; it is actually infinite; independently of things external to it, its inherent parts maintain the same eternal and immutable order; all true 'positions, distances, and local motions' have

---

[38] Ibid.
[39] Newton, *De gravitatione*, pp. 136-137.
   "Spatium est entis quaetenus ens affectio. Nullum ens existit vel potest existere quod non aliquo modo ad spatium refertur. Deus est ubique, mentes creatae sunt alicubi, et corpus in spatio quod implet, et quicquid nec ubique nec ullibi est id non est. Et hinc sequitur quod spatium sit entis primario existentis effectus emanativus, quia posito quolibet ente ponitur spatium. Deque Duratione similia possunt affirmari: scilicet ambae sunt entis affectiones sive attributa secundum quae quantitas existentiae cujuslibet individui quoad amplitudinem praesentiae et perseverationem in suo esse denominatur. Sic quantitas existentiae Dei secundum durationem aeterna fuit, et secundum spatium cui adest, infinita; et quantitas existentiae rei creatae secundum durationem tanta fuit quanta duratio ab inita existentia, et secundum amplitudinem praesentiae tanta ac spatium cui adest."

reference to these parts alone; and lastly, space is passively receptive to being occupied, and can offer no resistance to motion. It is clear that Newton considers space to be an individual in the sense of being the proper subject of certain sorts of predications.... Newton believes space to be a positive individual embodying real properties which derive from its actual extension in three dimensions. As such, it is...an individual fully actual at all time, and endowed with a rich inherent structure.[40]

Space and time thus have a sort of quasi-substantival status for Newton.

What becomes clear from this historical excursus is that the classical, Newtonian concept of time is rooted in a theistic metaphysic. Far from being after-the-fact theological reflections on concepts arising solely from physical considerations, the views of the General Scholium on time and space are the fruit of a long study which from start to finish was metaphysical and theological in character.[41] James Force rightly concludes,

> Newton would have been astonished to learn that some of his interpreters, following Hume's lead, have claimed that theology, metaphysics, and epistemology have no necessary, integrated, synthetic relationship in themselves, much less that he himself has been placed into this school. Newton's own thought is in fact a seamless unity composed of theology, metaphysics, and epistemology all mixed together because, at their base, is the Lord God of supreme dominion.[42]

Our exposition of Newton's views on God, space, and time makes it evident that when Newton speaks of divine eternity, he does not, like scholastic theologians in the Augustinian tradition, mean a state of timelessness, but rather infinite and everlasting temporal duration. In a preliminary draft of the General Scholium,

---

[40] J. E. McGuire, "Space, Infinity and Indivisibility: Newton on the Creation of Matter," in *Contemporary Newton Research*, p. 163.

[41] Westfall states,
"Composed virtually at the end of his active life, the General Scholium contained a vigorous reassertion of those principles which Newton had adopted in his rebellion against the perceived dangers of Cartesian mechanical philosophy. The same principles had continued to govern his scientific career as he followed the consequences of his rebellion into a new natural philosophy and a new conception of science" (Westfall, *Never at Rest*, p. 749).

[42] James E. Force, "Newton's God of Dominion: The Unity of Newton's Theological, Scientific, and Political Thought," in *Essays on the Context, Nature, and Influence of Isaac Newton's Theology*, ed. James E. Force and Richard H. Popkin, International Archives of the History of Ideas 128 (Dordrecht: Kluwer Academic Publishers, 1990), p. 90. Cf. Bechler's comment, "The view of physics and metaphysics as one indivisible block was inevitable to him, for physics was the inquiry into the nature of bodies in space and time, but since space and time were the abode of God, there was no possible way to demonstrate separate compartments of inquiry within all this" (Bechler, "Introduction," p. 14); similarly McGuire: "...his ontology of space and time cannot be understood without fully appreciating how it relates to the nature of divine existence" (McGuire, "Existence, Actuality, and Necessity: Newton on Space and Time," *Annals of Science* 35 [1978]: 463); and Brooke, "Newton himself provides one of the most spectacular examples of the integration of scientific and religious interests in one and the same mind" (John Hedley Brooke, "Science and Religion," in *Companion to the History of Modern Science*, p. 775). Elsewhere Brooke observes, "If theistic discourse constitutes part of natural philosophy, it becomes anachronistic to ask how Newton *reconciled* his science with his religion. The right question is more likely to be: *how did his distinctive view of God's dominion, in both nature and history, affect his interpretation of nature?*" (John Brooke, "The God of Isaac Newton," in *Let Newton Be!*, ed. John Fauvel, Raymond Flood, Michael Shortland, and Robin Wilson [Oxford: Oxford University Press, 1988], pp. 171-172. So also Ivor Leclerc, "The Relation between Natural Science and Metaphysics," in *The World View of Contemporary Physics*, ed. with an Introduction by Richard F. Kitchener (Albany: State University of New York Press, 1988), pp. 26-27.

Newton had explicitly rejected the conception of God's eternity as an eternal now: "His duration is not a *nunc stans* without duration, nor is his presence nowhere."[43] Writing to Des Maiseaux in 1717, he elaborated,

> The schoolmen made a *Nunc stans* to be eternity & by consequence an attribute of God & eternal duration hath a better title to that name, though it be but a mode of his existence. For a *nunc stans* is a moment wch always is & yet never was nor will be, which is a contradiction in terms.[44]

Far from being atemporal, God's now or present is thus the present of absolute time. Since God is not "a dwarf-god" located at a place in space,[45] but is omnipresent, every indivisible moment of duration is everywhere, as we saw in the General Scholium. There is thus a worldwide moment which is absolutely present. Dieks effectively captures the implications of Newton's theism for our understanding of time when he writes,

> In classical mechanics absolute simultaneity can easily be interpreted in terms of the '*flowing now*' we know from introspection. Just like [sic] our personal '*now*' is connected with our progressing history, the universal 'now' determined by the absolute simultaneity relation may be regarded as related to one history, e.g. the history of the Universe as a whole. For Newton *God* endures, and thereby constitutes time. This naturally fits in with the conception that there must be a universal succession of moments, determining the existence of one entity...there is *one* history, with *one* '*now*,' dividing past and future.[46]

Newton's temporal theism thus provides the foundation for both absolute simultaneity and absolute becoming. These are features first and foremost of metaphysical time, God's time, and derivatively of measured or physical time.

NEWTONIAN TIME AND RELATIVISTIC TIME

Returning to Newton's Scholium to his Definitions, we find that Newton, after making further distinctions with regard to absolute and relative place and absolute and relative motion, proceeds to grant freely that due to Galilean relativity we cannot for the most part know whether something is at absolute rest or in absolute motion.

> But since these parts of space cannot be seen and cannot be distinguished from one another by our senses, we use sensible measures in their stead. For we define all places on the basis of the positions and distances of things from some body that we regard as immovable, and then we reckon all motions with respect to these places, insofar as we conceive of bodies as being changed in position with respect to them. Thus, instead of

---

[43] Cited in McGuire, "Predicates of Pure Existence: Newton on God's Space and Time," in *Philosophical Perspectives on Newtonian Science*, p. 93.
[44] Newton to Des Maiseaux, in *Unpublished Papers*, p. 357. See also his rejection of God's existing *totum simul* in "Place, Time, and God" (ms. add. 3965, sect. 13, f. 545r-546r, in McGuire, "Newton on Place, Time, and God," p. 121).
[45] Newton, "Place, Time, and God," in McGuire, "Newton on Place, Time, and God," p. 123.
[46] Dennis Dieks, "Newton's Conception of Time in Modern Physics and Philosophy," in *Newton's Scientific and Philosophical Legacy*, ed. P. B. Scheurer and G. Debrock (Dordrecht: Kluwer Academic Publishers, 1988), pp. 156-157.

absolute places and motions we use relative ones, which is not inappropriate in ordinary human affairs, although in philosophy abstraction from the senses is required.[47]

But these commonly used relative quantities should not be confused with their absolute counterparts:

> Relative quantities, therefore, are not the actual quantities whose names they bear but are those sensible measures of them (whether true or erroneous) that are commonly used instead of the quantities being measured. But if the meanings of words are to be defined by usage, then it is these sensible measures which should properly be understood by the terms 'time,' 'space,' 'place,' and 'motion,' and the manner of expression will be out of the ordinary and purely mathematical if the quantities being measured are understood here. Accordingly those who there interpret these words as referring to the quantities being measured do violence to the Scriptures. And they no less corrupt mathematics and philosophy who confuse true quantities with their relations and common measures.[48]

Newton believed that there were certain properties, causes, and effects which served to distinguish absolute motion and rest from relative motion and rest.[49] It is noteworthy, however, that some of these do not manifest themselves empirically. Others do, such as rotation, and Newton concluded his Scholium with the promise, "...but in what follows, a fuller explanation will be given of how to determine true motions from their causes, effects, and apparent differences, and, conversely, of how to determine from motions, whether true or apparent, their causes and effects. For this was the purpose for which I composed the following treatise."[50]

How, then, we may ask, did Newton fall short in comparison with Einstein's analysis of time? A little reflection reveals that the shortcoming of Newton's analysis lay not in his belief in a metaphysical time distinct from physical time, as is so often alleged, but in his not realizing that the accuracy of physical time in its approximation to metaphysical time depends on the relative motion of one's clocks. Though Newton labeled his distinction between time and the sensible measures thereof "absolute" versus "relative," his physical time was in fact as absolute as metaphysical time in the sense that it, too, was well-defined independent of any

---

[47] Newton, *Principia*, pp. 410-411.
"Verum quoniam hae Spatii partes videri nequeunt, & ab invicem per sensus nostros distingui; earum vice abhibemus mensuras sensibiles.... Sic vice locorum & motuum absolutorum relativis utimur; nec incommode in rebus humanis: in philosophicis autem abstrahendum est a sensibus."

[48] Ibid., pp. 413-414.
"Quantitates relativæ non sunt igitur eæ ipsæ quantitates, quarum nomina præ se serunt, sed sunt earum mensuræ illæ sensibiles (veræ an errantes) quibus vulgus loco quantitatum mensuratarum utitur. At si ex usu definiendæ sunt verborum significationes; per nomina illa Temporis, Spatii, Loci & Motûs propriè intelligendæ erunt hæ mensuræ sensibilis; & sermo erit insolens & purè mathematicus, si quantitates mensuratæ hic intelligantur. Proinde vim inferunt sacris literis, qui voces hasce de quantitatibus mensuratis ibi interpretantur. Neque minus contaminant Mathesin & Philosophiam, qui quantitates veras cum ipsarum relationibus & vulgaribus mensuris confudunt."

[49] Ibid., p. 411.

[50] Ibid., p. 415.
"Motus autem veros ex eorum causis, effectibus, & apparentibus differentiis colligere, & contra ex motibus seu veris seu apparentibus eorum causas & effectus, docebitur fusius in sequentibus. Hunc enim in finem tractatum sequentem composui."

reference frame apart from that of absolute space. The sense in which it was relative is that it was a more or less accurate approximation to metaphysical time. Because clocks may not be truly periodic, physical time may not register the true time:

> In astronomy, absolute time is distinguished from relative time by the equation of common time. For natural days, which are commonly considered equal for the purpose of measuring time, are actually unequal. Astronomers correct this inequality in order to measure celestial motions on the basis of a truer time. It is possible that there is no uniform motion by which time may have an exact measure. All motions can be accelerated and retarded, but the flow of absolute time cannot be changed. The duration or perseverance of the existence of things is the same, whether their motions are rapid or slow or null; accordingly, duration is rightly distinguished from its sensible measures and is gathered from them by means of an astronomical equation.[51]

What Newton did not realize, nor could he have suspected, is that physical time is not only *relative,* but also *relativistic,* that the approximation of physical time to metaphysical time depends not merely upon the regularity of one's clock, but also upon its motion. Unless a clock were at absolute rest, it would not accurately register the passage of metaphysical time. Moving clocks run slow. This truth, unknown to Newton, only intimated by Larmor and Lorentz in the concept of "local time," was finally grasped by Einstein.

Where Newton fell short, then, was not in his conception of absolute or metaphysical time—he had theological grounds for positing such a time—, but in his incomplete understanding of physical time. He assumed too readily that an ideal clock would give an accurate measure of metaphysical time independently of its motion. But the above quotation suggests that Newton, if confronted with relativistic evidence, would have welcomed this correction and seen therein no threat at all to his doctrine of metaphysical time. Lucas emphasizes, "The relativity that Newton rejected is not the relativity that Einstein propounded; and although the Special Theory of Relativity has shown Newton to be wrong in some respects,...it has not shown that time is relative in Newton's sense, and merely some numerical measure of process."[52] In short, relativity corrects Newton's concept of physical time, not his concept of metaphysical time.

Of course, it hardly needs to be said that there is a great deal of antipathy in modern physics and philosophy of space and time toward such metaphysical entities

---

[51] Ibid., p. 410.
"Tempus Absolutum a Relativo distinguitur in Astronomia per æquationem temporis vulgi. Inæquales enim sunt dies naturales, qui vulgo tanquam æquales pro mensura temporis habentur. Hanc inæqualitatem corrigunt Astronomi, ut ex veriore tempore mensurent motus coelestes. Possibile est, ut nullus sit motus æquabilis, quo tempus accurate mensuretur. Accelerari & retardari possunt motus omnes, sed fluxus temporis absoluti mutari nequit. Eadem est duratio seu perseverantia existentiæ rerum, sive motus sint celeres, sive tardi, sive nulli: proinde hæc a mensuris suis sensibilibus merito distinguitur, & ex iisdem colligitur per æquationem astronomicam."

[52] Lucas, *Treatise,* p. 90. Later he adds,
"It is often said that relativity refuted Newton. But it is a misleading oversimplification. There is no straight opposition between relativity theories and Newtonian, absolute, theories.... Theologically (and not only theologically) speaking we may assign a preferred frame of reference,...which is at rest. There is no reason why we should not—God may have looked, and seen that it was so, and told Newton. Only there is no *physical* reason why we should" (Ibid., pp. 197-198).

as Newtonian space and time, primarily because they are physically undetectable and serve no physical purpose. In their standard text *Gravitation*, Misner, Thorne, and Wheeler, for example, challenge Newton's metaphysic by asking, "But how does one give meaning to Newton's absolute space, find its cornerstones, mark out its straight lines?" Complaining that "his ideal geometry is beyond observation," they conclude, "Newton's absolute space is unobservable, non-existent."[53] Newton would have been singularly unimpressed with this verificationist challenge and the positivistic equation between unobservability and non-existence. From his predecessor Isaac Barrow, he had been taught, "'*We do not perceive it, therefore it does not exist*,' is a fallacious inference...."[54] The grounds for metaphysical space and time were not physical, but philosophical, or more precisely, theological, as Lucas explains:

> The critics of Newtonian space, Leibniz (on occasion), Mach, and Einstein, urge an epistemological approach: 'How do we tell whether something is moving or at rest?'...Newton did not feel the force of the epistemological criticism. He takes a 'God's-eye' view of the universe. God is present 'from infinity to infinity' and 'governs all things, and knows all things that are or can be done.' There are no epistemological problems for God. He is 'omnipresent, who...sees the things themselves intimately and thoroughly perceives them, and comprehends them wholly, by their immediate presence to Himself.' He knows, just knows, where everything is—more, Newton would say, He knows because He puts it there; God places each atom in its place by fiat of His will—and so knows where it is because He knows what He is doing, immediately and without any room for any epistemological problem to arise.[55]

Epistemological problems fail to worry Newton because, as Lucas nicely puts it, "He is thinking of an omniscient, omnipresent Deity whose characteristic relation with things and with space is expressed in the imperative mood."[56] Modern physical theories say nothing against the existence of such a God or the metaphysical space and time constituted, in Newton's thinking, by His eternity and omnipresence. What relativity theory did, in effect, was simply to cut God out of the picture and to substitute in His place a finite observer. "Thus," according to Holton, "the *RT* [relativity theory] merely shifted the focus of spacetime from the sensorium of Newton's God to the sensorium of Einstein's abstract *Gedanken*experimenter—as it were, the final secularization of physics."[57] But to a man like Newton, who wrote in his General Scholium, "to treat of God from phenomena is certainly a part of

---

[53] C. Misner, K. S. Thorne, and J. A. Wheeler, *Gravitation* (San Francisco: W. H. Freeman, 1973), p. 19.
[54] Isaac Barrow, *The Geometrical Lectures of Isaac Barrow*, trans. with Notes by J. M. Child (Chicago: Open Court, 1916), p. 36.
[55] Lucas, *Treatise*, p. 143.
[56] Ibid.
[57] Gerald Holton, "On the Origins of the Special Theory of Relativity," in *Thematic Origins of Scientific Thought: Kepler to Einstein* by Gerald Holton (Cambridge, Mass.: Harvard University Press, 1973), p. 171. Cf. Capek's remark that it was the "alliance of theology and cosmology" from the Middle Ages through the nineteenth century that provided the basis for the concept of an absolute frame of reference—an alliance broken in the twentieth century (Milic Capek, "Introduction," in *The Concepts of Space and Time*, ed. M. Capek, Boston Studies in the Philosophy of Science 22 [Dordrecht: D. Reidel, 1976], p. xxii).

'natural' philosophy,"⁵⁸ such a secular outlook impedes rather than advances our understanding of the nature of reality. And even if we do not go so far as Newton in including discourse about God in scientific theorizing, still it is clear that if we are prepared to draw metaphysical inferences about the nature of space, time, and spacetime on the basis of physical science, then we must also be ready to entertain theistic metaphysical hypotheses such as Newton deemed relevant. I have argued elsewhere that Newton was correct in thinking not only that God exists but also that God is (at least since the moment of creation) temporal.⁵⁹ It remains to be seen what impact such a hypothesis will have on our interpretation of relativity theory.

---

[58] Newton, *Principia*, p. 943. "Et haec de Deo; de quo utique ex Phaenomenis disserere, ad Philosophiam Naturalem pertinet."
[59] See William Lane Craig, *The Kalam Cosmological Argument*, Library of Philosophy of Religion (London: Macmillan, 1979); William Lane Craig and Quentin Smith, *Theism, Atheism, and Big Bang Cosmology* (Oxford: Clarendon Press, 1993); William Lane Craig, *God, Time, and Eternity* (Dordrecht: Kluwer Academic Publishers, forthcoming).

CHAPTER 7

THE POSITIVISTIC FOUNDATIONS

OF RELATIVITY THEORY

For over two hundred years the classical, Newtonian concept of time reigned in physics. But by the end of the nineteenth century, physicists had realized that there was something fundamentally wrong with Newton's analysis of physical time and space. The failure to detect the earth's motion through the aether, which constituted a relative or physical space at rest with respect to metaphysical space,[1] prompted a crisis in physics which compelled men like Lorentz, Larmor, and Poincaré to revise and then abandon the Galilean transformation equations in favor of the relativistic Lorentz transformations. In so doing, they had already sounded the death knell of Newtonian physics, for they had relativized the sensible measures of metaphysical time and space in a way undreamed of by Newton. But they did so without abandoning the notion that there really is a true time and a true space.

Einstein interrupted this research programme with a radically different approach. In our exposition of the Special Theory, we have seen that foundational to Einstein's approach was his denial of absolute space and his consequent re-definition of time and simultaneity so as to deny their absolute status as well. What Einstein did, in effect, was to shave away Newton's metaphysical time and space, and along with them the aether, thus leaving behind only their sensible measures, so that physical time became the only time there is and physical space the only space there is. Since these are relativized to inertial frames, one ends up with the relativity of simultaneity and of length.

What justification did Einstein have for so radical a move? How did he know that metaphysical time and space do not exist? The answer, in a word, is positivism. Although one rarely finds this discussed in textbook expositions of the theory or even in discussions of the philosophical foundations of the theory, nevertheless historians of science have demonstrated convincingly that at the philosophical roots of Einstein's theory lies an epistemological positivism of Machian provenance which issues in a verificationist analysis of the concepts of time and space.[2]

---

[1] Sir Oliver Lodge was the first to make this assumption explicit: "the whole of this subject [the aether problem] indicates that the aether is a physical standard of rest; and that motion relative to it, which is becoming cognisable by us, is in that sense an ascertained absolute motion" (O. J. Lodge, "Note on Mr. Sutherland's Objection to the Conclusiveness of the Michelson-Morley Aether Experiment," *Philosophical Magazine* 46 [1898]: 344). Even Lorentz did not make explicit the aether frame's privileged status until after Einstein's development of SR. For discussion see Tetu Hirosige, "The Ether Problem, the Mechanistic Worldview, and the Origins of the Theory of Relativity," *Historical Studies in the Physical Sciences* 7 (1976): 3-82; A. J. Knox, "Hendrik Antoon Lorentz, the Ether, and the General Theory of Relativity," *Archive for History of Exact Sciences* 38 (1988): 67-78.

[2] I am indebted to the writings of Gerald Holton; see in particular Gerald J. Holton, "Mach, Einstein and the Search for Reality," in *Ernst Mach: Physicist and Philosopher*, Boston Studies in the Philosophy

POSITIVIST INFLUENCES ON EINSTEIN

In 1905, when Einstein published his paper on the electrodynamics of moving bodies, and for several years thereafter, Einstein was a self-confessed epistemological pupil of the positivistic physicist Ernst Mach, and the epistemological analysis of space and time given in the opening section of that paper clearly displays this influence. According to Holton, "If one studies Einstein's earliest papers on special relativity from a philosophically critical point of view, one can discern the influence of many, partly contradictory, points of view.... But perhaps the most important of these influences was that of the empirico-critical positivism of the type of which the Austrian philosopher Ernst Mach was the foremost exponent."[3] Mach was a giant of nineteenth century German physics, of whom Einstein later remarked, "Even those who think of themselves as Mach's opponents hardly know how much of Mach's views they have, as it were, imbibed with their mother's milk."[4] Holton remarks, "The influence of Mach's point of view, particularly in German-speaking countries, was enormous—in physics, on physiology, on psychology, and on the fields of the history and the philosophy of science."[5]

Mach's philosophy of science was phenomenalist or sensationalist in character. In experience we are given various sensations, such as colors, sounds, pressures, and so forth, which Mach called "the elements", and the aim of scientific theorizing is to construct the simplest possible description of the connections among these sensations.[6] Mach had no use for theoretical entities or even for entities behind the sensations. Statements in theories were meaningful only if they were related directly to sensations. His attitude was militantly "anti-metaphysical," and in his *Die Mechanik in ihrer Entwicklung*, which Einstein studied carefully, Mach declared in his Preface to the work that "Its intention is...an enlightening one, or to put it more bluntly, an anti-metaphysical one."[7] In line with his phenomenalism,

---

of Science 6 (Dordrecht: D. Reidel, 1970), pp. 165-199; idem, "Where Is Reality? The Answers of Einstein," in *Science and Synthesis*, ed. UNESCO (Berlin: Springer-Verlag, 1971), pp. 45-69; the essays collected together in idem, *Thematic Origins of Scientific Thought: Kepler to Einstein* (Cambridge, Mass.: Harvard University Press, 1973).
[3] Holton, "Where Is Reality?" p. 45. It is interesting that Holton refers to Mach as a philosopher. Mach himself disclaimed that title, saying, "...I have already declared explicitly that *I am by no means a philosopher, but only a scientist*" (Ernst Mach, *Erkenntnis und Irrtum* 2d ed. [Leipzig: D. Reidel, 1906], pp. vii-viii).
[4] Albert Einstein, "Ernst Mach," *Physikalische Zeitschrift* 17 (1916): 101, reprinted in Ernst Mach, *Die Mechanik in ihrer Entwicklung historisch-kritisch dargestellt*, ed. Renate Wahsner and Horst-Heino Borzeszkowski (Berlin, DDR: Akademie-Verlag, 1988), pp. 683-689. Einstein also refers to Mach as "the greatest influence on the epistemological orientation of the natural scientists of our time."
[5] Holton, "Search for Reality," p. 168.
[6] See Philipp Frank, "The Importance of Ernst Mach's Philosophy of Science for our Times," [1917], rep. in *Ernst Mach: Physicist and Philosopher*, pp. 219-234.
[7] Ernst Mach, *Mechanik in ihrer Entwicklung*, p. 13 (Ernst Mach, *The Science of Mechanics: A Critical and Historical Account of Its Development*, trans. Thomas J. McCormack [LaSalle, Ill.: Open Court, 1960], p. xxii). Here we may discern the even deeper influence of the father of positivism, Auguste Comte. Comte analyzed the evolution of thought along three stages: the theological, the metaphysical, and the positive. Positivism involves the determination to have nothing to do with theology and metaphysics. The classical concept of time as a metaphysical reality rooted in theology was

Mach held that "Space and time are well-ordered systems of series of sensations," comparable to sensations of color, sounds, and smells.[8] In his *Mechanics*, Mach assailed Newton's concepts of absolute space and time, which in his preface to the final seventh edition (1912) he denounced as "blunders" and "conceptual monstrosities."[9] According to Mach,

> No one is competent to predicate things about absolute space and absolute motion; they are pure things of thought, pure mental constructs, that cannot be produced in experience. All our principles of mechanics are...experimental knowledge concerning the relative positions and motions of bodies.... No one is warranted in extending these principles beyond the boundaries of experience. In fact, such an extension is meaningless, as no one possesses the requisite knowledge to make use of it.[10]

In Newtonian mechanics, absolute space served no purpose, since Newton's laws of motion held good in all inertial frames, not simply in frames at rest with respect to absolute space. The elimination of absolute space would not in any way reduce the empirical content of classical dynamics. It was therefore "metaphysical," and Mach proposed to do away with it along with any distinction between inertial and non-inertial frames.

Similarly, in his *Prinzipien der Wärmelehre*, Mach drew applications from his critique of the concept of temperature to Newton's concept of absolute time. Of temperature, Mach wrote,

> It is remarkable how long a period elapsed before it definitely dawned upon inquirers that the designation of thermal states by numbers rests on a convention. Thermal states exist in nature, but the conception of temperature exists only by virtue of our arbitrary *definition*, which might very well have taken another form. Yet until very recently inquirers in this field appear more or less unconsciously to have sought after a *natural* measure of temperature, a real temperature, a sort of Platonic Idea of temperature of which the temperatures read from the thermometric scales were only the imperfect and inexact expression.[11]

Mach thought that the Newtonian concepts of absolute time and space were as misconceived as the concept of an absolute measure of temperature: "Newton's conceptions of 'absolute time,' 'absolute space,' *etc.*, which I have discussed in another place, originated in quite a similar manner. In our conceptions of time the sensation of duration plays the same part with regard to the various measures of time as the sensation of heat played in the instance just adduced. The situation is similar with respect to our conceptions of space."[12] The implication is that just as there is no absolute temperature behind our various conventional measures of our sensation of heat, so no absolute time exists behind the various conventional measures of our

---

anathema. Mach refers favorably to Comte's three stages in his *Erkenntnis und Irrtum*, p. 99. For discussion see J. Bradley, *Mach's Philosophy of Science* (London: Athlone Press, 1971), pp. 204-207.

[8] Mach, *Mechanik in ihrer Entwicklung*, p. 522 (*Science of Mechanics*, p. 611).
[9] Ibid. pp. 23-24.
[10] Ibid., p. 252 (*Science of Mechanics*, p. 280). In his obituary notice of Mach, it is to precisely these sections of the *Mechanics* (II. 6, 7) that Einstein draws attention (Einstein, "Mach," pp. 685-688).
[11] Ernst Mach, *Prinzipien der Wärmelehre* (Leipzig: J. A. Barth, 1896), p. 154 (Ernst Mach, *Principles of the Theory of Heat*, trans. T. J. McCormack, rev. and compl. P. E. B. Jourdain and A. E. Heath, with an Introduction by M. J. Klein, ed. Brian McGuiness, Vienna Circle Collection 17 [Dordrecht: D. Reidel, 1986], pp. 53-54).
[12] Ibid. Cf. Mach, *Mechanik in ihrer Entwicklung*, p. 249 (*Science of Mechanics*, p. 276).

sensation of duration. Schaffner concludes that Mach's methodology in the *Wärmelehre* "seems to have influenced Einstein in his analysis of the concept of time."[13]

Mach's influence upon Einstein was reinforced by positivist physicists whom Einstein read while a student at the Eidgenössische Technische Hochschule in Zürich. Holton describes the phenomenalistic and revisionist programme they pursued:

> Their positivism provided an epistemology for the new, phenomenologically based science of correlated observations, linking energetics and pure sensationism...'hypothetical' quantities such as atomic entities were omitted; instead, these authors claimed they were satisfied, as Herz wrote around 1904, with 'measuring such quantities as are presented directly in observation, such as energy, mass, pressure, volume, temperature, heat, electrical potential, etc., without reducing them to imaginary mechanisms or kinetic quantities'. As a consequence the introduction of such conceptions as the ether, with properties not accessible to direct observation, was condemned. Instead, these philosophers urged a reconsideration of the ultimate principles of all contemporary physical reasoning, notably the scope and validity of the Newtonian Laws of motion, the conception of force, and the conceptions of absolute and relative motion.[14]

Chief among the influences on the young Einstein was August Föppl, whose *Introduction to Maxwell's Theory of Electricity* (1894) employed a Machian analysis of absolute space in dealing with experiments with moving conductors.[15] In the fifth chapter, entitled "The Electrodynamics of Moving Conductors," Föppl discusses the case of the magnet and the conductor with which Einstein opens his 1905 paper. Föppl replaced Newton's absolute space with the aether because the former was "not at all subject to experience."[16] But he struggled with the fact that insofar as the moving conductor and magnet were concerned, it is only their relative motion which matters. If both are absolutely moving in tandem, no electric current is produced, since they are at rest relative to each other. According to Holton, Föppl, with his "antimetaphysical and self-conscious empiricism," helped Einstein to realize that "the fundamental problem to be cracked is how to achieve a new point of view on the conceptions of time and space."[17] Another of these positivist physicists was W. Ostwald, with whom Einstein corresponded and whose findings were employed by Einstein in a paper of 1901. Ostwald went further than Föppl in rejecting the aether altogether, declaring, "There is no need to inquire for a carrier of [energy] when we find it anywhere. This enables us to look upon radiant energy as independently existing in space"—a position which Einstein was to adopt.[18]

Einstein himself freely expressed his indebtedness to Mach with respect to the epistemological foundations of Special Relativity. It was his student friend and

---

[13] Kenneth F. Schaffner, "Einstein versus Lorentz: Research Programmes and the Logic of Comparative Theory Evaluation," *British Journal for the Philosophy of Science* 25 (1974): 59.
[14] Gerald Holton, "Influences on Einstein's Early Work," in *Thematic Origins*, pp. 45-46.
[15] See Holton, "Influences," pp. 198-212; Arthur I. Miller, *Albert Einstein's Special Theory of Relativity* (Reading, Mass.: Addison-Wesley,1981), pp. 150-154.
[16] August Föppl, *Einführung in die Maxwell'sche Theorie der Elektrizität* (Leipzig: Teubner, 1894), p. 307.
[17] Holton, "Influences," pp. 207, 212.
[18] Wilhelm Ostwald, *Chemische Energie: Lehrbuch der allgemeinen Chemie*, 2d ed. (Leipzig: Verlag von Wilhelm Engelman, 1893), 2:1016.

colleague at the Berne patent office, Michael Besso—an enthusiastic Machist and the only person acknowledged in Einstein's 1905 paper[19]—who first introduced Einstein to Mach. In a letter of April 8, 1952, Einstein wrote, "My attention was drawn to Ernst Mach's *History of Mechanics* by my friend Besso while a student, around the year 1897. The book exerted a deep and persisting impression upon me..., owing to its physical orientation toward fundamental concepts and fundamental laws."[20] On December 8, 1947, Besso recalled in a letter to Einstein, "Is it not true that this introduction [to Mach] fell into a phase of development of the young physicist [*i.e.*, Einstein] when the Machist style of thinking pointed decisively at observables—perhaps even, indirectly, to clocks and meter sticks?"[21] In a reply to Besso the following month, Einstein conceded, "As for the influence of Mach on my thinking, it has certainly been very great. I remember very well how, during my early years as a student, you directed my attention to his treatise on mechanics and to his theory of heat, and how these two works made a deep impression on me."[22]

Einstein's deep admiration for Mach is evident in their surviving correspondence between 1909 and 1913. In the second edition of his *Conservation of Energy* (1909), Mach declared, "I subscribe, then, to the principle of relativity, which is also firmly upheld in my *Mechanics* and *Theory of Heat*."[23] In August of that year Einstein wrote to Mach upon receipt of the book, "You have had such a strong influence upon the epistemological conceptions of the younger generation of physicists that even your opponents today, such as Planck, undoubtedly would have been called Mach-followers by physicists of the kind that was typical a few decades ago."[24] Eight days later, Einstein wrote again in reply to a lost letter from Mach, saying, "I am very glad that you are pleased with the relativity theory" and signing off "Your admiring pupil."[25] As Einstein progressed toward the General Theory of Relativity he still considered himself to be employing Mach's epistemology.[26] In fact, in the spring of 1911 Mach had helped to formulate a manifesto calling for the founding of a society for positivistic philosophy, and among the signatories of this

---

[19] He wrote, "In conclusion I wish to say that in working at the problem here dealt with, I have had the loyal assistance of my friend and colleague M. Besso, and that I am indebted to him for several valuable suggestions" (Albert Einstein, "On the Electrodynamics of Moving Bodies," trans. Arthur I. Miller, Appendix to Miller, *Einstein's Special Theory*, p. 415).
[20] Albert Einstein to Carl Seelig, April 8, 1952, unpublished letter in the Einstein Archives at Princeton cited by Holton, "Where Is Reality?" pp. 47-48.
[21] Michael Besso to Albert Einstein, December 8, 1947, in *Correspondance 1903-1955*, trans. with Notes and an Introduction by Pierre Speziali (Paris: Hermann, 1979), p. 228.
[22] Albert Einstein to Michael Besso, January 6, 1948, in *Correspondance*, p. 231.
[23] Ernst Mach, *Die Geschichte und die Wurzel des Satzes von der Erhaltung der Arbeit* (Prague: J. G. Calvé, 1872), in Ernst Mach, *Abhandlungen*, ed. J. Thiele (Amsterdam: E. J. Bonset, 1969), p. 60 (Ernst Mach, *History and Root of the Principle of the Conservation of Energy*, trans. Philip E. B. Jourdain [Chicago: Open Court, 1911], p. 95).
[24] Albert Einstein to Ernst Mach, August 9, 1909, in Mach, *Mechanik in ihrer Entwicklung*, p. 679 (Holton trans.).
[25] Albert Einstein to Ernst Mach, August 17, 1909, in Mach, *Mechanik in ihrer Entwicklung*, p. 680 (Holton trans.).
[26] See letters reprinted in Mach, *Mechanik in ihrer Entwicklung*, pp. 680-681 and trans. in Holton, "Where Is Reality?" pp. 51-52.

document, such as David Hilbert, Sigmund Freud, and J. Petzoldt, was Albert Einstein.[27]

Eventually Mach came to repudiate relativity theory,[28] and Einstein threw off Mach's phenomenalism for a critical realism; but the founder of relativity theory continued to acknowledge Mach's influence on him during his early years. Writing off Mach's rejection of relativity theory to the intransigence of old age, Einstein insisted that "the whole direction of thought of this theory conforms with Mach's...."[29] While Mach's positivistic philosophy proved itself unfruitful in positive theory construction, still it served its purpose in eliminating unwanted metaphysical entities such as absolute time and space: "It cannot give birth to anything living, it can only exterminate harmful vermin."[30] Though converted into a "believing rationalist" through his work on gravitation, Einstein conceded that in his earlier work on the Special Theory he was "coming from sceptical empiricism of somewhat the kind of Mach's...."[31] In his later "Autobiographical Notes," Einstein connects his denial of absolute simultaneity with the critical reasoning of Ernst Mach and mentions the influence of another sceptical empiricist in this connection:

> Today everyone knows, of course, that all attempts to clarify this paradox satisfactorily were condemned to failure as long as the axiom of the absolute character of time, viz., of simultaneity, unrecognizedly was anchored in the unconscious. Clearly to recognize this axiom and its arbitrary character really implies already the solution of the problem. The type of critical reasoning which was required for the discovery of this central point was decisively furthered, in my case, especially by the reading of David Hume's and Ernst Mach's philosophical writings.[32]

A similar mention is found in Einstein's memorial notice for Mach in 1916: referring to Mach's critique of Newton's absolute space and time, Einstein asserts, "No one can take it away from the epistemologists that here they paved the way for this development [of relativity theory]; for my own part at least I know that I have been greatly aided, directly and indirectly, through especially Hume and Mach."[33] In his letter to Besso, already quoted, which was written about the time of "Autobiographical Notes," he asserts that Hume's influence on his thinking was

---

[27] See Friedrich Herneck, "Nochmals über Einstein und Mach," *Physikalische Blätter* 17 (1961): 276.

[28] Or so it has traditionally been believed. According to Brown, Gerion Wolters has shown Mach's rejection of relativity to have been a forgery by his son Ludwig (James Robert Brown, "Einstein's Brand of Verificationism," *International Studies in the Philosophy of Science* 2 [1987]: 36).

[29] Albert Einstein to Armin Weiner, September 18, 1930, unpublished letter from the Archives of the Burndy Library in Norwalk, Connecticut, cited by Holton, "Where Is Reality?" p. 55. In his obituary notice "Ernst Mach," Einstein goes so far as to say, "It is not improbable that Mach would have discovered the theory of relativity, if, at the time when his mind was still young and susceptible, the problem of the constancy of the speed of light had been discussed among physicists" (Einstein, "Mach," p. 688 [Holton trans.]).

[30] Albert Einstein to Michael Besso, May 13, 1917, in *Correspondance*, p. 68.

[31] Albert Einstein to C. Lanczos, January 24, 1938, unpublished letter in the Einstein Archives at Princeton cited in Holton, "Where Is Reality?" p. 64.

[32] Albert Einstein, "Autobiographical Notes," in *Albert Einstein: Philosopher-Scientist*, ed. P. A. Schilpp, Library of Living Philosophers 7 (LaSalle, Ill.: Open Court, 1949), p. 53. The paradox to which he refers is the question which occurred to him at the age of sixteen: if one pursues a light beam at the velocity c, one should observe it as an electromagnetic field at rest; but there seems to be no such thing. On the force of this reasoning, see Adolf Grünbaum, *Philosophical Problems of Space and Time*, 2d ed. Boston Studies in the Philosophy of Science 12 (Dordrecht: D. Reidel, 1973), pp. 73-75.

[33] Einstein, "Mach," p. 686.

even *greater* than Mach's. Referring to the *Mechanics* and *Theory of Heat*, Einstein reflected, "The extent to which they influenced my own work is, to say the truth, not clear to me. As far as I am conscious of it, the immediate influence of Hume on me was greater.... But, as I said, I am not able to analyze that which lies anchored in unconscious thought."³⁴

It is not too difficult to imagine what aspect of Hume's thought proved influential on Einstein at this point. In Hume's view, ideas which are not derived from sense impressions are empty of content and meaningless.

> When we entertain, therefore, any suspicion that a philosophical term is employed without any meaning or idea (as is but too frequent), we need but enquire, *from what impression is that supposed idea derived?* And if it be impossible to assign any, this will serve to confirm our suspicion.³⁵

Propositions expressing analytic relations of ideas "are discoverable by the mere operation of thought, without dependence on what is anywhere existent in the universe."³⁶ Such *a priori* reasoning is valid only in the realm of mathematics; "...all attempts to extend this more perfect species of knowledge beyond these bounds are mere sophistry and illusion."³⁷ Propositions expressing matters of fact cannot be known or demonstrated *a priori* but can only be known on the basis of sense experience. "The existence, therefore, of any being can only be proved by arguments from its cause or its effect; and these arguments are founded entirely on experience. If we reason *a priori,* anything may appear able to produce anything."³⁸ According to Hume, unless we are able to furnish arguments based on sense experience for a purported matter of fact, that supposed fact is revealed to be vacuous.

> When we run over libraries, persuaded of these principles, what havoc must we make? If we take in our hand any volume; of divinity or school metaphysics, for instance; let us ask, *Does it contain any abstract reasoning concerning quantity or number?* No. *Does it contain any experimental reasoning concerning matter of fact and existence?* No. Commit it then to the flames: for it can contain nothing but sophistry and illusion.³⁹

Although Hume's anti-metaphysical remarks were directed principally at scholastic philosophizing, Einstein would have seen in his words an equal indictment of Newton's metaphysical concepts of absolute time and space.⁴⁰

---

³⁴ Einstein to Besso, January 8, 1948, in *Correspondance*, p. 231.
³⁵ David Hume, *An Enquiry concerning Human Understanding*, in *Enquiries concerning Human Understanding and concerning the Principles of Morals* [1777], ed. with Introduction by L. A. Selby-Bigge, 3d ed., rev. with Notes by P. H. Nidditch (Oxford: Clarendon Press, 1975), II.17 (p. 22).
³⁶ Ibid., IV.i.20 (p. 25).
³⁷ Ibid., XII.iii.131 (p. 163).
³⁸ Ibid., XII.iii.132 (p. 164).
³⁹ Ibid. (p. 165).
⁴⁰ Cf. Hume's phenomenalistic analysis of space and time in his *A Treatise of Human Nature* [1739-1740], ed. L. A. Selby-Bigge, rev. with Notes by P. H. Nidditch (Oxford: Clarendon Press, 1978), I.ii.3 (pp. 33-39), though Hume's analysis of time in terms of the succession of mental impressions does not succumb to Einstein's physicalism. See also Hirosige, "Origins of the Theory of Relativity," p. 58. Later in life Einstein blamed Hume for having fostered "a fateful 'fear of metaphysics'" which became "a malady of contemporary empirisistic philosophizing" (Albert Einstein, "Remarks on Bertrand Russell's Theory of Knowledge," in *The Philosophy of Bertrand Russell*, 3d ed., ed. P. A. Schilpp, Library of Living Philosophers [New York: Tudor Publishing, 1951], p. 289); but even so he urges that a system of

A positivistic philosophy of science was also mediated to Einstein through his study of the writings of Poincaré. Einstein recalled, "In Bern I had regular philosophical reading and discussion evenings, together with K. Habicht and Solovine, during which we were mainly concerned with Hume.... The reading of Hume, along with Poincaré and Mach, had some influence on my development."[41] Einstein's biographer Abraham Pais comments, "I stress that Einstein and his friends did much more than just browse through Poincaré's writings. Solovine has left us a detailed list of books which the Akademie members read together. Of these he singles out one and only one, *La Science et l'hypothèse*, for the following comment: '[this] book profoundly impressed us and kept us breathless for weeks on end!'"[42] In *Science and Hypothesis* Poincaré repudiated an absolute metric of time and objective simultaneity relations on the basis of a verificationist theory of meaning:

> There is no absolute time; to say two durations are equal is an assertion which has by itself no meaning and which can acquire one only by convention.... Not only have we no direct intuition of the equality of two durations, but we have not even direct intuition of the simultaneity of two events occurring in different places; this I have explained in an article entitled 'La mesure du temps.'[43]

Poincaré's positivism issues not only in his metric conventionalism (which is compatible with there being a privileged time) but also points to the conventionality of simultaneity. In his "Measure of Time"—Pais says it is "virtually certain" that Einstein also knew it[44]—Poincaré deals with two questions: (1) can we transform psychological time, which is qualitative, into quantitative time, and (2) can we reduce to one and the same measure facts which transpire in different worlds? The first is the problem of time's metric; the second is the question of distant simultaneity. The approach to both questions is thoroughly positivistic: with respect to (1), since we have no direct intuition of the equality of temporal intervals, the affirmation that they are equal has no meaning at all, and so we must conventionally define a measure of time such that the equations of mechanics may be as simple as possible; with respect to (2) affirmations of the simultaneity of distant events "have by themselves no meaning" and we should ask ourselves how we came up with the idea of "putting into the same frame" events so far removed from one another.[45] It is not hard to discern here possible inspiration for Einstein's operational definition of time and the frame-dependence of simultaneity, but far more important, because unquestioningly presupposed, is Poincaré's positivistic epistemological approach to issues of space and time, a presupposition which underlay Einstein's own SR.

---

concepts will not degenerate into metaphysics so long as enough propositions are connected with sensory experiences!

[41] Albert Einstein to Michael Besso, March 6, 1952, in *Correspondance*, p. 464. The translation is Pais's.

[42] Abraham Pais, *'Subtle is the Lord...': The Science and Life of Albert Einstein* (Oxford: Oxford University Press, 1982), pp. 133-134. The citation is from A. Einstein, *Lettres à Maurice Solovine* (Paris: Gauthiers-Villars, 1956), p. viii.

[43] Henri Poincaré, *Science and Hypothesis*, in *The Foundations of Science* by H. Poincaré (Science Press, 1913; rep. ed.: Washington, D. C.: University Press of America, 1982), p. 92.

[44] Pais, *'Subtle is the Lord,'* p. 133.

[45] Henri Poincaré, "The Measure of Time," *Revue de métaphysique et de morale* 6 (1898): 1ff., rep. in H. Poincaré, *The Value of Science* (1905) in *The Foundations of Science*, trans. G. B. Halstead, p. 228.

CHAPTER 7

EINSTEIN'S POSITIVISM

When we turn to Einstein's 1905 paper, we discover that the introductory sections of the paper offer a verificationist analysis of space and time based on Machian positivism. Einstein's verificationism comes through most clearly in his operationalist re-definition of key concepts. Holton comments:

> ...the Machist component...shows up prominently in two related respects: first, by Einstein's insistence from the beginning of his relativity paper that the fundamental problems of physics cannot be understood until an epistemological analysis is carried out, particularly so with respect to the meaning of the conceptions of space and time; and second, by Einstein's identification of reality with what is given by sensations, the 'events,' rather than putting reality on a plane beyond or behind sense experience.
>
> From the outset, the instrumentalist, and hence sensationist, views of measurement and of the concepts of space and time are strikingly evident.
>
> ...We can say that just as the *time* of an event assumes meaning only when it connects with our consciousness through sense experience (that is, when it is subjected to measurement-in-principle by means of a clock present at the same place), so also is the *place*, or space co-ordinate, of an event meaningful only if it enters our sensory experience while being subjected to measurement-in-principle (that is, by means of meter sticks present on that occasion at the same time).[46]

Holton's insistence on the influence of Mach's sensationism is exaggerated; but his pinpointing Einstein's verificationist analysis of time and space as exhibiting the influence of positivism is accurate.[47] Time and space are defined by Einstein in terms of physical operations. Newton's distinction between time and the sensible measures thereof is quietly abolished. Einstein does not even attempt to justify his operationalism; it is just presupposed. He asserts,

> Now we must bear carefully in mind that a mathematical description of this kind has no physical meaning unless we are quite clear as to what we will understand by 'time'. We have to take into account that all our judgments in which time plays a role are always judgments of *simultaneous events*. If, for instance, I say, 'The train arrives here at 7 o'clock,' I mean something like this: 'The pointing of the small hand of my watch to 7 and the arrival of the train are simultaneous events.'[48]

It is taken for granted that *all* our judgements in which time plays a role must have a physical meaning. When it comes to judgements concerning the simultaneity of distant events, the concern is to find a "practical arrangement" to compare clock times.[49] In order to "define" a common time for spatially separated clocks, we assume that the time light takes to travel from A to B equals the time it takes to travel from B to A—a definition which, as we have seen, *presupposes* that absolute space does not exist. Thus, time is reduced to physical time (clock readings) and

---

[46] Holton, "Search for Reality," pp. 169-170.
[47] See T. A. Ryckman, "'P(oint)-C(oincidence) Thinking': The Ironical Attachment of Logical Empiricism to General Relativity (and some lingering Consequences)," *Studies in History and Philosophy of Science* 23 (1992): 473, who observes that Holton's portrayal of Einstein as a convert from sensationism to arch-realism is no longer considered to be accurate, for Einstein was never a sensationist and he remained a Machean or Humean up to the end with respect to his verificationism.
[48] Einstein, "Electrodynamics of Moving Bodies," p. 393.
[49] Ibid. Frank comments, "It is easy to see which lines of Mach's thought have been particularly helpful to Einstein. The definition of simultaneity in the special theory of relativity is based on Mach's requirement that every statement in physics has to state relations between observable quantities" (Philipp Frank, "Einstein, Mach, and Logical Positivism," in *Einstein: Philosopher-Scientist*, p. 272).

space to physical space (readings of measuring rods) and both of these are relativized to local frames. Simultaneity is defined in terms of clock synchronization via light signals. All of this is done by mere stipulation. With admirable bluntness, Heinz Pagels comments,

> These definitions, with their appeal to measurement, cut through all the excess philosophical baggage that the ideas of space and time had carried for centuries. The positivist insists that we talk only about what we can know through direct operations like a measurement. Physical reality is defined by actual empirical operations, not by fantasies in our head.[50]

Through Einstein's operational definitions of time and space, Mach's positivism triumphs in the Special Theory of Relativity. Reality is reduced to what our measurements read; Newton's metaphysical time and space, which transcend operational definitions, are implied to be mere figments of our imagination. "Thus," concludes Gutting, "Einstein's seemingly innocuous requirement that simultaneity be operationally defined led to the rejection of a basic concept of classical physics, the concept of an absolute time valid in all coordinate systems."[51]

In Einstein's other early papers on relativity, his verificationist theory of meaning comes even more explicitly to the fore. Concepts which cannot be given empirical content and assertions which cannot be empirically verified in principle are discarded as meaningless. In his article in the *Jahrbuch der Radioaktivität und Elektronik* of 1907, after giving his operational definitions for time and simultaneity, he states, "According to the definition of time given in section 1, a specific time has sense only in relation to a reference frame having a particular state of motion."[52] To

---

[50] Heinz Pagels, *The Cosmic Code* (London: Michael Joseph, 1982), p. 59. Less polemically, Katsumori makes the point:
> "In Einstein's view the concepts of space and time can by no means be defined prior to possible experience. Rather the meaning of any physical concept is inseparable from the procedure of measurement of the corresponding phenomena; the concepts of space and time, in particular, are endowed with physical meaning only through the prescription of measurement using rigid measuring-rods (*starre Maßstäbe*) and clocks (*Uhren*)" (Makoto Katsumori, "The Theories of Relativity and Einstein's Philosophical Turn," *Studies in History and Philosophy of Science* 23 [1992]: 560).

Tonnelat comments,
> "The notion of distant simultaneity—of simultaneity which is fundamental, universal, and independent of the movement of bodies—has always appeared, until now, to be a concept endowed with an evident meaning.
> For the physicist this notion must be capable of being experimentally defined in such a way as to permit the application of physical criteria. Any other idea of absolute simultaneity, however intuitive it may be, has no chance of being an object of experience if it does not at some moment enter into realizable operations.
> In other words, the simultaneity of two distant events and, in general, the time which an observer must ascribe to a distant phenomenon must be defined by *operations* which could be effectively accomplished, even if it is a matter of a process which can only be approximately realized" (Marie-Antoinette Tonnelat, *Histoire du principe de relativité*, Nouvelle Bibliothèque Scientifique [Paris: Flammarion, 1971], p. 142).

[51] Gary Gutting, "Einstein's Discovery of Special Relativity," *Philosophy of Science* 59 (1972): 60. On Gutting's view, Einstein's two fundamental principles were the principle of relativity and the principle of operational definition. "Therefore, our assessment of the nature of Einstein's discovery will, to an important extent, depend on our view of the status of these principles" (Ibid., p. 63).

[52] A. Einstein, "Über das Relativitätsprinzip und die aus demselben gezogenen Folgerungen," *Jahrbuch der Radioaktivität und Elektronik* 4 (1907): 417.

refer to the time of an event without reference to its inertial frame thus has no sense (*Sinn*). In his piece in the *Physikalische Zeitschrift* of 1909, he muses on the apparent contradiction between the Relativity Principle and Lorentz's conviction that light travels at constant velocity in the aether, and comments,

> The theorem of addition of velocities rests, however, on the arbitrary presupposition that time specifications, as well as statements about the shape of moving bodies, have a meaning independent of the state of motion of the relevant coordinate system. But one is convinced that the introduction of clocks which are at rest relative to the relevant coordinate system is required for a definition of time and the shape of moving bodies.[53]

Here Einstein asserts that statements about the time of an event have no meaning (*Bedeutung*) unless one refers to clocks at rest in the relevant inertial system. In his summary paper, "Die Relativitäts-Theorie," published in 1911, Einstein expresses himself at greater length concerning the meaning of statements about time and space. After laying out the "dreadful dilemma" confronting late nineteenth century physicists concerning the apparent incompatibility of the Relativity Principle and the Principle of the Constancy of the Velocity of Light, Einstein writes,

> Now it turns out that Nature is completely innocent with regard to this dilemma, but that this dilemma follows from the fact that in our thinking...we have silently and arbitrarily made presuppositions which must be dropped in order to arrive at a consistent and simple conception of things.
> ...The first and most important of these arbitrary presuppositions concerns the concept of time, and I shall try to describe wherein this arbitrariness consists. In order to do this well, I shall first deal with space, in order to put time into parallel with it. When we want to express the position of a point in space, that is, the position of a point relative to a coordinate system k, we give its right angle coordinates x, y, z. The meaning of these coordinates is the following: one constructs according to familiar procedures perpendiculars on the coordinate planes and sees how many times a given unit measure can be laid against these perpendiculars. The results of this counting are the coordinates. A spatial specification in coordinates is thus the result of certain manipulations. The coordinates which I assign have accordingly a very precise physical meaning; one can verify if a certain given point really has the assigned coordinates or not.
> How do things stand in this connection with time? There we shall see that we are not in such a good position. Until now one was always content to say: time is the independent variable of the event. With such a definition the measurement of the time value of an actually present event can never be founded. We must therefore try to so define time that time measurements are possible on the basis of this definition.[54]

Einstein then proceeds to describe the familiar light signal synchronization procedure. In so doing, he asserts that since we are unable to measure the one-way velocity of light without making "arbitrary stipulations," we are therefore justified in making such stipulations. "We now stipulate that the velocity of propagation of light in a vacuum on the way from point A to point B is the same as the velocity of propagation of a light beam from B to A."[55] Having described the synchronization procedure, he concludes, "When we have fulfilled this prescription, we have thus

---

[53] A. Einstein, "Über die Entwicklung unserer Anschauungen über das Wesen und die Konstitution der Strahlung," *Physikalische Zeitschrift* 10 (1909): 819.
[54] A. Einstein, "Die Relativitäts-Theorie," *Vierteljahrsschrift der naturforschenden Gesellschaft in Zürich* 56 (1911): 6-7.
[55] Ibid., p. 8.

attained a time determination from the standpoint of the measuring physicist. The time of an event is, namely, equal to the reading of these clocks regulated by the above prescription which are at the place of the event."[56] Einstein thinks this all sounds so self-evident, but what is noteworthy, he states, is that "in order to arrive at time specifications of a very precise sense" we use a prescription that relates to clocks which are relative to a certain coordinate system k. We have not gained simply a time, but a time relative to a coordinate system. "It is not said that time has an absolute...meaning. That is an arbitrary element which was contained in our kinematics."[57]

We now come, Einstein proceeds, to the second arbitrary element in kinematics. "We speak of the shape of a body, for example, the length of a rod, and believe that we know exactly what its length is, even when it is in motion with respect to a reference system from which we are describing appearances."[58] But a little reflection shows that the length of a rod is not so simple a concept as we instinctively imagine. "We now ask: how long is this rod? This question can have only the meaning: what experiments must we carry out in order to discover how long the rod is?"[59] Einstein then proceeds to describe length measurement of a moving rod by means of synchronized clocks.

By abandoning the presuppositions of absolute time and space and substituting in their stead operational definitions, Einstein reduces time and space to our measurements of them. He concludes, "Since we have in a precise way physically defined coordinates and time, every relation between spatial and temporal entities will have a very precise physical content."[60] Statements about spatial or temporal relations which are metaphysical in character, that is, are independent of clocks, rods, or reference frames, are nonsense.

It is frequently asserted that as Einstein labored on the General Theory, he came to see the bankruptcy of Mach's positivism.[61] But this claim needs to be carefully qualified. When one reads Einstein's 1916 paper laying the foundations of General Relativity, Mach's verificationism fairly shouts at one in section 2 of the article. Einstein presents here a "weighty argument from the theory of knowledge" in favor of the extension of the Postulate of Relativity to non-inertial frames of reference.[62] The purpose of this argument is to remedy "an inherent epistemological defect" in both classical mechanics and SR "which was...clearly pointed out by Ernst Mach."[63] The defect is that if we imagine two spatially isolated fluid bodies S_1 and S_2, so situated that observers at rest relative to each respectively will judge that the other is in rotation about their common axis, then there still emerges an absolute effect, namely, one of the bodies will be, say, spherical, while the other is ellipsoidal.

---

[56] Ibid.
[57] Ibid., p. 9.
[58] Ibid.
[59] Ibid.
[60] Ibid., p. 11.
[61] See Holton, "Search for Reality," pp. 185-186.
[62] A. Einstein, "The Foundations of General Relativity Theory," in *General Theory of Relativity*, ed. C. W. Kilmister, Selected Readings in Physics (Oxford: Pergamon Press, 1973), p. 144. The original paper appeared in *Annalen der Physik* 49 (1916): 769.
[63] Ibid., p. 143.

"What is the reason for this difference in the two bodies? No answer can be admitted as epistemologically satisfactory, unless the reason given is an *observable fact of experience*."[64] In Newtonian mechanics, such rotational motion is said to reveal the existence of absolute space, but Einstein objects to such an account on the basis of the positivistic principle enunciated above. "But the privileged space R_1 of Galileo, thus introduced, is a merely *factitious* cause, and not a thing that can be observed."[65] This leads Einstein to embrace the rather outlandish view of Mach that "the mechanical behavior of S_1 and S_2 is partly conditioned, in quite essential respects, by distant masses which we have not included in the system under consideration."[66] According to Einstein, "These distant masses and their motions relative to S_1 and S_2 must then be regarded as the seat of the causes (which must be susceptible to observation) of the different behavior of our two bodies S_1 and S_2. They take over the role of the factitious cause R_1."[67] We cannot say that R_1 is privileged "without reviving the above-mentioned epistemological objection."[68] Einstein's defense of Mach's Principle (as this point of view is called) is clearly based on a verificationist epistemology. As Earman points out, Mach's rejection of absolute space, like his much criticized rejection of atomism, stems from "his highly positivistic and operationalistic philosophy."[69] Fortunately, Mach's Principle does not belong essentially to the foundations of GR, and most textbook expositions of that theory pass over Einstein's argument in benign neglect, focusing instead on his subsidiary argument in section 2 on the equivalence of acceleration and gravitational fields.[70]

What Einstein's labor on General Relativity made clear to him was the inadequacy of Mach's phenomenalism. Scientific theorizing is not the mere linking of observation statements but involves a creative exercise of the mind, which is free to postulate theoretical entities not directly given in observation. Nevertheless, even after GR he continued to regard such theoretical terms as meaningless unless they could be somehow linked to observation statements. In his *Relativity: the Special and General Theory* (1920), for example, he asks his reader what sense (*Sinn*) there is in asserting that two lightning bolts simultaneously strike at two separate places. He explicitly rejects the answer that the meaning (*Bedeutung*) of the assertion is in and of itself clear, even if it might be difficult to show by means of observations that the two events occurred simultaneously. According to Einstein, the concept of simultaneity does not exist for the physicist until he can in a concrete situation discover whether the concept holds or not.

> We thus require a definition of simultaneity such that this definition supplies us with the means by which, in the present case, he can decide by experiment whether both lightning strokes occurred simultaneously. As long as this requirement is not satisfied, I

---

[64] Ibid.
[65] Ibid., pp. 143-144.
[66] Ibid., p. 144.
[67] Ibid.
[68] Ibid.
[69] John Earman, "Who's Afraid of Absolute Space?" *Australasian Journal of Philosophy* 48 (1970): 298.
[70] On GR's failure to render S_1 and S_2 indistinguishable see Michael Friedman, *Foundations of Spacetime Theories* (Princeton: Princeton University Press, 1983), pp. 211, 230-232.

allow myself to be deceived as a physicist (and of course the same applies if I am not a physicist) when I imagine that I am able to attach a meaning to the statement of simultaneity.[71]

For physicist and non-physicist alike the statement that two events occur simultaneously is meaningless unless an operational definition can be given for that concept.[72] Thus, he continued to cling to his rejection of metaphysical time and space:

> The only justification for our concepts and system of concepts is that they serve to represent the complex of our experiences; beyond this they have no legitimacy. I am convinced that the philosophers have had a harmful effect upon the progress of scientific thinking in removing certain fundamental concepts from the domain of empiricism, where they are under our control, to the intangible heights of the *a priori*. For even if it should appear that the universe of ideas cannot be deduced from experience by logical means, but is, in a sense, a creation of the human mind, without which no science is possible, nevertheless this universe of ideas is just as little independent of the nature of our experiences as clothes are of the form of the human body. This is particularly true of our concepts of time and space, which physicists have been obliged by the facts to bring down from the Olympus of the *a priori* in order to adjust them and put them in a serviceable condition.[73]

The Machist-Humean tones of this line of thinking are strikingly evident.

Even as late as his "Autobiographical Notes," it is clear that the incubus of Machism had not been exorcised from Einstein's soul. He had become a critical realist and emphasized the necessity of logical, mathematical, and other non-observation terms in scientific theory construction;[74] still the old positivism hung on. In stating his "epistemological credo," Einstein wrote,

> I see on the one side the totality of sense experiences, and, on the other, the totality of concepts and propositions which are laid down in books. The relations between the concepts and propositions among themselves and each other are of a logical nature, and the business of logical thinking is strictly limited to the achievement of the connection between concepts and propositions among each other according to firmly laid down rules, which are the concern of logic. The concepts and propositions get 'meaning,' viz., 'content,' only through their connection with sense-experiences. The connection of the latter with the former is purely intuitive, not itself of a logical nature. The degree

---

[71] Albert Einstein, *Relativity, the Special and the General Theory*, trans. Robert W. Lauren (London: Methuen, 1920), p. 26.
[72] That Einstein was a verificationist outside the realm of physics is also evident from his remarks in an interview with Max Wertheimer that when someone uses the word "hunchback," "If this concept is to have any clear meaning, there must be some way of finding out whether or not a man has a hunched back. If I would conceive of no possibility of reaching such a decision, the word would have no real meaning for me" (Max Wertheimer, *Productive Thinking*, ed. Michael Wertheimer, enlar. ed. [London: Tavistock, 1961], p. 220).
[73] Albert Einstein, *The Meaning of Relativity*, 6th ed. (1922; rep. ed.: London: Chapman and Hall, 1967), p. 2. Cf. his "Fundamental Ideas and Problems of the Theory of Relativity," [1923], in *Nobel Lectures, Physics: 1901-1921* (New York: Elsevier, 1967), pp. 479-490, where he lays down a postulate called "the stipulation of meaning," which requires that concepts and distinctions are only admissible to the extent that observable facts can be assigned to them without ambiguity. He considers this postulate to be of "fundamental importance" epistemologically.
[74] See his Herbert Spencer lecture of 1933, "On the Methods of Theoretical Physics," in *The World as I See It* (New York: Covici-Friede, 1934), pp. 30-40.

> of certainty with which this relation, viz., intuitive connection, can be undertaken, and nothing else, differentiates empty phantasy from scientific 'truth.'[75]

It seems evident that Einstein was still holding to a relaxed verificationist theory of meaning. It is extremely instructive to observe that when Einstein several pages later blasts Mach and Ostwald for their "positivistic philosophical attitude" which led them to deny the reality of atoms, he equates positivism with the view "that facts by themselves can and should yield scientific knowledge without free conceptual construction"[76]—a view which Einstein had long rejected. But as Frank points out in the same volume, Einstein's requirement that there must be some observational statements to give meaning to the body of concepts and propositions is positivistic.[77] In Victor Lenzen's essay, "Einstein's Theory of Knowledge," which Einstein hails as "convincing and correct" in everything it says,[78] Lenzen explains that "The concept of objective time is introduced through the intermediary of space. Just as the rigid body is a basis for space, so is the clock for time."[79] "Objective local time is based on the correlation of the temporal course of experience with the indications of a 'clock'.... Objective extended time is based upon synchronization of distant clocks by signals."[80] Lenzen goes on to contrast these concepts to the "hypostatization" of time and space in classical theory. These re-definitions of time and space in operational terms, however, were and remain positivistic in character.

In any case, Einstein's later views remain somewhat beside the point for our purposes. What is relevant are his early epistemological views, and these are solidly Machist. In his "Autobiographical Notes," Einstein admits that his epistemological credo "actually evolved only much later and very slowly and does not correspond with the point of view I held in younger years."[81] In those years, it was Mach who held sway: "I see Mach's greatness in his incorruptible skepticism and independence; in my younger years, however, Mach's epistemological position also influenced me very greatly, a position which today appears to me to be essentially untenable."[82] Holton concludes, "Taking the early papers as a whole, and in the context of the physics of the day, we find that Einstein's philosophical pilgrimage

---

[75] Einstein, "Autobiographical Notes," pp. 11-13. Cf. his statement, "Concepts can only acquire content when they are connected, however indirectly, with sensible experience" (A. Einstein, "The Problem of Space, Ether, and the Field in Physics," in *Ideas and Opinions*, trans. Sonja Bargmann [New York: Crown Publishers, 1954], p. 277).

[76] Ibid., p. 49.

[77] Frank, "Einstein, Mach, and Logical Positivism," p. 274. Frank concludes that the difference between Einstein and later logical positivists is purely verbal (Ibid., p. 282). Cf. Ernst Nagel, "Relativity and Twentieth-Century Intellectual Life," in *Some Strangeness in the Proportion*, ed. Harry Woolf (Reading, Mass.: Addison-Wesley, 1980), p. 41, who claims that logical positivists came to embrace a moderate operationalism like Einstein's, an operationalism which came to dominate the philosophy of science during the second half of the century.

[78] Albert Einstein, "Reply to Criticism," in *Einstein: Philosopher-Scientist*, p. 683.

[79] Victor F. Lenzen, "Einstein's Theory of Knowledge," in *Einstein: Philosopher-Scientist*, pp. 369-370.

[80] Ibid., p. 370.

[81] Einstein, "Autobiographical Notes," p. 11.

[82] Ibid., p. 21.

did start on the historic ground of positivism. Moreover, Einstein thought so himself and confessed as much...."[83]

POSITIVIST RECEPTION OF SR

Positivistic philosophers and physicists were quick to recognize in SR a kindred spirit and embraced the theory eagerly. According to Holton, "This was the kind of operationalist message which, for most of his readers, overshadowed all other philosophical arguments in Einstein's paper. His work was enthusiastically embraced by the groups who saw themselves as philosophical heirs of Mach, the Vienna Circle of neopositivists and its predecessors and related followers, providing a tremendous boost for the philosophy that had initially helped to nurture it.[84] In fact Friedman believes that SR helped to spawn the logical positivism of 1925-35, particularly its theoretical/ observational distinction.[85] The earliest adherents to the theory, such as Planck, Wien, and Minkowski, embraced SR because they found it "congenial," stressing that no empirical facts stood *against* it.[86] Planck noted in particular the "devastating results of [Einstein's] new interpretation of the time concept, which have repercussions on the whole of physics, and above all on mechanics and thence deep into epistemology."[87] Walter Kaufmann, who thought his experimental data falsified SR, maintained that the only advantage of Einstein's theory was "epistemological" in that it did not involve unknown velocities.[88] Max von Laue opined that one can hardly avoid preferring the relativity principle to the aether, since "it contradicts all epistemological principles to ascribe to a body physical reality when it can never be detected."[89] At the first meeting of the *Gesellschaft für positivistische Philosophie* in 1912, Petzoldt declared, "Lorentz's theory is, at its conceptual center, pure metaphysics, nothing else than Schelling or Hegel's philosophy of nature," but Einstein's theory represents "the victory over the metaphysics of absolutes in the conceptions of space and time...a mighty impulse for the development of the philosophical point of view in our time."[90] In an Appendix entitled "The Relationship of Mach's Worldview to Relativity Theory" to the eighth edition of Mach's *Mechanik*, Petzoldt points to Mach's section on Newton as responsible for creating the atmosphere "without which Einstein's relativity theory would not have been possible."[91] He dismisses Newton's absolute

---

[83] Holton, "Search for Reality," p. 171. Einstein's pilgrimage began with "a philosophy of science in which positivism was the center" (Holton, "Where Is Reality?" p. 45).
[84] Holton, "Search for Reality," pp. 170-171.
[85] Friedman, *Foundations of Spacetime Theories*, pp. 24-25.
[86] Gerald Holton, "Einstein's Scientific Program: The Formative Years," in *Strangeness in the Proportion*, p. 58.
[87] Max Planck, cited in C. Seelig, *Albert Einstein: A Documentary Biography* (London: Staples Press, 1956), pp. 144-145.
[88] W. Kaufmann, "Über die Konstitution des Elektrons," *Annalen der Physik* 19 (1906): 487-553.
[89] M. Laue, *Das Relativitätsprinzip*, Sammlung Naturwissenschaftlicher und mathematischer Monographien 38 (Braunschweig: Friedrich Vieweg & Sohn, 1911), p. 33.
[90] Cited in Holton, "Search for Reality," pp. 171.
[91] Joseph Petzoldt, "Das Verhältnis der Machschen Gedankenwelt zur Relativitätstheorie," in *Die Mechanik in ihrer Entwicklung*, 8th ed., by Ernst Mach (Leipzig: F. A. Brockhaus, 1921), p. 494. Petzoldt records Einstein's agreement with this judgement.

space and absolute motion as having "no physical meaning" and concludes that relativity theory "fits in fully with Mach's worldview" and is "a golden fruit of his deeply rooted and powerfully expansive tree of thought."[92]

P. W. Bridgeman, the father of operationalism, extolled Einstein's procedure:

> ...he recognized that the meaning of a term is to be sought in the operations employed in making application of the term. If the term is one which is applicable to concrete physical situations, as 'length' or 'simultaneity,' then the meaning is to be sought in the operations by which the length of concrete physical objects is determined, or in the operations by which one determines whether two concrete physical events are simultaneous or not.[93]

Reichenbach recognized the positivism underlying such a procedure:

> ...he saw that certain physical problems could not be solved unless the solutions were preceded by a logical analysis of the fundamentals of space and time, and he saw that this analysis, in turn, presupposed a philosophic readjustment of certain familiar conceptions of knowledge. The physicist who wanted to understand the Michelson experiment had to commit himself to a philosophy for which the meaning of a statement is reducible to its verifiability, that is, he had to adopt the verifiability theory of meaning if he wanted to escape a maze of ambiguous questions and gratuitous complications. It is this positivist, or let me rather say, empiricist commitment which determines the philosophical position of Einstein.[94]

Ernst Cassirer, in his essay "Einstein's Theory of Relativity considered from the Epistemological Standpoint," clearly recognized the crucial role played by positivism in SR and GR. All historical examples of "the real inner connection between epistemological problems and physical problems," he says, are outdone by relativity theory; "from its very beginning" in its analysis of the concept of time, the

---

[92] Ibid., pp. 495, 516-517.

[93] P. W. Bridgeman, "Einstein's Theories and the Operational Point of View," in *Einstein: Philosopher-Scientist*, p. 335. Gaston Bachelard adds, "This *operational* definition of simultaneity dissolves the notion of *absolute* time. Since simultaneity is linked to physical experiments which occur in space, the temporal contexture (*contexture*) is one with spatial contexture. Since there is no absolute space, there is no absolute time" (Gaston Bachelard, "The Philosophic Dialectic of the Concepts of Relativity," in *Einstein: Philosopher-Scientist*, p. 571). According to Prokhovnik, "Perhaps Einstein's most radical innovation was his introduction of precise operational definitions (or 'conventions') for synchronising separated clocks and for determining the space and time co-ordinates of an event. By implication he was thereby asserting that there was no other meaning for these co-ordinates apart from their determination in terms of agreed measurements—he had no need for any abstract notions of space and time" (S. J. Prokhovnik, *Light in Einstein's Universe* [Dordrecht: D. Reidel, 1985], p. 33). Epstein observed, "The theory of relativity had an immense educational influence on physicists and other scientists by teaching them the approach to scientific concepts that has been more recently named *the operational point of view*;" the main philosophic value of relativity is that it has divested concepts of space and time "of their former metaphysical connotations" (Paul S. Epstein, "The Time Concept in Restricted Relativity," *American Journal of Physics* 10 [1942]: 1).

[94] Hans Reichenbach, "The Philosophical Significance of the Theory of Relativity," in *Einstein: Philosopher-Scientist*, pp. 290-291. Norton observes that when in the 1905 paper Einstein stated that he had no need to introduce the ether, "Einstein was not consciously applying the verifiability principle when he wrote these words. The criterion was not formulated until over twenty years later. However, if we are to identify any justifiable, scientific application of the criterion, then this case surely would be included" (John D. Norton, "Philosophy of Space and Time," in *Introduction to the Philosophy of Science* [New Jersey: Prentice-Hall, 1992], p. 183). Since the state of absolute rest had no observable consequences, "We are thereby enjoined to despise the entity or state as an idle metaphysical conception and to banish it from our discourse" (Ibid.).

theory extends into the field of epistemology.[95] SR's advantage over competing explanations, such as Lorentz's contraction hypothesis, "is based not so much on its empirical material as on its pure logical form, not so much on its physical as on its general *systematic* value."[96] He explains,

> The Lorentzian hypothesis...was sufficient to give a complete explanation of all known observations. An experimental decision between Lorentz and Einstein's theories was thus not possible; it was seen that between them there could fundamentally be no *experimentum crucis*. The advocates of the new doctrine accordingly had to appeal—an unusual spectacle in the history of physics—to general philosophical grounds, to the advantages over the assumption of Lorentz which the new doctrine possessed in a systematic and epistemological respect.[97]

Cassirer goes on to explain that Lorentz's approach was deemed "epistemologically unsatisfactory" because it ascribed to the aether definite effects which preclude the aether's ever being an object of possible observation. What in the last analysis counted against Lorentz's theory was not an empirical, but a methodological defect:

> It conflicted most sharply with a general principle, to which Leibniz has appealed in his struggle against the Newtonian concepts of absolute space and time, and which he formulated as the 'principle of observability' *(principe de l'observabilité.)* When Clarke, as the representative of Newton, referred to the possibility that the universe in its motion relatively to absolute space might undergo retardation or acceleration which would not be discoverable by our means of measurement, Leibniz answered that nothing fundamentally outside the sphere of observation possessed 'being' in the physical sense: *quand il n'y a point de changement observable, il n'y a point de changement du tout.* It is precisely this principle of 'observability' which Einstein applied at an important and decisive place in his theory, at the transition from the special to the general theory of relativity, and which he has attempted to give a necessary connection with the general principle of causality.[98]

From start to finish, then, relativity theory was seen to be based in a positivistic epistemology which won for it its principal advantage over its competitors.

POSITIVISM AND THE REJECTION OF METAPHYSICS

Under the influence of positivism and the verificationist criterion of meaning, physicists and philosophers of space and time during the first half of this century exhibited an abhorrence for what was called "metaphysics." An instructive piece is a 1941 article by Henry Margenau ostensibly defending metaphysical elements in physics. Observing that "our time appears to be distinguished by its *taboos*, among which there is to be found the broad convention that the word *metaphysics* must never be used in polite scientific society," Margenau counters that there not only are,

---

[95] Ernst Cassirer, *Substance and Function and Einstein's Theory of Relativity*, trans. W. C. Swabey and M. C. Swabey (1923; rep ed.: Dover, 1953), pp. 354-355.
[96] Ibid., p. 354.
[97] Ibid., pp. 375-376.
[98] Ibid., pp. 376-377. Cassirer's analysis can mislead. The debate between Clarke and Leibniz concerned substantivalism vs. relationalism. Leibniz was just as non-relativistic as Clarke, and contemporary relativists are often substantivalists. Moreover, Einstein's positivism tended to deny that statements about an absolute frame had any meaning, so that his principle resembled a verification criterion of meaning in a way Leibniz's did not. Still, both critiques of the classical concept of time dispute that concept based on verificationist arguments.

but ought to be, metaphysical elements in physical science.[99] But then he emasculates this bold contention by explaining that he means thereby that we must have "epistemology"; but as physicists "we reject ontology."[100] He reduces metaphysics to what he calls the methodology of science, and insists that we must not relax our standards here, lest the "obnoxious ontological elements" find their way back into science.[101] What these elements are he leaves in no doubt: the luminiferous aether and simultaneity in different Lorentz frames are classed along with the external world and the *Ding-an-sich* and dismissed as "ultra-perceptory and hence meaningless."[102] This judgement is based on the "positivistic criticism" that propositions not verifiable in principle are meaningless, a criterion which elicits Margenau's ringing endorsement: "this recognition should be one of the premises of philosophy of science; it enjoys, indeed, almost universal consent."[103]

It is not difficult to find examples of this anti-metaphysical bent in treatments of relativity theory. Complaining that "it is hardly possible to open a textbook on the theory of relativity—even if written by an otherwise competent physicist—without coming upon sentences of an entirely metaphysical character," Frank asserted that such sentences are wholly meaningless in physics.[104] By a metaphysical sentence, he meant one which asserts a fact not verifiable by observation. All such sentences form a system of sentences which is coherent and isolated from the verifiable sentences. "One usually calls these isolated sentences 'metaphysical' sentences which state something about a real world. Being non-verifiable sentences they are meaningless as far as science is concerned."[105] Relativity theory, through its operational definitions of space and time, relegates sentences about metaphysical space and time to a state of "splendid isolation."[106]

Similarly, Eddington endorsed a verificationist analysis of meaning. He wrote, "If we happen to make a deduction which could not conceivably be corroborated or disproved by these diligent measures, there is no criterion of its truth or falsehood and it is thereby a meaningless deduction."[107] With the advent of relativity theory, he explains, concepts of space and time are defined by pointer readings rather than

---

[99] H. Margenau, "Metaphysical Elements in Physics," *Reviews of Modern Physics* 13 (1941): 176.
[100] Ibid., p. 177.
[101] Ibid.
[102] Ibid., p. 178.
[103] Ibid.
[104] Frank, *Interpretations and Misinterpretations of Modern Physics*, Actualités Scientifiques et Industrielles 587: Exposés de Philosophie Scientifique 2 (Paris: Hermann & Cie, 1938), p. 34. Cf. his later remark, "It can be said without exaggeration that there is no philosophical congress, no philosophical textbook, not even an issue of a philosophical journal, where we do not encounter examples of attempts to draw arguments in favor of metaphysical opinions from the statements of the theory of relativity" (Ibid., p. 46). He gives as an example James Jeans's attempt to deduce fatalism from SR.
[105] Ibid, p. 38.
[106] Ibid., p. 39. For an example of this tendency, see Ehrenfest's characterization of SR as the combination of the following assertions: (1) Light sources send us light signals as independent phenomena through empty space. (2) The measured velocities of light rays from a source moving towards us end from a source at rest would be observed to be the same. (3) We declare that we are satisfied with the combination of these two statements. (Paul Ehrenfest, *Zur Krise der Lichthypothese* [Berlin: Springer Verlag, 1913], p. 19).
[107] Arthur Eddington, *The Nature of the Physical World*, with an Introductory Note by Sir Edmund Whittaker, Everyman's Library (1928; rep. ed.: London: J. M. Dent and Sons, 1964), p. 223.

having the metaphysical significance we might have expected.[108] Thus, in an imaginary dialogue between a physicist and a relativist we find this exchange concerning time dilation:

> *Phys.* I think your unit of time would change according to the motion of your 'clock' through the aether.
> *Rel.* Then you are comparing it with some notion of absolute time. I have no notion of time except as the result of measurement with some kind of clock.[109]

Again, when discussing the possibility that the aether could serve to partition spacetime into privileged, successive three-dimensional slices, Eddington demurs, "It seems an abuse of language to speak of a division *existing*, when nothing has ever been found to pay any attention to the division."[110] He opined, "the off-chance that some future generation may discover a significance in our utterances is scarcely an excuse for making meaningless noises."[111]

Similarly, Max Born discarded absolute time with the words, "But what exists 'without reference to any external object whatsoever' is not ascertainable, and is not a fact" and absolute space with the same aplomb: "A conception has physical reality only when there is something ascertainable by measurement corresponding to it in the world of phenomena."[112] With respect to distant simultaneity he advises, "the quantitative physicist...sees no meaning in the statement that an event at A and an event at B are simultaneous, since he has no means of deciding the truth or the incorrectness of this assertion."[113] Since Einstein showed that there is no means of determining absolute time, "this signifies that absolute time has no physical reality."[114] Generalizing, Born draws attention to the particular emphasis of his exposition that "only ascertainable facts have physical reality."[115]

When the contemporary student of physics reads such anti-metaphysical polemics, he must feel as though he were peering into a different world. For it is now widely recognized that the boundaries of science are impossible to fix with precision, and during the last few decades theoretical physics has become characterized precisely by its metaphysical, speculative character. In various fields such as quantum mechanics, classical cosmology, and quantum cosmology, debates

---

[108] Ibid., p. 247.
[109] Arthur Eddington, *Space, Time and Gravitation*, Cambridge Science Classics (1920: rep. ed.: Cambridge: Cambridge University Press, 1987), p. 15.
[110] Ibid., p. 39.
[111] A. S. Eddington, *The Mathematical Theory of Relativity*, 2d ed. (Cambridge: Cambridge University Press, 1930), p. 8.
[112] Max Born, *Einstein's Theory of Relativity*, trans. Henry L. Brose (New York: E. P. Dutton, n. d.), pp. 51, 61.
[113] Ibid., p. 193. Cf. the strictures of d'Abro:
> "For we may speak of simultaneity throughout space, of an instantaneous space, as much as we please, but until we know enough about these evasive concepts to be able to distinguish events that are simultaneous from those which are not, we cannot claim to have any definite idea of what we are talking about....
> ...if two events take place in different places, we can no longer attribute any universal meaning to the opinion that these two events have taken place at one and the same instant of time" (A. d'Abro, *The Evolution of Scientific Thought*, [1927] 2d rev. ed. [n. p.: Dover Publications, 1950], pp. 186, 201).

[114] Born, *Einstein's Theory*, p. 210.
[115] Ibid., p. 252.

rage over overtly metaphysical issues. Take quantum physics, for example. In 1935 Max Born wrote, "For what lies within the limits is knowable, and will become known; it is the world of experience, wide, rich enough in changing hues and patterns to allure us to explore it in all directions. What lies beyond, the dry tracts of metaphysics, we willingly leave to speculative philosophy."[116] Commenting on this statement, Euan Squires marvels, "how far from the truth such an attitude really is."[117] In Squires's opinion, "In an effort to understand the quantum world, we are led beyond physics, certainly into philosophy and maybe even into cosmology, psychology and theology."[118] Is there a quantum world at all, or is there only, as at least one version of the Copenhagen Interpretation would have it, an abstract quantum physical description which does not at all purport to find out how nature really is? Or do quantum realities come into being, as other Copenhagen Interpreters maintain, by and upon the occasion of measurement with a classical apparatus? How and when does the wave function of a quantum entity collapse? Since any physical measuring apparatus can itself be described in quantum physical terms, are we forced to regard observation by a conscious being as the final link in the chain, thus resulting in an observer-dependent reality? But since human consciousnesses are linked to physical brains, which can be described quantum mechanically, would we not be led to posit some transcendent Consciousness, or God, who is finally responsible for the collapse of all wavefunctions in the physical universe? But perhaps the wavefunction of quantum entities never collapses. Perhaps we should adopt the Many Worlds Interpretation, according to which the universe splits at every quantum measurement into parallel worlds each similar except for the differing value of the measurement in each respective world—Or, even more radically, the Many Minds Interpretation, which posits an infinite number of selves associated with any mind which respectively apprehend the different values of the wave function. Or perhaps Einstein was right after all: a non-local hidden variables theory along the lines of the de Broglie-Bohm pilot wave is mathematically consistent and complete and would explain all empirical observations. Then again, maybe the wave function is an *ens fictum*; maybe quantum physics describes only the behavior of ensembles of particles, not individual quantum entities. All of these theories take up metaphysical positions and are part and parcel of contemporary debate in physics.

Or take the field of classical cosmology. George Gale has observed that most cosmologists in the twentieth century remarked with relief the disengagement of modern cosmology from its origin in philosophy. But such relief was short-lived. "Cosmology, even as practiced today," says Gale, "is science done at the limit: at the limit of our concepts, of our mathematical methods, of our instruments, indeed, of our very imaginations."[119] In an article in *Astronomy*, Rothman and Ellis pose the question, "Has astronomy become metaphysical?" and answer that it has.[120] Think,

---

[116] Max Born, *Atomic Physics*, 5th ed., trans. J. Douglas (London: Blackie, 1951), p. 306.
[117] Euan Squires, *The Mystery of the Quantum World* (Bristol: Adam Hilger, 1986), p. 4.
[118] Ibid.
[119] George Gale, "Cosmos and Conflict," paper presented at the conference "The Origin of the Universe," Colorado State University, Ft. Collins, Colorado, 22-25 September, 1988.
[120] Tony Rothman and George Ellis, "Has Astronomy Become Metaphysical?" *Astronomy* (February 1987): 7.

for example, about the origin of the universe. Einstein's field equations for GR (minus the cosmological constant) predicted the expansion of the universe, which was confirmed in 1927 by Hubble's observation of the galactic red-shift, which indicated a universal and isotropic expansion. In 1928 Eddington wrote of the beginning of the universe predicted by the Big Bang model, "It is one of those conclusions from which we can see no logical escape—only it suffers from the drawback that it is incredible. As a scientist I simply do not believe that the present order of things started off with a bang.... But I can make no suggestion to avoid the deadlock."[121] Since then the Steady State Theory has come and gone. Today, a standard text reports, "No problem of cosmology digs more deeply into the foundations of physics than the question of what 'preceded' the 'initial state' of infinite (or near infinite) density, pressure, and temperature."[122] Did matter and energy, space and time, all of physical reality come into being out of nothing at a point in the finite past? How can this be understood? Does it point, as Milne and Whittaker thought, to an extra-mundane Creator of the universe? Or should we prefer naturalistic, but equally metaphysical, explanations, such as an oscillating universe? —Or perhaps adjust the standard model by adding some sort of inflationary scenario, according to which the observable universe is but a bubble in a sea of foam? Then again, there is the staggeringly improbable complexity of our universe to be dealt with. The old design argument, thought to have been banished by Hume and Darwin, has come roaring back in the debate surrounding the Anthropic Principle. The delicate balance of physical constants and quantities in the initial conditions of the universe necessary for the universe to be life-permitting cries out for explanation. But what sort of explanation? A cosmic Intelligence, who designed the world, as John Leslie argues? Or are we forced to posit an infinite and exhaustively random World Ensemble, so as to give purchase to the Anthropic Principle? Or should we insist that no explanation is needed for the life-permitting conditions in the first place? These questions, hotly debated in physical journals, are metaphysical in character. Gale concludes, "Although both sides—Inflationists as well as Anthropic Big Bangers—accuse the other of being 'metaphysical,' and just insofar as that, unscientific, it is clear that metaphysics continues to play an honorable role in cosmology. And, to the extent that it is an honorable role, it is no dishonor to use metaphysics in one's cosmologizing."[123]

Physics becomes most metaphysical in the budding field of quantum cosmology. Relativistic physicists in the first half of this century exulted in the four-dimensional spacetime view of reality which relativity theory had given them. Now quantum theory threatens to undo the fabric of their world. At distances of the order of the Planck length, 1.6×10^{-33} cm and less, quantum fluctuations take place in the geometry of space. Quantum geometrodynamics, as this field of study is called, posits a superspace as the dynamical arena in which all three-dimensional geometries exist. The various three-geometries are assigned probability wave functions which make it possible only to approximate a classical spacetime as a leaf

---

[121] Eddington, *Nature of the Physical World*, pp. 90-91.
[122] C. Misner, K. S. Thorne, and J. A. Wheeler, *Gravitation* (San Francisco: W. H. Freeman, 1973), p. 769.
[123] Gale, "Cosmos and Conflict."

of superspace. *"No prediction of spacetime, therefore no meaning for spacetime*, is the verdict of the quantum principle. That object which is central to all of classical general relativity, the four-dimensional spacetime geometry, simply does not exist, except in a classical approximation."[124] If this vision of reality did not raise metaphysical questions enough, some theorists like Wheeler go further, proposing to analyze superspace in terms of a "pre-geometry," mathematical objects which are the constituents that make up superspace. Quantum geometrodynamics becomes relevant to cosmology in the on-going quest of a quantum theory of gravity. Proponents of quantum cosmology assign a wave function to the universe as a whole. But if one adopts some version of the Copenhagen Interpretation, then one is forced, it seems, to posit some sort of Ultimate Observer who transcends the universe and collapses its wave-function.[125] In order to avoid this implication, which smacks of theism, quantum cosmologists often embrace a Many Worlds Interpretation, the worlds being all the three-geometries in superspace. One can then employ quantum mechanical methods to calculate the probability of our universe. In the most celebrated of such models, the Hartle-Hawking model, the further metaphysical position must be adopted that time is imaginary (that is, the time coordinate is an imaginary number), if one is to avoid intractable infinities in performing the quantum calculations. But clearly, the notion that our universe exists in imaginary time raises profound metaphysical difficulties.

Metaphysical questions, hypotheses, and difficulties are abundant in these and other fields of modern physics—for example, particle physics, which Victor Weisskopf mischievously accuses of suffering from the defect of developing "too many new unproven ideas per unit time."[126] Like it or not, theoretical physics has become thoroughly impregnated with metaphysics.

But we have forgotten one of the prime examples of the role of metaphysics in scientific theorizing: relativity theory itself! As Sklar points out, those who are

---

[124] Misner, Thorne, and Wheeler, *Gravitation*, p. 1183.

[125] John Barrow writes that the program of quantum cosmology "immediately faces an impasse that can only be overcome by coming to grips with the meaning of quantum observership. If we only ascribe reality to what is observed, who observes the Universe?" Unless we adopt the Many Worlds Interpretation, "...we are left asking the question 'who or what collapses the wave function of the universe?'—some 'ultimate Observer' at the world's end, or outside the Universe of space and time altogether?" The Many Worlds Interpretation "is adopted by quantum cosmologists because it does not require the Universe to be observed" (John Barrow, *The World within the World* [Oxford: Clarendon Press, 1988], pp. 156, 363).

[126] Victor Weisskopf, comment on Steven Weinberg, "Elementary Particle Physics in the Very Early Universe," in *Astrophysical Cosmology*, ed. H. A. Brück, G. V. Coyne, M. S. Longair, Pontificiae Academiae Scientiarvm Scripta Varia 48 (Vatican City: Pontificia Academia Scientiarvm, 1982), p. 527. He thinks that particle physics has become "even more uncertain and hypothetical than cosmology"! (Ibid., p. 528). For a discussion of the metaphysical issues raised by this field, see George Gale, "Some Metaphysical Perplexities in Contemporary Physics," paper presented at the 36th Annual Meeting of the Metaphysical Society of America, Vanderbilt University, March 14-16, 1985. He asserts that "...although quarks apparently satisfy the classical metaphysical desiderata of the atomistic research program, they do so at enormous cost: the sacrifice of the epistemological and methodological criteria which explicitly define modern science," since they are "*in principle* unobservable." Of course, the same may be said about string theory. See Michio Kaku, *Hyperspace* (New York: Oxford University Press, 1994); Michio Kaku, *Introduction to Superstrings and M-Theory*, 2d ed., Graduate Texts in Contemporary Physics (New York: Springer, 1999); Brian Greene, *The Elegant Universe* (New York: W. W. Norton, 1999).

realists about problems of space and time reject the positivist doctrine that observational equivalence is a sufficient condition of the full equivalence of two theories.[127] Realists often hold that in order to be equivalent two theories must not only be empirically indistinguishable but also have some structural isomorphism at the non-observational level. For example, in response to the sort of conventionalism propounded by Poincaré and Reichenbach, according to which there is no factual difference between a curved spacetime and a flat spacetime conjoined with compensating universal forces, the realist will insist, in Sklar's words, that

> Two theories might have all the same observational predictions but be so radically different in their structure at the theoretical level that one ought to take them as attributing (realistically) quite different explanatory structures to the world. Only commonality of structure at the level of the theoretical ontology introduced by the theories to explain the commonly predicted observational results is enough for us to say that the two accounts are genuinely, realistically, equivalent to one another.[128]

Sklar quite rightly observes that this response raises the further question as to what then justifies the choice of one theory as true over a non-equivalent, but empirically indistinguishable alternative. But notice that this is, indeed, a *further* question. At worst, we are left with a sort of watered-down, epistemic conventionalism, such as Clark Glymour espouses:

> Cases do arise, however, where it is unreasonable to think that two theories can both be true, yet we cannot decide between them on empirical grounds. It is with such cases that conventions in physics have a real importance, and that importance is epistemic. The significant cases of conventional topologies are those in which the world is one way or the other, not both, but things are so arranged that we cannot discover which way the world is.[129]

This epistemic conventionalism goes no distance toward justifying a positivistic, ontological conventionalism. As Nerlich emphasizes, "Maybe we could never know, even in principle, whether the earth or our whole space was Euclidean or spherical or what. But we are not concerned with the problems of finding out about spaces. We are worried about the spaces themselves. Our question was whether the space was determinate in its global topology, not whether our knowledge was."[130]

Now with respect to special relativity, we have three competing approaches or theories which are on all hands admitted to be observationally equivalent but which differ fundamentally in terms of their theoretical ontology, namely, the aether compensatory approach of Lorentz, the relativity approach of Einstein and the spacetime approach of Minkowski. The central difference between these three approaches lies in the structures posited at the theoretical level, which yield quite different accounts of the predicted phenomena. Contrasting Einstein's approach to Lorentz's, Kilmister argues that what makes relativity theory interesting is the

---

[127] Lawrence Sklar, *Philosophy and Spacetime Physics* (Berkeley: University of California Press, 1985), p. 6.
[128] Ibid., p. 56.
[129] Clark Glymour, "Topology, Cosmology, and Convention," in *Space, Time, and Geometry*, ed. Patrick Suppes, Synthèse Library (Dordrecht: D. Reidel, 1973), pp. 213-214.
[130] Graham Nerlich, *The Shape of Space* (Cambridge: Cambridge University Press, 1976), p. 146.

"major change in ontology" wrought by Einstein *vis à vis* nineteenth century aether compensatory theories.[131] René Dugas points out,

> Lorentz's theory already embodies the essential results of special relativity, namely, the transformation (T) of the coordinates of space and time, the law (C) of the transformation of electrical and magnetic fields, the law (M) of the variation of the mass with the velocity. To these results, Einstein's first paper was to add only the law (V) of the composition of the velocities and the formula relating the mass to the energy.
>
> But in the passage of Lorentz's theory to that of Einstein, what is essential is the *novelty of the Einsteinian point of view*. Lorentz's theory is not relativistic, in this sense that a system at absolute rest in relation to the ether continues to enjoy special properties there.[132]

In Lorentz's theory there is a preferred rest frame and temporal parameter and, hence, absolute length and absolute simultaneity. Relativistic phenomena are the effects of dynamic causes operative as a result of motion relative to the rest frame. Einstein denies all of this theoretical structure. Early interpreters of SR, no doubt under the influence of positivism, tended to paper over these differences, often referring to SR as the Lorentz-Einstein theory. But these theoretical differences are so stark that in time Einstein's relativistic interpretation came to be distinguished from and preferred to the aether compensatory approach. This involved positivist thinkers in a strange inconsistency: in order to reject Lorentz's theory in favor of Einstein's they appealed to a verifiability criterion of meaning to eliminate the aether frame, absolute simultaneity, and so on; but in so doing they undermined the thesis of the equivalence of observationally indistinguishable theories, thereby exposing the metaphysical elements in science.

Moreover, we have seen that SR itself is patient of two radically different interpretations, namely, Einstein's original relativity interpretation and Minkowski's spacetime interpretation. Again, these two interpretations are observationally equivalent but are fundamentally different at the theoretical level. The spacetime realist posits a radically different, four-dimensional ontology than the three-dimensional ontology (with its reference frames, light signals, and so forth) of the

---

[131] C. W. Kilmister, "Why is Relativity Interesting?" *British Journal for the Philosophy of Science* 42 (1991): 414. Dieks makes the point well:

"The philosophical difference between Lorentz's electron theory and the special theory of relativity is exactly the different ontology of the two theories. Whereas an adherent of the electron theory works with classical spacetime relations, and therefore feels obliged to *explain* deviations from these relations which occur if moving rods are used, the relativist will, following Einstein's postulates take it for granted that the world is relativistic, and not classical" (Dennis Dieks, "The 'Reality' of the Lorentz Contraction," *Zeitschrift für allgemeine Wissenschaftstheorie* 15/2 [1984]: 341).

According to Dieks, the Relativity Principle thus assumes a central explanatory role within SR which can be compared to the role of the classical ontology of space and time in the Lorentz theory.

[132] René Dugas, *A History of Mechanics*, with a Foreword by Louis de Broglie, trans. J. R. Maddox (New York: Central Book Company, 1955), p. 489. Cf. the comment by David Bohm, *The Special Theory of Relativity*, Lecture Notes and Supplements in Physics (New York: W. A. Benjamin, 1965), pp. vii-viii: "Einstein's basic contribution was less in the proposal of new formulas than in the introduction of fundamental changes in our basic notions of space, time, matter, and movement." See further ibid., p. 71. See also Carlo Rovelli, "Halfway through the Woods: Contemporary Research on Space and Time," in *The Cosmos of Science*, ed. J. Norton and J. Earman (Pittsburgh: University of Pittsburgh Press, 1998), p. 196, who observes "Einstein's contribution to special relativity was the *interpretation* of the theory, not its formalism, since the formalism existed already."

space and time relativist.[133] Thus, unless we are to collapse the Lorentzian aether compensatory theory, the Einsteinian relativity interpretation of SR, and the Minkowskian spacetime interpretation of SR into the same theory, despite the structural differences between them at the theoretical level, then we have no choice but to recognize in SR itself the metaphysical currents which breach the dike of positivism. John Earman concludes that when it comes to questions about the nature of space and time, there is simply no way to justify an empirical/philosophical dichotomy; the appropriate term for the study is the old one: Natural Philosophy.[134]

In a recent essay, "Is Physics at the Threshold of a New Stage of Evolution?" Rompe and Treder echo Planck's question of 1908 and answer in the affirmative: "For several decades physics *'wächst über sich selbst hinaus'* (increases beyond its own limits."[135] Gale, in surveying some of "the metaphysical perplexities abounding in today's physics," contends, "...we are entering a phase of scientific activity during which the physicist has out-run his philosophical base camp, and, finding himself cut off from conceptual supplies, he is ready and waiting for some relief from his philosophical comrades-in-arms."[136] Noting that in recent years such "metaphysical conundrums" as *creatio ex nihilo* "have entered the mainstream of scientific discussions," John Barrow remarks, "Traditional dogmas as to what criteria must be met by a body of ideas for it to qualify as a 'science' now seem curiously inappropriate in the face of problems and studies far removed from the human enterprise."[137] The point is that the positivistic, anti-metaphysical view of physics which dominated the first two-thirds of the twentieth century is simply outmoded in light of contemporary theoretical physics.

In summary, we have seen that the key justificatory factor in Einstein's rejection of Newton's absolute or metaphysical time and space and consequently his reduction of time and space to measured or physical time and space was Machian

---

[133] The metaphysical element in relativity theory becomes most apparent when it is given a spacetime formulation, in which case it becomes impossible justifiably to reject Newton's absolute time and space on the grounds that they are metaphysical. Arthur comments,
> "Mach himself, of course, is no longer a force to be reckoned with in the philosophy of space and time, the positivist position which he inaugurated having come in for its own share of debunking in recent years. This applies particularly to his attack on Newton's conception of an enduring absolute space, a conception which the latest generation of spacetime theorists has done much to rehabilitate" (Richard T. W. Arthur, "Newton's Fluxions and Equably Flowing Time," *Studies in History and Philosophy of Science* 26 [1995]: 323).

The question remains whether the preference on the part of these theorists for a four-dimensional spacetime metaphysic over a metaphysic featuring an absolute space enduring through time is not itself grounded in a positivist critique of the Newtonian concepts. See also Carl Hoefer and Nancy Cartwright, "Substantivalism and the Hole Argument," in *Philosophical Problems of the Internal and External Worlds*, ed. John Earman, Alan I. Janis, Gerald J. Massey, and Nicholas Rescher, Pittsburgh-Konstanz Series in the Philosophy and History of Science (Pittsburgh: University of Pittsburgh Press, 1993), pp. 23-43.

[134] John Earman, "Whose Afraid of Absolute Space?" *Australasian Journal of Philosophy* 48 (1970): 317.

[135] R. Rompe and H.-J. Treder, "Is Physics at the Threshold of a New Stage of Evolution?" in *Quantum, Space and Time—The Quest Continues*, ed. Asim O. Barut, Alwyn van der Merwe, and Jean-Pierre Vigier, Cambridge Monographs on Physics (Cambridge: Cambridge University Press, 1984), p. 608.

[136] Gale, "Metaphysical Perplexities."

[137] Barrow, *World within the World*, pp. 2, vii-viii.

positivism, which eschewed such metaphysical entities as absolute time and space as literally meaningless and which in Einstein's case issued in a verificationist analysis of time and space along operationalist lines. Despite the triumph of Einstein's Special Theory, contemporary physics has burst the old wineskins of positivism and verificationism, as metaphysical elements have increasingly entered the mainstream of modern physical discussions. Therefore, as we enter a new century of scientific exploration, freed from the blinders of positivism, the time is ripe for a reassessment of the metaphysical question of the existence of absolute time and space.

CHAPTER 8

THE ELIMINATION OF ABSOLUTE TIME

In light of what we have seen in our previous chapter, what can be said of SR's putative elimination of metaphysical time and space?

POSITIVISM'S ESSENTIAL ROLE IN SR

The first thing to be noticed is that the positivism which characterized the historical formulation of SR belongs essentially to the philosophical foundations of the theory. The relativity of length depends upon the relativity of simultaneity, which in turn rests upon Einstein's re-definition of simultaneity in terms of clock synchronization by light signals. But that re-definition assumes necessarily that the time light takes to travel between two relatively stationary observers A and B is the same from A to B as from B to A. That assumption presupposes that A and B are not both in absolute motion, or in other words that neither metaphysical space nor a privileged rest frame exists. The only justification for that assumption is that it is empirically impossible to distinguish uniform motion from rest relative to such a frame. But if metaphysical space and absolute motion or rest are undetectable empirically, therefore they do not exist (and may even be said to be meaningless). Such an inference is clearly verificationist, and therefore positivistic.

In a clear-sighted analysis of the epistemological foundations of SR, Lawrence Sklar underlines the essential role played by this verificationism: "Certainly the original arguments in favor of the relativistic viewpoint were rife with verificationist presuppositions about meaning, etc. And despite Einstein's later disavowal of the verificationist point of view, no one to my knowledge has provided an adequate account of the foundations of relativity which isn't verificationist in essence."[1] It would be desirable to do so, muses Sklar, but "what I don't know is either how to formulate a coherent underpinning for relativity which isn't verificationist to begin with, or how, once begun, to find a natural stopping point for verificationist claims of under-determination and conventionality."[2]

POSITIVISM'S UNTENABILITY

But if positivism belongs essentially to the foundations of SR, the next thing to be noted is that positivism has proved to be completely untenable. This fact is so universally acknowledged that it will not be necessary to rehearse the objections

---

[1] Lawrence Sklar, "Time, reality and relativity," in *Reduction, Time and Reality*, ed. Richard Healey (Cambridge: Cambridge University Press, 1981), p. 141.
[2] Ibid.

against positivism here.[3] In a recent review of the literature, Tyler Burge has remarked that "the central event" in philosophy during the last half century has been "the downfall of positivism and the re-opening of discussion of virtually all the traditional problems in philosophy."[4] This is the case not only in philosophy of language and of mind. Healey observes that "...positivism has come under such sustained attack that opposition to it has become almost orthodoxy in the philosophy of science."[5] The rug has thus been pulled from beneath the positivists' anti-metaphysical polemics. Richard Kitchener comments,

> As anyone familiar with the recent history of philosophy knows, the once popular view of logical empiricists that 'metaphysics is meaningless' is no longer tenable; indeed, in retrospect it seems clear that the logical empiricists had a metaphysics and that what they were opposed to was a metaphysics that was *transcendent*, one that made claims about a supernatural (or 'super-empirical') reality that could in no way be checked empirically.[6]

Positivism provides absolutely no justification for thinking that Newton erred, for example, in holding that God exists in a temporal series which transcends our physical measures of it and may or may not be accurately registered by them. It matters not a whit whether we finite creatures know what time it is in God's metaphysical time; God knows, and that is enough.

With the failure of positivism, the A-theorist, especially if he is a theist, is quite free to make, with Newton, a distinction between physical time and space (clock and rod measurements) and metaphysical time and space (ontological time and space independent of physical measures thereof). SR is a theory about physical time and space and says nothing about the nature of metaphysical time and space. Questions dealing with the latter are philosophical in nature and must be dealt with as such. In an unusual appreciation of this point in relativity discussions, Max von Laue wrote in the *Physikalische Zeitschrift* of 1912:

> In addition it seems to me that the whole question of the existence of the aether and of absolute time could without harm be banned from physical discussion; so long as entirely new facts, like, say, experimental proof of the existence of superluminal velocities or a contradiction between the appearances of gravitation and the Relativity Principle, do not permit us to decide physically between both of the mentioned viewpoints. They deliver exactly the same statements about all the quantities susceptible to measurement. That is not to say that the question has no interest, on the contrary, it seems to me to be of great philosophical significance. But precisely because of that, it should remain reserved for treatment with philosophical methods.[7]

How tragic that this quite sensible advice was ignored by the succeeding generation of positivist philosophers and physicists, who misinterpreted the relativization of

---

[3] See the excellent survey in Frederick Suppe, "The Search for Philosophic Understanding of Scientific Theories," in *The Structure of Scientific Theories*, 2d ed., ed. F. Suppe (Urbana, Ill.: University of Illinois Press, 1977), pp. 3-118.
[4] Tyler Burge, "Philosophy of Language and Mind," *Philosophical Review* 101 (1992): 49.
[5] Richard Healey, "Introduction," in *Reduction, Time and Reality*, p. vii.
[6] Richard F. Kitchener, "Introduction: The World View of Contemporary Physics: Does It Need a New Metaphysics?" in *The World View of Contemporary Physics*, ed. Richard F. Kitchener (Albany: State University of New York Press, 1988), p. 5.
[7] M. Laue, "Zwei Einwände gegen die Relativitätstheorie und ihre Widerlegung," *Physikalische Zeitschrift* 13 (1912): 120.

physical time as the destruction of metaphysical time! Richard Swinburne does not exaggerate when he complains that "scientific talk about space and time has been influenced by verificationist presuppositions for the past century, perhaps more than any other scientific talk."[8] In an interesting survey article, John Norton concurs that it is in the philosophy of space and time that one finds some of the "clearest applications" of the ideas of logical positivism, including

1. An application of the verifiability criterion in Einstein's SR, its target being the absolute state of rest of Newton's theory of space and time,

2. Anti-realist claims about the geometry of space and the simultaneity of distant events (conventionality claims),

3. A reduction of spatio-temporal relations to causal relations in the causal theory of time.[9]

What is striking about this list is that applications (2) and (3), namely, conventionalism and causal theories of time, are now widely rejected, whereas (1) continues to be just as widely accepted, as if immune to the collapse of the epistemology essential to its underpinning. This constitutes an untenable situation. In his presidential address to the Royal Astronomical Society, E. A. Milne declared, "It is a good rule that however well a theory appears to fit the facts, it cannot be accepted as satisfactory if it is not *philosophically satisfactory.*"[10] The contemporary philosopher should insist that an exclusively physical methodology is simply inadequate to deal with the problems of time and space. P. J. Zwart concurs,

> According to modern physics,...the duration of a certain interval of time is dependent upon the state of motion of the observer or his clock. However, strictly speaking this is not a philosophical standpoint, but a purely scientific one, since it only relates to the *meaning* of time and not to its nature. Only when adopting the positivistic view that time is nothing more than what is measured by a clock, could the relativistic theory of time be considered a philosophical one.... However, when one is of the opinion that the scientific description of time as a quantity applied in certain formulas and measured by means of a clock is far from complete, one should restrict oneself to the statement that clocks (and processes in general) in systems in motion run more slowly than in systems at rest.[11]

SR may provide a perfectly adequate account of physical phenomena without giving us any philosophical insight into the nature of time. Decrying the "illegitimate passage from the results of physics to metaphysics," Mary Cleugh

---

[8] Richard Swinburne, "Verificationism and Theories of Spacetime," in *Space, Time, and Causality*, ed. Richard Swinburne, Synthèse Library 157 (Dordrecht: D. Reidel, 1983), p. 63. He aptly remarks, "A satisfactory science ought to reveal the extent of our ignorance, not pretend that what is not knowable is not true" (Ibid., p. 74). Stating that "...I know of no good argument for verificationism," Swinburne contends that there is "no need to follow Einstein in his verificationism" (Richard Swinburne, *Space and Time*, 2d ed. [London: Macmillan, 1981], pp. 6, 201).

[9] John D. Norton, "Philosophy of Space and Time," in *Introduction to the Philosophy of Science*, ed. Merilee Salmon (New Jersey: Prentice-Hall, 1992), p. 179.

[10] E. A. Milne, "Presidential Address to the Royal Astronomical Society," *Monthly Notices of the Royal Astronomical Society* 104 (1944): 121.

[11] P. J. Zwart, "The Flow of Time," *Synthèse* 24 (1972): 134.

asserts, "...physicists have a perfect right to use what concepts they please: and criticism can enter only when one of these is labeled 'Time'. Then it is fair to ask whether those characteristics of Time which are neglected and even implicitly negated are not important. If they are, it is reasonable to hold that, however useful 't' might be for physics, its *complete* identification with Time is fallacious."[12] Take absolute simultaneity, for example. It is granted by all parties that we have an intuitive sense that on the planet Neptune, for example, there are events occurring right now, simply simultaneously with events here on Earth. To point out that observers in different inertial frames synchronizing their clocks by reflected light signals will calculate different events on Neptune to be simultaneous with events here is to raise an irrelevancy. We are not talking about light signals or clocks, we are talking about what is happening at this moment of time. It is only by a positivistic collapse of ontological time into measured time that absolute simultaneity is drawn into question. As Sir James Jeans put it, SR "deals with measures of things, and not with things themselves, and so can never tell us anything about the nature of the things with the measure of which it is concerned. In particular it can tell us nothing as to the nature of space and time."[13] Thus, when Ralph Baierlein asserts, "...the fault is not with our clocks. It is with our intuition. We expect all observers to measure the same time interval between a pair of events, but space and time are not that way,"[14] one is tempted to reply, "*Au contraire*, the fault is not with the clocks or with our intuition. It is with our definitions. We are told that the time interval between a pair of events is the same as what observers measure, but space and time are not that way." Mary Cleugh rightly concludes, "Relativity physics, then, in raising the question as to the precise significance which temporal relations such as simultaneity have *for it*, is not necessarily saying anything which is relevant for the metaphysician, still less is it laying down dicta to which metaphysics must conform."[15] "It cannot be too often emphasized that physics is concerned with the measurement of time, rather than with the essentially metaphysical question as to its nature."[16]

What is astounding is the degree to which contemporary treatments of time, despite the demise of positivism, still cling to the old verificationist perspective. For example, Bondi declares,

---

[12] Mary F. Cleugh, *Time and its Importance in Modern Thought*, with a Foreword by L. Susan Stebbing (London: Methuen, 1937), pp. 29-30, 61.

[13] James Jeans, *Physics and Philosophy* (Cambridge: University Press, 1942), p. 68; cf. p. 66. Some theorists have disputed the reduction of time to its measures but have gratuitously assumed that time is that quantity described by relativity theory. See, for example, the critique of operational definitions of time by Mario Bunge, "Physical Time: The Objective and Relational Theory," *Philosophy of Science* 35 (1968): 355-388, though Bunge's concept of time is still physical, not metaphysical, in that it is analyzed as a map from sets of events, reference frames, and chronometric scales to the real numbers; see also Henry Mehlberg, *Time, Causality, and the Quantum Theory*, 2 vols., ed. Robert S. Cohen, Boston Studies in the Philosophy of Science 19 (Dordrecht: D. Reidel, 1980), 1: 1, 189-190, 251, who thinks that physical time is but an aspect of universal time and yet seems to assume that universal time is relativistic.

[14] Ralph Baierlein, *Newton to Einstein: The Trail of Light* (Cambridge: Cambridge University Press, 1992), p. 215.

[15] Cleugh, *Time*, p. 61.

[16] Ibid., p. 51.

> Time is that which is measured by a clock. This is a sound way of looking at things. A quantity like time, or any other physical measurement, does not exist in a completely abstract way. We find no sense in talking about something unless we specify how we measure it. It is the definition by the method of measuring a quantity that is the one sure way of avoiding talking nonsense about this kind of thing.[17]

A verificationist theory of meaning clearly underlies this thinking, and it never occurs to Bondi that collapsing time to measured time might be one sure way of distorting reality. Verificationism also underlies Møller's claim that "The concept of simultaneity between two events in different places obviously has no exact objective meaning at all, since we cannot give any experimental method by which this simultaneity could be ascertained."[18] Like Bondi, Pagels makes the simple inference: clocks run slow in a gravitational field; time is what clocks measure; so time runs slow![19] Concerning a related phenomenon, Clifford Will advises, "Both descriptions are physically equivalent.... The observable phenomenon is unambiguous: the received signal is blue-shifted. To ask for more is to ask questions without observational meaning. This is a key aspect of relativity, indeed of much of modern physics: we focus only on observable, operationally defined quantities, and avoid unanswerable questions."[20] Capek asserts that to admit the "intrinsic unobservability" of certain events in spacetime and yet still to insist on their existence "would contradict the most elementary rules of scientific methodology."[21] In line with this thinking, Barbour declares that because absolute motion in space is undetectable, therefore it does not exist.[22] Freundlich in his discussion of temporal asymmetries repeatedly asserts that in order for a statement about the physical world to be meaningful, it must make claims about the "ways we are appeared to"; two physical statements which do not make different claims about the ways we are appeared to do not differ in meaning.[23] This same sort of reasoning appears to underlie the claim of persons like Duffy that the Lorentzian and Einsteinian approaches to relativity, since they are empirically equivalent, are in fact the *same* theory, interpreted in different ways.[24] According to d'Inverno, the

---

[17] Hermann Bondi, *Relativity and Common Sense* (New York: Dover Publications, 1964), p. 65.
[18] C. Møller, *The Theory of Relativity*, 2d ed. (Oxford: Clarendon Press, 1972), p. 31.
[19] Heinz Pagels, *The Cosmic Code* (London: Michael Joseph, 1982), p. 50.
[20] Clifford M. Will, *Was Einstein Right?* (New York: Basic Books, 1986), pp. 49-50.
[21] M. Capek, "The Inclusion of Becoming in the Physical World," in *The Concepts of Space and Time*, ed. Milic Capek, Boston Studies in the Philosophy of Science 22 (Dordrecht: D. Reidel, 1976), p. 520; cf. his verificationist remarks that "the succession of causally unrelated events" is devoid of physical meaning, so that the simultaneity of distant events and the succession of causally independent events *simply do not exist* (Ibid., pp. 514-515).
[22] Julian B. Barbour, *Absolute or Relative Motion?*, vol. 1: *The Discovery of Dynamics* (Cambridge: Cambridge University Press, 1989), pp. 8-9.
[23] Yehudah Freundlich, "'Becoming' and the Asymmetries of Time," *Philosophy of Science* 40 (1973): 497-498.
[24] Michael C. Duffy, "The Modified Vortex Sponge: a Classical Analogue for General Relativity," paper presented at the International Conference of the British Society for the Philosophy of Science, "Physical Interpretations of Relativity Theory," Imperial College of Science and Technology, London, 16-19 September, 1988. He quotes with approval H. P. Robertson's remark, "Ives had in fact set up a theory completely equivalent in substance to the special theory of relativity...but I was never able to convince him that since what he had was in fact indistinguishable in its predictions from the relativity theory within the domain of physics, it was...the same theory...." Ives was apparently the keener epistemologist of the two.

Lorentz-FitzGerald hypothesis "had the philosophical defect that its fundamental assumptions were unverifiable," whereas Einstein's derivation of the Lorentz transformations "was physically meaningful."[25]

All too often the conflation of time with its measures has led to even faulty theological inferences. For example, P. C. W. Davies, observing that if a spacetime singularity did occur at the Big Bang, as predicted by Friedmann models, then it will be impossible "to continue physics, or physical reasoning, through it to an earlier stage of the universe,"[26] goes on to conclude from this that it is "meaningless" to speak of God's creating the universe.[27] For a cause must precede its effect temporally, but there is no temporal moment before the Big Bang. Therefore, the Big Bang can have no cause.[28] But (leaving aside the faulty premiss that a cause must temporally precede its effect) if we draw a distinction between metaphysical time and physical time as Newton did, it is quite evident that a beginning of the latter does not imply a beginning of the former. God in metaphysical time could be quite active prior to creation (perhaps creating angelic realms) and could bring physical space and time into being after having existed without their being co-existent with Him.

In discussions of quantum theory, too, one frequently encounters extravagant theological inferences based on positivistic analyses of time. For physical time disintegrates at the elementary level due to quantum fluctuations in the geometry of physical space. Explaining that we must forego the view that every event occupies a position in spacetime with the Einstein interval from each event to its neighbor eternally established, Wheeler roundly proclaims, "There is no spacetime, there is no time, there is no before, there is no after. The question of what happens 'next' is without meaning."[29] Such inferences are justified only if one is presupposing a positivistic reductionism of the time concept. When one connects quantum geometrodynamics with cosmology, these fluctuations also become important because in the earliest stages of the universe prior to the Planck time the observable universe becomes so small that its scale factor is of sub-atomic proportions, where

---

[25] Ray d'Inverno, *Introducing Einstein's Relativity* (Oxford: Clarendon Press, 1992), p. 16.
[26] P. C. W. Davies, *Space and Time in the Modern Universe* (Cambridge: Cambridge University Press, 1977), p. 160.
[27] P. C. W. Davies, *God and the New Physics* (New York: Simon & Schuster, 1983), pp. 38-39.
[28] He asks,
> "But what does it mean to say that God *caused* time to come into existence, when by our usual understanding of causation a cause must precede its effect? Causation is a temporal activity. Time must already exist before anything can be caused. The naive image of God existing *before* the universe is clearly absurd if time did not exist—if there was no 'before'" (Ibid., p. 44).

More recently, in his acceptance speech of the Templeton Prize for Progress in Religion, Davies enunciates his belief that the quantum origin of spacetime "has illuminated the ancient theological debate" over divine eternity because the contingency of time reveals that the ultimate Being must be atemporal (Paul Davies, "The Acceptance Speech of Professor Paul Davies," Westminster Abbey, London, 3 May 1995). Davies naively applies physical time concepts to a metaphysical question.
[29] C. Misner, K. S. Thorne, and J. A. Wheeler, *Gravitation* (San Francisco: W. H. Freeman, 1973), p. 1183. Cf. Wheeler's declaration, "*There is no such thing as spacetime in the real world of quantum physics*...superspace leaves us space but not spacetime and therefore not time. With time gone the very ideas of 'before' and 'after' also lose their meaning" (J. A. Wheeler, "From Relativity to Mutability," in *The Physicist's Conception of Nature*, ed. J. Mehra [Dordrecht: D. Reidel, 1973], p. 227).

quantum physics predominates. The uncertainty this introduces into the time coordinate is exploited by Banks to draw a marvelous metaphysical inference:

> ...as we enter this regime [prior to the Planck time] the intuitive concept of time loses all meaning. There is no content in the question of what happened before the big bang, not because the universe becomes singular, but because quantum fluctuations invalidate the notion of 'absolute time....'[9]

[9] I. Newton, *Philosophiae naturalis principia mathematica* (1687)....[30]

It is not explained, of course, how the indeterminacy of physical time is supposed to invalidate Newton's absolute time, which, based in God's eternity, ought, to recall Newton's words, to be "rightly distinguished from its sensible measures." Sometimes the metaphysical conclusions proclaimed on the basis the positivistic analysis of time in quantum theory can be quite ludicrous. For example, appealing to the invariance of quantum field theories under consecutive reversals of time, charge, and space (TCP invariance), Henry Mehlberg states,

> If all natural laws are time reversed invariant and no irreversible processes occur in the physical universe, then there is no inherent, intrinsically meaningful difference between past and future—just as there is no such difference between 'to the left of' and 'to the right of.' If this is actually the case, then all mankind's major religions which preach a creation of the universe (by a supernatural agency) and imply, accordingly, a differentiation between the past and the future, i.e., an intrinsic difference between both, would have to make an appropriate readjustment of man's major religious and 'creationist' creeds and the scientific findings.[31]

This solemn and ridiculous pronouncement clearly rests on the identification of God's time with physical time, a reduction which is positivistic in character. In general, the whole debate surrounding TCP invariance and the asymmetry or anisotropy of time is predicated on the verificationist assumption that time is identical with physical time, so that the direction(s) or arrow(s) of time must be found exclusively in physical processes like entropy increase, the expansion of space, or other irreversible processes. But if we reject the verificationist identification of time with its physical measures, then these irreversible processes are merely *evidence* of the asymmetry of time, not *constitutive* of it. An A-theorist of time sees the distinction between the past and future as rooted in the fact of absolute (that is, ontological, mind-independent) becoming. Events of which any observer in the metaphysical present can have traces have occurred and are therefore past, events in the actual world (tenselessly speaking of the possible world that obtains) of which no observer in the metaphysical present can have traces have not yet happened and are therefore future. Since becoming is real, metaphysical time is intrinsically anisotropic.

In fact, when one reflects on some of the counter-intuitive situations into which one is forced by the reduction of time to physical time, it becomes all the more evident that the two are to be distinguished. The dissolution of time in quantum physics would be just one example.[32] Another example would be GR's permitting

---

[30] T. Banks, "*TCP*, Quantum Gravity, the Cosmological Constant, and All That...," *Nuclear Physics B* 249 (1985): 340.
[31] Henry Mehlberg, "Philosophical Aspects of Physical Time," *Monist* 53 (1969): 363.
[32] To quote Wheeler again:

time travel in universes having a global rotation of matter.[33] By following a wide enough curve, a spaceship can travel into its own past (or future) and meet up with itself before its journey (Figure 8.1).

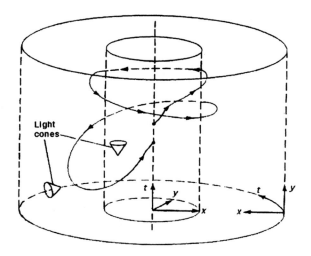

Figure 8.1. Time travel in a Gödel space-time. Space-time is divided into two parts, an internal cylinder with a vertical temporal axis and the surrounding space-time, whose temporal dimension is circular. The world-line of the rocket ship begins in the central axis, moves upward and outward through the cylinder boundary, spirals down outside the cylinder, re-enters it, and then moves back toward the center, ending at a point earlier than its origin.

"'Before' and 'after' don't rule everywhere, as witness quantum fluctuations in the geometry of space at the scale of the Planck distance. Therefore, 'before' and 'after' cannot legalistically rule anywhere. Even at the classical level, Einstein's standard closed-space cosmology denies all meaning to 'before the big bang' and 'after the big crunch.' Time cannot be an ultimate category in the description of nature. We cannot expect to understand genesis until we rise to an outlook that transcends time" (J. A. Wheeler, *Frontiers of Time* [Amsterdam: North-Holland, 1979], p. 20).
Wheeler should have said, "transcends *physical* time," *i.e.*, to God's outlook in metaphysical time in which "before" and "after" exist regardless of any breakdown in physical time concepts.
[33] On such models see Kurt Gödel, "A Remark about the Relationship between Relativity Theory and Idealistic Philosophy," in *Albert Einstein: Philosopher-Scientist*, ed. P. A. Schilpp, Library of Living Philosophers 7 (LaSalle, Ill.: Open Court, 1949), pp. 555-562. Gödel also recognized that such models were incompatible with objective becoming, but he opted for the static models, since "...no reason can be given why an objective lapse of time should be assumed at all" (Ibid., p. 561). He hoped to forestall the paradoxes of time travel by practical considerations like excessive fuel requirements. For discussion see Paul Horwich, *Asymmetries in Time* (Cambridge, Mass.: MIT Press, 1987), chap. 7 and William Lane Craig, *Divine Foreknowledge and Human Freedom* (Leiden: E. J. Brill, 1990), chap. 6.

The only restrictions on such scenarios' coming to pass are technological: excessive fuel requirements, for example, for journeys into the recent past. But with the right fuel, GR permits such a trip. This, however, generates all sorts of difficulties. For example, suppose the space traveller as a child gets the formula for the super-efficient, light-weight fuel from his future self, while the future self is on his journey into the past. As he grows up, he becomes a space engineer and directs the construction of a spacecraft designed to use the fuel specified in the secret formula. He then embarks on his journey through space and time and delivers the formula to his earlier self. So where does the knowledge for the formula come from? Such circular causation seems evidently absurd, and the best way to break it is to deny that time and physical time are identical. In metaphysical time such a journey is impossible, given an A-Theory, because what is metaphysically present is not affected by motion through space, and, therefore, model universes permitting time-travel are engaging mathematical constructs which cannot be descriptive of reality.

Or consider what happens to physical time in connection with an object undergoing gravitational self-collapse. When the object collapses to its so-called Schwarzschild radius at which $r = 2M$, where r is its gravitational radius and M its mass,—a process which elapses in a finite proper time but takes infinite coordinate time as reckoned by an external observer—, then *the time and space coordinates switch*, that is to say, the spacelike coordinate r becomes at that point the temporal coordinate and the time coordinate t becomes spatial. The further decrease of r to zero represents the passage of time. But unless we distinguish metaphysical from physical time, so that the same metaphysical quantity is being represented by different coordinates, then this seems to be a blatant example of spatializing time, indeed, of actually turning time into space at a certain juncture. This is surely absurd, and their failure to distinguish metaphysical and physical time lead Misner, Thorne, and Wheeler into the following *mélange:*

> The unseen power of the world which drags everyone forward willy-nilly from age twenty to forty and from forty to eighty also drags the rocket in from time coordinate $r = 2M$ to the later value of the time coordinate $r = 0$. No human act of will, no engine, no rocket, no force...can make time stand still.[34]

In this description of a rocket's being sucked into a black hole from its Schwarzschild radius, we have a colorful mix of physical and metaphysical time. The flow of Newtonian absolute time is represented as a power pulling people along which nothing can arrest. But then this is identified as the same power that pulls the rocket into the black hole, namely, gravitation or spacetime curvature! Finally, the radius of the black hole is identified with time, so that one's powerlessness to halt the gravitational collapse of the black hole is identified with one's inability to stop the flow of time! Surely this is an egregious example of the confusion which inevitably results when time is identified with its measures.

Or consider the attempt on the part of Hartle and Hawking to avoid the initial cosmological singularity by having the time co-ordinate assume imaginary values at an early epoch of the universe, which again has the effect of converting time into

---

[34] Misner, Thorne, and Wheeler, *Gravitation*, p. 823.

space.³⁵ According to Hawking, "...there is no difference between the time direction and directions in space...time is imaginary and is indistinguishable from directions in space."³⁶ As Isham points out, the Hartle-Hawking model depends crucially on (i) the legitimacy of defining "time" phenomenologically in terms of the gravitational and/or material content of the universe (for example, its radius) and (ii) employing what in special relativistic contexts is a mere mathematical trick, namely, the introduction of imaginary values for the time co-ordinate, in general relativistic contexts, where its use entails radical differences for the nature of spacetime.³⁷ Taken literally, the Hartle-Hawking model is metaphysically absurd, since time is distinct from space, so that it must be regarded as the manipulation of physical time and its quantities only, not as a temporal ontology of the universe.

A final *Gedankenexperiment* suggested by Grünbaum will illustrate my point.³⁸ He invites us to imagine a particle moving circularly on a platform without observer and with no light. Each time the particle "returns" to point A, it is in an identical state. Therefore, given the Identity of Indiscernibles, it would really be the *same* time. There would be no basis in the particle's attributes or in the relation of events for saying that the same kind of event is recurring eternally. Rather we should have a closed time. If the motion of the particle were pendulum-like, then terminii could be established which are different from points on the path, and so time would be open. But on a circular path, in the absence of light or contemplation, the particle would exist in closed time. If it is said that time is by nature open, then the same thing could be said about space, which would be false, as non-Euclidean geometry shows.

Now if we reduce time to its empirical measures, then Grünbaum is quite correct. But the conclusion is obviously wrong: a particle circling in the dark would return to the same point again and again, regardless of physical observation of it. The fallacy lies in the positivistic interpretation of the Identity of Indiscernibles. For the particle's being at A obtains at different moments of metaphysical time, so that these successive states are discernible, having different temporal properties, even if they are empirically indistinguishable. Arthur Pap asserts that the proposition "No event precedes itself" is a synthetic *a priori* truth;³⁹ becoming is intrinsic to metaphysical time. Grünbaum's analogy to non-Euclidean space evinces the tendency of B-theorists to spatialize time; as Capek notes, if we believe that time is adequately represented by a geometrical line, then there is no reason why this line should not be curved or closed.⁴⁰ But the A-theorist considers such a representation

---

[35] James Hartle and Stephen W. Hawking, "Wave Function of the Universe," *Physical Review D* 28 (1983): 2960-2975.
[36] Stephen W. Hawking, *A Brief History of Time: From the Big Bang to Black Holes*, with an Introduction by Carl Sagan (New York: Bantam Books, 1988), pp. 14, 35.
[37] C. J. Isham, "Creation of the Universe as a Quantum Process," in *Physics, Philosophy, and Theology: A Common Quest for Understanding*, ed. R. J. Russell, W. R. Stoeger, and G. V. Coyne (Vatican City: Vatican Observatory, 1988), pp. 376, 391, 399, 402-403.
[38] Adolf Grünbaum, *Philosophical Problems of Space and Time*, 2d ed., Boston Studies in the Philosophy of Science 12 (Dordrecht: D. Reidel, 1973), pp. 197-201.
[39] Arthur Pap, *Introduction to the Philosophy of Science* (London: Eyre & Spottiswoode, 1963), pp. 97-98.
[40] M. Capek, "Introduction," in *The Concepts of Space and Time*, ed. M. Capek, Boston Studies in the Philosophy of Science 22 (Dordrecht: D. Reidel, 1976), p. xxx.

inadequate. Even if no basis exists for distinguishing physical times, it is evident in this case that different metaphysical times are involved.

In a fascinating review of the time concept in various fields of physics alone, Carlo Rovelli has emphasized how unlike the intuitive notion of time physical time concepts are and how diverse they are when compared among themselves.[41] He lists eight characteristics commonly associated with time:

1. One dimensional: time can be thought of as a collection of instants which can be arranged in a one-dimensional line
2. Metric: time intervals can be measured such that two intervals can be said to have equal duration
3. Temporally global: the real variable t which we use to denote the measure of time goes through every real value from −infinity to +infinity
4. Spatially global: the time variable can be uniquely defined at all space points
5. External: the flow of time is independent of the specific dynamics of the objects moving in time
6. Unique: there are not many times, but just the time
7. Directional: it is possible to distinguish the past from the future direction of the time-line
8. Present: there always exists an ontologically preferred instant of time, the Now

Rovelli then provides the following chart to illustrate the diversity of the physical time concept (Figure 8.2).

The Notion of Time Used in:	has properties:
Natural language	1, 2, 3, 4, 5, 6, 7, 8
Thermodynamics	1, 2, 3, 4, 5, 6, 7
Newtonian mechanics	1, 2, 3, 4, 5, 6
SR	1, 2, 3, 4, 5
Cosmology	1, 2, 3, 4
GR - proper time	1, 2, 3, 5
GR - coordinate time	1, 3, 4
GR - clock times	1, 2
Quantum gravity	none

Figure 8.2. The concept of time in various fields of physics in comparison with the customary concept.

---

[41] Carlo Rovelli, "What Does Present Days [sic] Physics Tell Us about Time and Space?" Lecture presented at the Annual Series Lectures of the Center for Philosophy of Science of the University of Pittsburgh, 17 September 1993.

Even if one is disposed to dispute some of the specifics, there is, I think, no gainsaying Rovelli's point that physical time is both very different from our intuitive notion of time and, moreover, that because time is differently defined in different fields there is no unitary notion of physical time. It is difficult to resist the conclusion that all of these operationally defined "times" are not really time at all, but just various measures of time suitable for their respective fields of inquiry. The *reductio ad absurdum* of the positivistic reduction of time to its measures is surely the conclusion of someone like Barbour that because time in quantum gravity disappears, therefore time does not exist![42]

In sum, I think that in light of the collapse of positivism a re-appraisal of the time concept(s) in physics and in relativity theory in particular is long overdue. The demise of positivism should be neither ignored nor mourned; on the contrary, its lingering shadow over certain discussions of the time concept in contemporary physics has obfuscated the philosophical issues involved and sometimes led to quite unjustified metaphysical and theological conclusions. The above examples well bear out the verdict of Jonathan Powers: "In the end positivism itself passed into history: its ambitions had proved to be a mirage. But in the slogans of the New Physics, its ghost goes marching on."[43]

Certain thinkers have, we may be thankful, been more careful and discriminating than their above cited colleagues in their discussions of the concept of time. Such discrimination is already evident in the title of Peter Kroes's *Time: Its Structure and Role in Physical Theories*, for example. In his preface, Kroes makes it quite clear that the concept handled in his book is the concept of physical time, that is, time as used by physicists in their theories. He explicitly limits his discussion so as not to pronounce upon metaphysical time, commenting, "nor is it my intention to examine the metaphysical status of time. I shall restrict myself to an analysis of the structure of physical time and of the role time plays in physical theories."[44] That he takes this distinction seriously is evident when one comes to his discussion of the objectivity of temporal becoming. He construes the controversy to concern whether it is possible to introduce the notion of the flow of time into the physicist's conception of nature.[45] He concludes, "Whether or not it is in principle impossible for physics to incorporate the flow of time in its description of physical reality, is still an open question. Up to the present, all attempts to capture this mysterious *but essential aspect of time* in the language of physics have failed."[46] Kroes carefully differentiates between physical time, into which the notions of past, present, and future have yet in his opinion to be successfully introduced, and what he calls "real time," for which these notions are essential. Positivistic philosophers would have concluded from the absence of becoming in physical time that objective becoming

---

[42] Julian B. Barbour, "The Timelessness of Quantum Gravity: I & II," *Classical and Quantum Gravity* 11 (1994): 2853-2873, 2875-2897. Barbour holds that instants exist, but these are analyzed as three-dimensional relative configurations of the universe in superspace.

[43] Jonathan Powers, *Philosophy and the New Physics*, Ideas (London: Methuen, 1982), p. 12.

[44] Peter Kroes, *Time: Its Structure and Role in Physical Theories*, Synthèse Library 179 (Dordrecht: D. Reidel, 1985), p. xiii; cf. p. viii.

[45] Ibid., p. 196.

[46] Ibid., p. xxiv [my emphasis]; cf. p. 209.

and temporal transience are therefore unreal. But with the failure of positivism, so rash an inference without further justification would only betray a lack of care.[47]

It must be admitted, too, that even some of the positivist thinkers I have cited above employed a verificationist analysis of time with respect to natural science only but showed themselves open to a more full-orbed concept of time when discussing issues of a metaphysical character. For example, Margenau, after dismissing absolute simultaneity and other notions as ultra-perceptory and, hence, meaningless, adds the qualifier that he is not saying that these notions are meaningless in the sense intended by the logical positivist, but only that "...they cease to be the physicist's professional business."[48] Similarly, Frank's claim concerning metaphysical sentences was that "...they are meaningless *as far as science is concerned.*"[49] But when we begin to ask questions that are broader in scope, the meaningfulness of such sentences may change considerably:

> Our judgement about the usefulness of such expressions may change considerably if we consider not only the realm of physical facts in the narrower sense (*e.g.*, the motion of planets) but ask also for a general picture of the world and include the phenomena of human behavior as facts to be represented.[50]

Frank even opines that here religious ideas may enter the picture. Finally, Eddington also limited his verificationism to physical science and seemed open to wider realms of reality. In his imaginary dialogue between a physicist and a relativist we find the following exchange:

> *Phys.* What you have shown is that we have not sufficient knowledge to determine in practice which are simultaneous events on the Earth and Arturus. It does not follow that there is no definite simultaneity.
> *Rel.* That is true, but it is at least possible that the reason why we are unable to determine simultaneity in practice...in spite of many brilliant attempts, is that there is no such thing as absolute simultaneity of distant events. It is better therefore not to base our physics on this notion of absolute simultaneity, which may turn out not to exist, and is in any case out of reach at present.[51]

Here the relativist admits that the relativity of simultaneity may be merely an epistemic matter; but because we cannot establish absolute simultaneity and because it may not exist, we ought not *to base our physics* on it. Such a contention, even if correct, does absolutely nothing to undermine Newton's metaphysical time. In another place, Eddington distinguishes two questions that may be asked concerning time: (i) what is the true nature of time, and (ii) what is the nature of that quantity which has under the name of time become a fundamental part of the structure of classical physics? Einstein's theory, says Eddington, cleared up the *second* of these

---

[47] See also Herbert Dingle, "Time in Philosophy and Physics," *Philosophy* 54 (1979): 99-104. Though overly generous to Einstein in stating that "'time,' the ever-rolling stream, never entered his thoughts" in the 1905 paper, Dingle is correct that Einstein's theory had only to do with clock-times and proved nothing about the nature of time itself. Cf. idem, *Science at the Crossroads* (London: Martin Brian & O'Keefe, 1972), p. 137.
[48] H. Margenau, "Metaphysical Elements in Physics," *Reviews of Modern Physics* 13 (1941): 183.
[49] Philipp Frank, *Interpretations and Misinterpretations of Modern Physics*, Actualités Scientifiques et Industrielles 587: Exposés de Philosophie Scientifique 2 (Paris: Hermann & Cie., 1938), p. 38.
[50] Philipp Frank, *Philosophy of Science* (Englewood Cliffs, N. J.: Prentice-Hall, 1957), p. 144.
[51] Arthur Eddington, *Space, Time and Gravitation*, Cambridge Science Classics (1920; rep. ed.: Cambridge: Cambridge University Press, 1987), p. 17.

questions.[52] But since this distinction is precisely that drawn by Newton between absolute and relative time, Eddington in effect agrees that the contribution made by SR is the relativization of relative or physical time and that the theory does not even speak to the question of the nature of absolute or metaphysical time.

These thinkers erred in believing that positivism could be successful even as a philosophy of science, but at least they did not use verificationism as a proofstone for all conceptions of reality. Their restricted positivism was quite compatible with Newton's metaphysical time. Unless one is a full-blooded positivist, then one cannot support the claim that SR abolished notions of absolute simultaneity, absolute length, and so on. And full-blown positivism is widely recognized today as simply indefensible.

ARBITRARINESS OF SR'S SIMULTANEITY CONCEPT

The third thing to be noticed about SR's time concept is that it is predicated upon a definition of simultaneity which we are under no obligation to adopt. Contrary to the widespread, popular impression that Einstein proved the relativity of simultaneity, in fact all SR does, as should be evident from our exposition, is *re-define* simultaneity. The operational definitions Einstein gives to define simultaneity, and, hence, time, are completely arbitrary for someone who has not yet accepted the theory. Dingle remarks, "It is necessary to emphasize that this definition is entirely arbitrary.... Prejudices die hard, and there are still physicists who have not entirely rid their minds of the idea that Einstein was *discovering* the right way of timing a distant event, instead of *freely inventing* something to which he gave that name."[53] Why should we regard two separated events as simultaneous, just because a light signal is inferred to reach the reflection event at the time of an event midway between the signal's emission and reception? Why is this what we mean by "simultaneity?" As Cleugh points out, "It is one thing to say that we cannot make judgments of simultaneity with regard to events at some distance from each other without the help of light signals: it is quite another to *define* simultaneity as depending on light signals."[54]

Einstein himself freely confessed that his operational definitions were conventional. He commented,

> The theory of relativity is often criticized for giving, without justification, a central theoretical role to the propagation of light, in that it founds the concept of time upon the law of the propagation of light. The situation, however, is somewhat as follows. In order to give physical significance to the concept of time, processes of some kind are

---

[52] Arthur Eddington, *The Nature of the Physical World*, with an Introductory Note by Sir Edmund Whittaker, Everyman's Library (1928; rep. ed.: London: J. M. Dent & Sons, 1964), p.47; so also S. C. Tiwari, "Fresh Look on Relativistic Time and Lifetime of an Unstable Particle," paper presented at the International Conference of the British Society for the Philosophy of Science, "Physical Interpretations of Relativity Theory," Imperial College of Science and Technology, London, 16-19 September, 1988.
[53] Herbert Dingle, "Time in Relativity Theory: Measurement or Coordinate?" in *The Voices of Time*, 2d ed., ed. with a new Introduction by J. T. Fraser (Amherst, Mass.: University of Massachusetts Press, 1981), p. 462.
[54] Cleugh, *Time*, p. 57.

required which enable relations to be established between different places. It is
immaterial what kind of processes one chooses for such a definition of time.[55]

Two things are evident in these remarks: (i) Einstein's concern was to define time in such a way as to give it physical significance, and (ii) the choice of the processes used to establish temporal relations is arbitrary. One who has not already adopted the theory is thus under no obligation at all to regard as simultaneous events calculated to be simultaneous by Einstein's clock synchronization prescription. One could have used bullets instead of light signals to synchronize distant clocks. One could with equal justification (or lack thereof) have defined simultaneity by an exchange of signals in which the outbound signal is a light beam and the return signal is sound waves. For in the absence of any absolute time we do not know how distant simultaneity is to be defined.

What we mean intuitively by simultaneity is "the property of occurring at the same moment of time." If we feel bound to give operational definitions, then they ought to preserve the intuitive content of the concept or else be used to define a new and different concept. We ought to ask at the outset whether there may not be some operational method of defining distant simultaneity which is in tune with our intuitive beliefs. Continuity with the intuitive concept of simultaneity is preserved by T. Sjödin's proposal to define simultaneity operationally by using the concept of infinitely fast signals.[56] He emphasizes that "The *concept* 'infinitely fast signal' exists even if there is an upper limit to the velocity of causal propagations in our actual world. Consequently, also the *concept* 'simultaneity' exists even if there is no possibility for man to know if two events are simultaneous, or not."[57] He offers the following definition of simultaneity:

> *Definition of simultaneity:* Two events E_1 and E_2 at the points P_1 and P_2 are *simultaneous*, if no signal, independently how fast, may depart from P_1 after the event E_i and arrive at P_j before the event E_j ($i, j = 1, 2$).

While intuitively on target, this definition needs to be tightened up a bit. In the first place, it is really a proposed definition of distant simultaneity, not simultaneity *simpliciter*. Moreover, it states *prima facie* only a sufficient condition of distant simultaneity and so cannot constitute a definition. Finally, it allows that a signal might depart from P_1 after E_1 and arrive at P_2 at E_2 or that a signal might depart from P_1 at E_1 and arrive at P_2 before E_2, so that it fails to give even a sufficient condition of distant simultaneity. We may revise Sjödin's definition accordingly:

---

[55] Albert Einstein, *The Meaning of Relativity*, 6th ed., (1922; rep. ed.: London: Chapman & Hall, 1967), p. 27.
[56] T. Sjödin, "On the One-Way Velocity of Light and its Possible Measurability," paper presented at "Physical Interpretations of Relativity Theory." See also Richard Gale, "Human Time: Introduction," in *The Philosophy of Time: a Collection of Essays*, ed. R. M. Gale (New York: Humanities Press, 1968), p. 299; M. Capek, *The Philosophical Impact of Contemporary Physics* (Princeton: D. Van Nostrand, 1961), p. 172. It is noteworthy that the Galilean transformations follow automatically from the Lorentz equations if we substitute for c an infinite value.
[57] Sjödin, "One-Way Velocity of Light."

Definition of distant simultaneity: Two events E_1 and E_2 at the points P_1 and P_2 are simultaneous iff no signal, independently how fast, may depart from P_1 after E_1 and arrive at P_2 at or before E_2.

In other words, the time of E_1 is the latest that a signal, no matter how fast, can depart from P_1 and arrive at P_2 by the time of E_2. If signals travel infinitely fast, their moment of reception at a distance is the same as the moment of their emission. E_1 and E_2 must therefore occur at the same moment, since they are connectable only by an infinitely fast signal.

Sjödin's definition of simultaneity makes no use of clocks. But on the basis of his definition, he is prepared to define absolute synchronization as well:

Definition of absolute synchronisation: The synchronisation of two identical clocks at rest in some inertial frame is *absolute* if they show the same time simultaneously (according to the definition above).

Sjödin recognizes that because we are dependent on finite signals, we are obliged to choose from among the infinitely many possibilities of synchronizing clocks that all give different (but equivalent) descriptions of the physical world. Of all these different choices the standard Einstein synchronization procedure is of special practical importance, but it enjoys no conceptual importance in defining simultaneity.

Sjödin's definitions are so perspicuous and in harmony with the intuitive notion of simultaneity that they seem vastly preferable to Einstein's definitions (if one feels the need for operational definitions at all). What is especially intriguing about these definitions is that relativity theory does not prohibit the existence of superluminal velocities, even arbitrarily high velocities, so that Sjödin's prescription describes a nomologically possible procedure.[58] There is an intriguing body of literature surrounding the nature of and the search for particles that travel at superluminal velocities (dubbed "tachyons").[59] Grünbaum, who touts the conventionality of simultaneity, acknowledges that if we admit superluminal velocities then relativistic time order relations are not "ontologically sacrosanct."[60] Referring to a universal tachyonic background as constitutive of a special reference frame, Grünbaum poses the provocative question, "Perhaps this is tantamount to an ether reference frame whose time qualifies as physically preferred?" This inference is usually disputed on the grounds that infinite velocity tachyons, while sufficing to eliminate the conventionality of simultaneity within a single reference frame, would not serve to establish absolute simultaneity unless their velocity were also invariant.[61] For since E_1 and E_2, being connected by an infinite velocity tachyon beam in frame R, are spacelike separated, SR requires that in some other frame R', E_2 occurs later than E_1.

---

[58] SR prohibits the acceleration of particles from subluminal to luminal or superluminal speeds, not particles that always travel at such speeds. SR may also be incompatible with superluminal *signals* (controllable causal connections), but nothing in Sjödin's definition depends on the use of signals as opposed to raw tachyon beams occurring naturally.
[59] See the discussion and literature listed in Craig, *Divine Foreknowledge and Human Freedom*.
[60] Grünbaum, *Space and Time*, p. 827.
[61] Michael Friedman, *Foundations of Spacetime Theories* (Princeton: Princeton University Press, 1983), p. 167.

Thus, while a tachyon beam connecting E_1 and E_2 would possess an infinite velocity in R, it would possess only a finite velocity in another frame R'. In R' there is no reason to take E_1 and E_2 as simultaneous, since they are connected by a merely finite velocity signal, and faster tachyons in R' could serve to establish the non-simultaneity of E_1 and E_2. Worse, SR requires that in some frame R^* E_1 occurs later than E_2, so that the tachyon beam literally travels backward in time (and possesses negative energy). Now this bizarre result might provide just the motivation which is needed for regarding R as the preferred reference frame and for treating solutions in which E_1 and E_2 are not simultaneous as purely mathematical and non-physical.[62] Bilaniuk and Sudarshan, two pioneers of tachyon research, insist that this would not serve to overthrow SR. They ask,

> After all, is not the exclusion of a preferred frame what relativity is all about? *No, it is not.* The postulates of special relativity require the laws of physics, including the speed of light, to be the same in all inertial frames. They do not preclude the existence of cosmological boundary conditions that permit us to single out a particular local frame as a preferred reference system.[63]

Bilaniuk and Sudarshan are being somewhat disingenuous here, since the admission of a preferred frame would require us to adopt a neo-Lorentzian rather than Einsteinian relativity theory. Loathe to admit such a privileged frame, relativity theorists frequently invoke instead a so-called Reinterpretation Principle in order to rule out back-travelling tachyons.[64] According to this principle, a negative energy tachyon traveling backward in time may be reinterpreted as a positive energy tachyon traveling forward in time. Thus, what in one frame is the emission of a tachyon beam at E_1 and its reception at E_2 is in another frame the emission of the beam at E_2 and its reception at E_1. This Reinterpretation Principle, however, seems clearly to be just one more vestige within the philosophy of space and time of precisely that defunct positivism which we have been decrying. Because observers cannot tell the difference between a negative energy particle traveling backward in time and a positive energy particle travelling forward in time, it is declared that there is no objective, factual difference between the two. Apart from other objections,[65] the untenability of the Reinterpretation Principle is evident from the simple consideration that the observer associated with E_2 might not even have a tachyon transmitter but only a receiver, so that E_2 could not possibly be reinterpreted as the emission of a tachyon beam. Thus, resort to the Reinterpretation Principle to preclude back-traveling tachyons is a desperate and ultimately futile expedient. Rather what needs to be done, in Tim Maudlin's words, is to posit the extra structure

---

[62] See Wesley C. Salmon, *Space, Time, and Motion* (Encino, Calif.: Dickenson Publishing, 1975), pp. 119-122.
[63] Olexa-Myron Bilaniuk, *et al.*, "More about Tachyons," *Physics Today*, (December 1969): 52.
[64] See, for example, Michael Redhead, "Nonlocality and Peaceful Coexistence," in *Space, Time and Causality*, ed. Richard Swinburne, Synthèse Library 157 (Dordrecht: D. Reidel, 1983), pp. 167-179; idem, "The Conventionality of Simultaneity," in *Philosophical Problems of the Internal and External Worlds*, ed. John Earman, Alan I. Janis, Gerald J. Massey, and Nicholas Rescher, Pittsburgh-Konstanz Series in the Philosophy and History of Science (Pittsburgh: University of Pittsburgh Press, 1993), pp. 116-125; Graham Nerlich, *What Spacetime Explains* (Cambridge: Cambridge University Press, 1994), pp. 68-71.
[65] See Craig, *Divine Foreknowledge and Human Freedom*, pp. 122-127.

to spacetime requisite to preclude them, structure "which looks suspiciously like an absolute notion of simultaneity."[66] If there is a privileged frame in which E_1 and E_2 are simultaneous in virtue of their being connected by an infinite velocity tachyon beam, then E_1 and E_2 are non-conventionally and absolutely simultaneous.

In any case, Sjödin's definitions are unproblematic because they do not define simultaneity in terms of connectability by infinite velocity signals. His definitions are independent of reference frames and rule out back-traveling signals. Since his operational definitions are intuitively continuous with the common notion of simultaneity, they are preferable to Einstein's definitions.

Of course, one might complain that an operational definition presupposing the possibility of infinitely fast tachyonic signals is of no practical significance, since we have no such signals. But, in the first place, it should be understood that Einstein's own definitions are impracticable as well. Holton underscores this point:

> That Newton's absolute space and absolute time were not meaningful concepts in the sense of laboratory operations was, of course, not the original discovery of Mach; rather it was freely acknowledged by Newton himself. But Einstein was also quite explicit that in replacing absolute Newtonian space and time with an infinite ensemble of rigid meter sticks and ideal clocks he was not proposing a laboratory-operational definition. He stated it could be realized only to some degree, 'not even with arbitrary approximation' and that the fundamental role of the whole conception, both on factual and logical grounds, 'can be attacked with a certain right.'[20]
>
> [20] Albert Einstein, *Les prix Nobels en 1921-1922* (Stockholm, 1923), p. 2.[67]

It is at this juncture that Holton suggests that Einstein has merely substituted an abstract *Gedankenexperimenter* for God. There seems to be no in principle reason why our abstract *Gedankenexperimenter* should not employ tachyon beams rather than light rays in his synchronization procedure. In any case, issues of practicality are irrelevant for purposes of *defining* the concept of simultaneity. Even if Einstein's procedure were completely practicable, that does nothing to commend it as defining the concept of simultaneity. On the contrary, Einstein's prescription seems manifestly counter-intuitive in comparison with Sjödin's suggestion.

The arbitrariness of Einstein's operational definitions and their lack of conceptual continuity with the intuitive (and classical) conception of simultaneity provokes a critical problem with regard to the claim that relativity theory has superseded Newtonian theory, for it appears that the two theories are not even talking about the same thing. Cleugh cautions, "'Simultaneity' is now a dangerously ambiguous word, standing for at least two quite different notions."[68]

---

[66] Tim Maudlin, *Quantum Non-Locality and Relativity*, Aristotelian Society Series 13 (Oxford: Blackwell, 1994), pp. 115-116; cf. pp. 73-75.

[67] Gerald Holton, "On the Origins of the Special Theory of Relativity," in *Thematic Origins of Scientific Thought: Kepler to Einstein*, by Gerald Holton (Cambridge, Mass.: Harvard University Press, 1973), pp. 170-171. Dugas comments, "We shall not go as far to complain, as certain critics do, that Einstein made use of magical clocks and enchanted measuring rods in order to return, in the last analysis, to Lorentz's transformation (T). But there is no doubt that those ideal clocks and rods can only be used in idealised experiments" (René Dugas, *A History of Mechanics*, with a Foreword by Louis de Broglie, trans. J. R. Maddox [New York: Central Book Company, 1955], p. 490).

[68] Cleugh, *Time*, p. 59. She asserts, "...the simultaneity with which Einstein deals is only a very distant cousin of the simultaneity with which the plain man and the metaphysician are concerned...," and, she might have added, the classical physicist as well (p. 58).

On this account, the Newtonian physicist and the relativist do not mean the same thing in asserting, "Events e_1 and e_2 are simultaneous in frame F." Therefore, their theories do not contradict each other, since the conceptions of simultaneity are diverse. The Newtonian would concede quite happily that two events which are simultaneous in absolute time are not simultaneous according to Einstein's clock synchronization procedure.

In analyzing this problem, Friedman asserts that if we view the referents of theoretical terms as determined by the theoretical principles within which they occur, then we cannot say Newtonian mechanics represents a false view of time and simultaneity, for its terms have no referents, and so the theoretical principles involving those terms are not false, but truth-valueless.[69] But this verdict seems to presuppose the legitimacy of Einstein's operational definitions; if we choose Sjödin's, then the Newtonian concept of simultaneity does have a referent independent of the behavior of clocks. In this case, the Newtonian assertion that "Events e_1 and e_2 are simultaneous in frame F" and the Einsteinian assertion "Events e_1 and e_2 are not simultaneous in frame F" can both be true because the operational definitions of what it is to be "simultaneous" are different. Friedman proposes to untangle the muddle by adopting a causal theory of reference instead, according to which what a theoretical term refers to is a matter of which actual entities have the right sort of historical connection with the use of the term. The use of "time" and "simultaneous" in Newtonian mechanics provides Friedman with an illustration of the proper procedure in determining the meaning of theoretical terms. Given the entities postulated by our best current theory, we look for some among these which: (i) give a plausible distribution of truth values for the sentences involving "time" and "simultaneous" used by Newtonian physicists, (ii) are actually responsible for the phenomena explained by Newtonian mechanics, and (iii) are actually measured by the measuring procedures used to test Newtonian mechanics. One then shows how Newtonian physicists were referring to time and simultaneity as these concepts play a role in SR or else that these terms had no referents.

Unfortunately, Friedman still leaves us with a disjunction: either the Newtonian referent of simultaneity was the same as in SR or there was no such referent. Since different operational definitions could provide such a referent, and since such a referent's theoretical term could with equal, if not greater, plausibility also be claimed to stand in historical continuity with the use of "time" and "simultaneity" in Newtonian theory (after all, Poincaré's use of the light signal synchronization procedure was only valid in the rest frame of the aether, and he remained uncompromising on this in the face of SR), one reaches the conclusion that the terms "time" and "simultaneity" in Newtonian theory were referring either to the referents mentioned in Einstein's definitions or in, say, Sjödin's definitions. But since these are mutually incompatible, the referents of Newtonian theory cannot be identical with the referents of both.

Perhaps the difficulty here only serves to reveal the ineptness of operational definitions.[70] On an operationalist view, what is intuitively a single concept can be

---

[69] Friedman, "Simultaneity," pp. 411-415.
[70] As Suppe explains,

operationally defined in a multiplicity of ways, so that the resulting referents of the terms so defined are not after all identical. In the case at hand, it seems evident that "simultaneity" is a unitary, intuitive concept and that what ought to be at issue in the operational procedures is not *defining* simultaneity, but *determining* or *discovering* it. All of the definitions discussed share the common, intuitive core meaning of "simultaneous" as "occurring at the same time," but they differ in how to determine or discover physically which events occur at the same time. Einstein adopts the convention of using light signals and clocks to determine which events are simultaneous. Adopting this prescription leads in turn to the relativity of simultaneity. But another prescription is possible, whose adoption does not lead to the conventionality or relativity of simultaneity. The proponent of this other prescription will agree that the behavior of light rays, clocks, and measuring rods is such that "simultaneity" according to the Einsteinian prescription is, indeed, relative; but he will regard this as a pseudo-simultaneity, not a genuine simultaneity. Genuine simultaneity is discovered, not constituted, by the relevant operations and is therefore non-conventional and absolute.

Einstein's definitions, then, are admittedly conventional and, arguably, not very plausible intuitively. It is interesting to read the defense of one relativist of Einstein's re-definition of time:

> ...the primitive concept of absolute time, which has been discarded, is so deep and has been so successful for so long that it has become part of our language. It is a concept so much a part of our thinking that it is difficult to change—to accept a definition of time that is different for different observers. And the mind always struggles, the question is always raised, 'Is it necessary? Is this what the experiments tell us?' The answer is 'No, it is not necessary.' And the fact that it is not necessary makes Einstein's achievement the more brilliant. For he has not shown us truth. Better, he has shown us a way to regard our experience as so perfect that, having seen his canvas, we find it difficult to look at the world another way.[71]

"Often there is more than one experimental procedure used to determine, for example, an object's mass. Since concepts or properties are identified with unique combinations of operations, each different experimental procedure defines a distinct concept, and so there are as many distinct concepts of *mass* as there are procedures for determining it. In actual scientific practice, however, these different procedures are taken as measuring the same thing, *mass*; thus operational definition is unsatisfactory as an analysis of the meaning (hence the cognitive significance) of theoretical terms" (Suppe, "Search for Philosophic Understanding," p. 19).

[71] Leon N. Cooper, *An Introduction to the Meaning and Structure of Physics* (New York: Harper & Row, 1968), p. 401. Cf. the curious intransigence of Rovelli, who after recognizing that "most metaphysicians underestimate the effect of their own epistemological prejudices," goes on to provide as examples of "discoveries that are forever" the following: "That the Earth is not the center of the universe, that simultaneity is relative. That we do not get rain by dancing. These are steps humanity takes, and does not take back." (Carlo Rovelli, "Quantum Spacetime: What Do We Know?" in *Physics Meets Philosophy at the Planck Scale*, ed. C. Callender and N. Hugger [Cambridge: Cambridge University Press, forthcoming], pp. 2, 15). See also the comment by Henry Margenau and Richard A. Mould, "Relativity: an Epistemological Appraisal," *Philosophy of Science* 24 (1957): 303:

> "...it does not make sense to speak of settling the matter between Einstein and Lorentz by experiment. What is significant about Einstein's principle is that it *can* be maintained, not that it *must* be maintained. In rejecting it, Lorentz did not commit any technical error, but he did forego its conceptual and heuristic advantages."

According to this most peculiar and interesting analysis, Einstein's theory is not required by experimental evidence but is based upon arbitrary and counter-intuitive re-definitions, so that what the theory asserts cannot even be claimed to be true. Nevertheless, it gives us a view of the world which is so compelling that we prefer it to alternative viewpoints! Engaging as the relativistic view of time may be, however, the empirical content of the theory is not incompatible with absolute simultaneity, so that the discarding of absolute time is unwarranted. Geoffery Builder concludes, "...the measurement procedures prescribed by the restricted theory are conventional; failure to recognize this has led many exponents of the restricted theory to assert, without sufficient justification, that these procedures demand a complete revision of the older concepts of space and time."[72] Revisions in Newtonian mechanics are necessary, of course; Newton did not realize that physical time—the time kept by clocks— needed to be relativized. But, as Frank explains, "The 'relativization' of space and time consists actually in the introduction of new operational definitions that are better adapted to the actual needs of the scientist. This 'relativization' of space and time is an advance in semantics and not, as has been frequently said, an advance in metaphysics or ontology."[73]

SUMMARY

We have thus seen that positivism belongs essentially to the philosophical foundations of SR, that such an epistemological outlook has been justifiably and universally rejected as untenable, and that SR is predicated upon arbitrary and counter-intuitive re-definitions of the key concepts of time and distant simultaneity. Lawrence Sklar concludes,

> The original Einstein papers on special relativity are founded, as is well known, on a verificationist critique of earlier theories.... Now it might be argued that Einstein's verificationism was a misfortune, to be encountered not with a rejection of special relativity, but with an acceptance of the theory now to be understood on better epistemological grounds....
> But I don't think a position of this kind will work in the present case. I can see no way of rejecting the old aether-compensatory theories, originally invoked to explain the Michelson-Morley results, without invoking a verificationist critique of some kind or other. And I know of no way to defend the move to a relativized notion of simultaneity, so essential to special relativity, without first offering a critique, in the same vein as Einstein's, of the pre-relativistic absolutist notion, and then continuing to observe that even the relativistic replacement for this older notion is itself, insofar as it outruns the 'hard data' of experiment, infected with a high degree of conventionality.[74]

It seems to me, therefore, that SR's elimination of absolute time is a miscarriage, resting as it does on defective epistemological foundations. We both can and should distinguish between metaphysical time and physical time and maintain that the

---

We have already seen that Einstein's theory is conceptually hobbled in comparison with Lorentz's, being based on an arbitrary and counter-intuitive re-definition of distant simultaneity motivated by an untenable epistemology; as for heuristic advantages, see discussion in the sequel.
[72] G. Builder, "The Constancy of the Velocity of Light," *Australian Journal of Physics* 11 (1958): 457-480, rep. in *Speculations in Science and Technology* 2 (1979): 421.
[73] Frank, *Philosophy of Science*, p. 143.
[74] Sklar, "Time, reality, and relativity," p. 132.

former is characterized by a universal and objective present and, hence, relations of absolute simultaneity. Once we examine critically the philosophical foundations of Einstein's Special Theory, we discover that there is nothing there that warrants the conclusion that metaphysical time, objective becoming, and absolute simultaneity do not exist. It is difficult, indeed, to understand how the authors quoted earlier can therefore speak of SR's "forcing" us to abandon the classical concepts of space and time or of its "destruction" of Newtonian absolute time. Sklar asserts, "One thing is certain. Acceptance of relativity cannot force one into the acceptance or rejection of any of the traditional views about the reality of the past and future."[75] The fatal shortcoming in the spacetime realist's objection to the objectivity of temporal becoming lies in his uncritical acceptance of Einstein's positivistic elimination of metaphysical time and, consequently, his unjustified assumption that the foundation of temporal becoming must be found in physical, measured time rather than in time itself. Of course, we are still left with Mellor's challenge of exactly how the relativity of physical time, which, as we have seen, has been well-established empirically, is to be reconciled with the objectivity of temporal becoming and the absolute simultaneity that the A-theorist postulates of metaphysical time. It is to that task that we now turn.

---

[75] Ibid., p. 140. In Gödel's words,
"the observational results by themselves really do not force us to abandon Newtonian time and space as objective realities, but only the observational results together with certain general principles, e.g., the principle that two states of affairs which cannot be distinguished by observers are also objectively equal" (unpublished *Nachlass* cited by Palle Yourgrau, *The Disappearance of Time* [Cambridge: Cambridge University Press, 1991], p. 54).

CHAPTER 9

ABSOLUTE TIME AND RELATIVISTIC TIME

NEO-LORENTZIAN RELATIVITY

We have seen that for Newton God's eternity and omnipresence were ontologically foundational for his views of time and space. I have elsewhere argued that it is plausible that a transcendent, personal Creator of the universe exists[1] and in yet another place that such a deity, in virtue of His real, causal relationship to the world, must (at least since the moment of creation) be temporal if an A-Theory of time is correct.[2] Unfortunately in our secular age physicists and philosophers of space and time rarely, if ever, give careful consideration to the difference God's existence makes for our conceptions of time and space. To borrow Sir Arthur Eddington's words, "physics has in the main contented itself with studying the abridged edition of the book of nature."[3] Such indifference was characteristic of Einstein himself. Only after 1930 did he begin to refer more frequently in his non-scientific writings to religious questions. But he was, in Holton's words, "quite unconcerned with religious matters during the period of his early scientific publications."[4] Einstein remarked that by the age of twelve he had lost his faith because of science and became a free-thinker.[5] Thus, he did not consider what difference theism might make to one's views of time and space.

But in a fascinating passage in his essay "La mesure de temps," Poincaré does briefly entertain the hypothesis of *"une intelligence infinie"* and considers the implications of such a hypothesis. Poincaré is reflecting on the problem of temporal succession. In consciousness, the temporal order of mental events is clear. But going outside consciousness, we confront various difficulties. One of these concerns how we can apply one and the same measure of time to events which transpire in "different worlds," that is, spatially distant events. What does it mean to say that two psychological phenomena in two consciousnesses happen simultaneously? Or what does it mean to say a supernova occurred before

---

[1] William Lane Craig, *The Kalam Cosmological Argument*, Library of Philosophy and Religion (London: Macmillan, 1979); William Lane Craig and Quentin Smith, *Theism, Atheism, and Big Bang Cosmology* (Oxford: Clarendon Press, 1993).
[2] William Lane Craig, *God, Time, and Eternity* (Dordrecht: Kluwer Academic Publishers, forthcoming).
[3] Arthur S. Eddington, "A Generalization of Weyl's Theory of the Electromagnetic and Gravitational Fields," *Proceedings of the Royal Society of London* A99 (1921): 108.
[4] Gerald Holton, "Mach, Einstein and the Search for Reality," in *Ernst Mach: Physicist and Philosopher*, Boston Studies in the Philosophy of Science 6 (Dordrecht: D. Reidel, 1970), p. 198; cf. p. 188.
[5] See also A. Einstein, "Wie ich die Welt sehe," in *Mein Weltbild*, ed. Carl Selig (Frankfurt: Ullstein Bücher, 1934), pp. 7-18 for Einstein's denial of personalistic theism.

Columbus saw the isle of Espanola? "All these affirmations," says Poincaré, "have by themselves no meaning."[6] Then he remarks,

> We should first ask ourselves how one could have had the idea of putting into the same frame so many worlds impenetrable to one another. We should like to represent to ourselves the external universe, and only by so doing could we feel that we understood it. We know we can never attain this representation: our weakness is too great. But at least we desire the ability to conceive an infinite intelligence for which this representation could be possible, a sort of great consciousness which should see all, and which should classify all *in its time*, as we classify, *in our time*, the little we see.
>
> This hypothesis is indeed crude and incomplete, because this supreme intelligence would be only a demigod; infinite in one sense, it would be limited in another, since it would have only an imperfect recollection of the past; it could have no other, since otherwise all recollections would be equally present to it and for it there would be no time. And yet when we speak of time, for all which happens outside of us, do we not unconsciously adopt this hypothesis; do we not put ourselves in the place of this imperfect god; and do not even the atheists put themselves in the place where God would be if he existed?
>
> What I have just said shows us, perhaps, why we have tried to put all physical phenomena into the same frame. But that cannot pass for a definition of simultaneity, since this hypothetical intelligence, even if it existed, would be for us impenetrable. It is therefore necessary to seek something else.[7]

Poincaré here suggests that, in considering the notion of simultaneity, we instinctively put ourselves in the place of God and classify events as past, present, or future according to His time. Poincaré does not deny that such a perspective would disclose to us true relations of simultaneity. But he rejects the hypothesis as yielding a definition of simultaneity because *we* could not know such relations; such knowledge would remain the exclusive possession of God Himself.

But clearly, Poincaré's misgivings are relevant to a definition of simultaneity only if one is presupposing some sort of verificationist theory of meaning, as he undoubtedly was. The fact remains that God knows the absolute simultaneity of events even if we grope in total darkness. Nor need we be concerned with Poincaré's argument that such an infinite intelligence would be a mere demigod, since it is a *non sequitur* that a being with perfect recollection of the past cannot be temporal. There is no conceptual difficulty in the idea of a being which knows all true past-tense propositions. That such a being would be temporal is evident from the fact that as events transpire, more and more past-tense propositions become true, so that the content of his knowledge is constantly changing.[8] Hence, it does not follow that if God is temporal, He cannot have perfect recollection of the past.

Poincaré's hypothesis suggests, therefore, that God's present is constitutive of relations of absolute simultaneity. J. N. Findlay was wrong when he said, "...the influence which harmonizes and connects all the world-lines is not God, not any featureless, inert, medium, but that living, active interchange called...Light,

---

[6] Henri Poincaré, "The Measure of Time," *Revue de métaphysique et de morale* 6 (1898): 1ff., rep. in H. Poincaré, *The Value of Science* (1905), trans. G. B. Halstead, in *The Foundations of Science*, by H. Poincaré (Science Press, 1913; rep. ed.: Washington, D. C.: University Press of America, 1982), p. 228.

[7] Ibid., pp. 228-229.

[8] If one takes propositions to be tenselessly true or God's knowledge to be non-propositional, it still follows that God's *de se* knowledge is constantly changing. For discussion, see my *Divine Foreknowledge and Human Freedom* (Leiden: E. J. Brill, 1990), Introduction.

offspring of Heaven firstborn."[9] On the contrary, the use of light signals to establish clock synchrony is a convention which finite and ignorant creatures have been obliged to adopt, but the living and active God, who knows all, is not so dependent. In God's temporal experience, there is a moment which is present in metaphysical time, wholly independently of physical clock times. He would know, without any dependence on clock synchronization procedures or any physical operations at all, which events were simultaneously present in metaphysical time. He would know this simply in virtue of His knowing at every such moment the unique set of present-tense propositions true at that moment, without any need of a *sensorium* or physical observation of the universe.

The question now presses: how, then, does God's metaphysical time relate to our physical time? It seems that God's existence in A-theoretic, metaphysical time and His real relation to the world would imply that a Lorentz-Poincaré theory of relativity is correct after all. For God in the "now" of metaphysical time would know which events in the universe are now being created by Him and are therefore absolutely simultaneous with each other and with His "now." The argument may be formulated as follows:

1. God exists.
2. An A-Theory of time is correct.
3. If God exists and an A-Theory of time is correct, then God is in time.
4. If God is in time, then a privileged reference frame exists.
5. If a privileged reference frame exists, then a Lorentz-Poincaré theory of relativity is correct.

I am taking (1) for granted, though I have tried to justify it elsewhere.[10] Similarly, I have argued at length for the truth of (2) in my companion volumes on the tensed *vs.* tenseless conceptions of time.[11] In my *God, Time, and Eternity*, I present two arguments on behalf of (3) which I find persuasive: the argument from God's real relation to the world and the argument from God's knowledge of tensed facts.[12] The truth of (4) is based on God's privileged status; the events which He knows to be occurring now can be associated with some hypothetical reference frame which is preferred precisely because it is God's frame. In speaking of God's frame, I mean merely that it is the reference frame on whose associated hyper-plane of simultaneity at the time of any event on the world line of a hypothetical observer at rest in that frame lie those events which God in the "now" of metaphysical time knows He is then creating.[13] Now it needs to be understood that we are speaking

---

[9] J. N. Findlay, "Time and Eternity," *Review of Metaphysics* 32 (1978-79): 6-7.
[10] See William Lane Craig, *The Kalam Cosmological Argument*; William Lane Craig and Quentin Smith, *Theism, Atheism and Big Bang Cosmology*.
[11] See my *The Tensed Theory of Time: a Critical Examination*; Synthèse Library (Dordrecht: Kluwer Academic Publishers, forthcoming) and my *The Tenseless Theory of Time: a Critical Examination*; Synthèse Library (Dordrecht: Kluwer Academic Publishers, forthcoming).
[12] See my *God, Time, and Eternity* chaps. 3, 4.
[13] I do not mean to endorse Newton's view that God exists spatially. See the sequel.

here in the context of SR, which posits a flat spacetime. When we come to GR, which allows a variable curvature due to the presence of mass-energy, it is possible that the events which God knows to be simultaneous cannot be gathered into a single inertial frame. In that case, as Lucas and Hodgson explain, the set of events known by God to be simultaneous would constitute a three-dimensional hyper-surface in spacetime, but not a hyper-plane:

> Only if it were a hyper-plane would it constitute the canon of simultaneity of some inertial frame of reference. Otherwise there would be no world-wide *inertial* frame of reference, but only different ones for different pairs of events. Instead of the flat spacetime of the Special Theory, we should be having to do with a curved one in which the significant hyper-surfaces were not hyper-planes. But that...is something we have been forced to countenance in the General Theory anyhow.[14]

A discussion of God's relation to time within the context of GR may be reserved for the sequel; our present concern is with the flat spacetime of SR. Within that context, the events which God knows He is now causing all lie on a three-dimensional hyper-plane which is associated with some inertial frame. In virtue of that frame's simultaneity relation's being God's, that frame is plausibly privileged. But if there is a privileged frame, then Lorentz and Poincaré were right, as (5) states. Einstein's SR presupposes that no such frame exists, as we have seen, and therefore cannot be correct. The postulation of a privileged frame is sufficient for some sort of "aether compensatory" theory, even if the so-called aether just is the three-dimensional privileged space. This privileged frame would be associated with a physical time which would constitute a sensible measure of God's metaphysical time.

Lorentz's understanding of relativity theory is well-suited to play the role ascribed to it by the above argument. It is one of the marks of Lorentz's genius that he stood out virtually alone among the major figures of early twentieth century physics in resisting the pervasive positivistic epistemology which lay at the root of Einstein's SR. Gracious almost to a fault, Lorentz always spoke appreciatively of Einstein's alternate approach and lectured sympathetically on both SR and GR, while remaining finally unconvinced that Einstein had abolished the classical conceptions of time and space. A major reason that Lorentz remained unconvinced was that he was not a positivist. In 1913 he wrote,

> According to Einstein it has no meaning to speak of motion relative to the aether. He likewise denies the existence of absolute simultaneity.
> It is certainly remarkable that these relativity concepts, also those concerning time, have found such a rapid acceptance.
> The acceptance of these concepts belongs mainly to epistemology.... It is certain, however, that it depends to a large extent on the way one is accustomed to think whether one is attracted to one or another interpretation. As far as this lecturer is concerned, he finds a certain satisfaction in the older interpretations, according to which the aether possesses at least some substantiality, space and time can be sharply separated, and simultaneity without further specification can be spoken of. In regard to this last point, one may perhaps appeal to our ability of imagining arbitrarily large velocities. In that way, one comes very close to the concept of absolute simultaneity.

---

[14] J. R. Lucas and P. E. Hodgson, *Spacetime and Electromagnetism* (Oxford: Clarendon Press, 1990), p. 120.

> Finally, it should be noted that the daring assertion that one can never observe velocities larger than the velocity of light contains a hypothetical restriction of what is accessible to us, [a restriction] which cannot be accepted without some reservation.[15]

Here Lorentz clearly discerns the crucial role played by Einstein's verificationist theory of meaning and rejects it. In defense of absolute simultaneity, he appeals to the use of arbitrarily fast signals, even though they are not presently observable. He quite rightly disregards the assumption that it is meaningless to speak of such unobservables. Elsewhere Lorentz affirms that it makes sense, if there is an aether, to speak of motion relative to it even if observers could not detect such motion.[16] He writes,

> But it needs to be clearly recognized that A could never assure himself of the immobility in the ether which we have attributed to him by supposition and that physicist B could with the same right, or rather with the same absence of right, claim that it is he who finds himself in these privileged circumstances. This incertitude, this impossibility of even disclosing a movement in relation to the ether, led Einstein and numerous other modern physicists to abandon completely the notion of an aether.
>
> There, it seems to me, is a question toward which each physicist must take a position which best accords with the manner of thinking to which he is accustomed.[17]

Lorentz clearly felt no obligation to abandon what seemed to him meaningful notions out of deference to a verificationist epistemology. The subsequent collapse of positivism has vindicated him with respect to this conviction and left his approach to problems of space and time, in contrast to Einstein's, unshaken by this epistemological revolution.

Second, Lorentz's conception of the aether was virtually equivalent to space itself. His aether was so dematerialized that Einstein, lecturing at the University of Leiden in 1920, could tease the Dutch physicist by declaring, "As regards the mechanical nature of Lorentz's aether one might say of it, with a touch of humor, that immobility was the only mechanical property which H. A. Lorentz left it."[18] Later Einstein would speak of Lorentz's "great discovery," namely, that all the phenomena of electromagnetism then known could be explained on the basis of two assumptions: that the aether is firmly fixed in space—that is to say, unable to move at all—and that electricity is firmly lodged in the mobile elementary particles.[19]

---

[15] H. A. Lorentz, A. Einstein, H. Minkowski *Das Relativitätsprinzip*, Fortschritte der mathematischen Wissenschaften 2, mit Anmerkungen von A. Sommerfeld und Vorwort von O. Blumenthal (Leipzig: B. G. Teubner, 1920), p. 23 (Pais translation).

[16] H. A. Lorentz, "Het relativiteitsbeginsel. Voordrachten in Teyler's Stichtung," *Archives Musée Teyler* 2 (1914): 26. See the outstanding historical study by Jozsef Illy, "Einstein Teaches Lorentz, Lorentz Teaches Einstein. Their Collaboration in General Relativity, 1913-1920," *Archive for History of Exact Sciences* 39 (1989): 272, who comments, "Lorentz distinguishes the existence of a state of motion from its observability—a distinction void of meaning according to Einstein."

[17] H. A. Lorentz, "Considérations élémentaires sur le principe de relativité," *Revue générale des Sciences* 25 (1914): 179ff, in H. A. Lorentz, *Collected Papers*, 9 vols., ed. P. Zeeman and A. D. Fokker (The Hague: Martinus Nijhoff, 1934-1939), 7: 165.

[18] A. Einstein, *Äther und Relativitätstheorie* (Berlin: Julius Springer Verlag, 1920), p. 7.

[19] A. Einstein, "The Problem of Space, Ether, and the Field in Physics" [*Mein Weltbild*, 1934], in *Ideas and Opinions*, trans. Sonja Bargmann (New York: Crown Publishers, 1954), p. 281. So also Tetu Hirosige, "Origins of Lorentz' Theory of Electrons and the Concept of the Electromagnetic Field," *Historical studies in the Physical Sciences* 1 (1969): 197; idem, "The Ether Problem, the Mechanistic

"Today his discovery may be expressed as follows: physical space and the ether are only different terms for the same thing: fields are physical states of space."[20] For Lorentz the aether just is the privileged spatial frame. He thus accepts a 3+1 ontology of spatial objects enduring through time, a metaphysic which is well-suited to an A-Theory of time.

Third, Lorentz distinguished true time, which was measured by a clock at rest in the aether frame, from merely local time and ascribed to events absolute simultaneity in true time. Writing in 1910, he contrasted his view with Einstein's:

> Assume there were an aether; then there would be among all systems x, y, z, t one singled out in that the coordinate axes as well as the clock are at rest in the aether. If one conjoins with this the idea...that space and time are something wholly different and that there is a 'true time' (simultaneity would then exist independently of location, in accord with the circumstance that it is possible for us to conceive of infinitely great velocities), then one easily sees that this true time would have to be indicated just by clocks which are at rest in the aether. If, then, the principle of relativity were generally valid in nature, then one would not be in a position to determine whether the coordinate system employed is that distinguished one. One thus comes to the same results as when one in agreement with Einstein and Minkowski denies the existence of the aether and the true time and treats all coordinate systems as equivalent. Which of the two modes of thought one may agree with is best left to the individual.[21]

Lorentz, realizing that his aether compensatory theory is empirically equivalent to the Einstein-Minkowski theory, leaves it up to the individual to choose which he shall adopt. But Lorentz preferred the classical conceptions of time and space on metaphysically intuitive grounds, as he made clear in his 1922 lectures at Cal Tech:

> All our theories help us form pictures, or images, of the world around us, and we try to do this in such a way that the phenomena may be coördinated as well as possible, and that we may see clearly the way in which they are connected. Now in forming these images we can use the notions of space and time that have always been familiar to us,

---

Worldview, and the Origins of the Theory of Relativity," *Historical Studies in the Physical Sciences* 7 (1976): 69.

[20] Einstein, "Space, Ether, and the Field," p. 281. I am skeptical that Lorentz would have agreed that fields are physical states of space rather than that physical fields exist in space. Lorentz held to the classical conceptions of time and space.

[21] H. A. Lorentz, "Alte und neue Fragen der Physik," *Physikalische Zeitschrift* 11 (1910): 1234ff, in *Collected Papers*, 7: 211. Notice that Lorentz eschews operational definitions of time and does not equate time with clock readings. This is particularly evident in his failure to ascribe at first physical reality to his local times. He recalled,

"I had the idea that there was an essential difference between the systems $x, y z, t$ and x', y', z', t'. In one one uses—such was my thought—coordinate axes which have a fixed position in the aether and which one can call the 'true' time; in the other system, on the other hand, it would be a matter of mere auxiliary quantities whose introduction is nothing but a mathematical artifice. In particular, the variable t' could not be called 'time' in the same sense as the variable t" (H. A. Lorentz, "Deux mémoires de Henri Poincaré sur la physique mathématique," *Acta mathematica* 38 (1914): 293ff, in *Collected Papers*, 7: 262.

Eventually Lorentz came to see the reality of clock retardation. But even the readings of a clock at rest in the aether are but a sensible measure of true time, not definitive of true time. Illy comments, "When he says that 'we make the readings of...clocks match our fundamental conception of time' he implies that time is not to be identified with the mere reading of clocks, in the way that Einstein defined time in his relativistic paper of 1905; consequently measurements cannot be interpreted without some previous agreement upon fundamental concepts; measurements may not be used for defining time (and space)" (Illy, "Einstein Teaches Lorentz," p. 275).

and which I, for my part, consider as perfectly clear and, moreover, as distinct from one another. My notion of time is so definite that I clearly distinguish in my picture what is simultaneous and what is not.[22]

Here Lorentz quite rightly refuses to jettison the intuitively obvious reality of absolute simultaneity among events in the world just because one cannot determine which spatially separated events are simultaneous or because Einstein's operationally re-defined notion of simultaneity is relative to reference frames. Moreover, he sees no good reason to scrap the intuitive distinctness of space and time in favor of Minkowski's unified reality, spacetime. It seems to me that Lorentz was quite justified in sticking with his intuitions and, moreover, that his intuitions were right on target. In any case, the ease with which his approach can be assimilated with our direction of thought is obvious.

Finally, Lorentz himself hinted at a theological application of his ideas. In January of 1915 Lorentz, convinced that Einstein's work on GR was quite compatible with the notion of a privileged aether frame, wrote to Einstein in response to the latter's paper "The Formal Foundations of the General Theory of Relativity."[23] In a passage redolent of the General Scholium and *Opticks* of Newton, Lorentz broached considerations whereby "I cross the borderland of physics":

> A 'World Spirit' who, not being bound to a specific place, permeated the entire system under consideration or 'in whom' this system existed and who could 'feel' immediately all events would naturally distinguish at once one of the systems U, U', etc. above the others.[24]

Such a being could "directly verify simultaneity."[25] Here we see from Lorentz's own hand the integratability of his approach to relativity theory with a theological perspective such as we have enunciated.

I do not mean to suggest, of course, that we should return to Lorentz's own, defunct electron theory. As we saw from Zahar's analysis,[26] that theory was doomed by the advent of quantum physics. But Lorentz's own attempted explanation of the phenomenon of length contraction is incidental to his basic approach. By 1908 Lorentz had already realized the incompatibility of his electron theory with Planck's quantum hypothesis, and by the 1911 Solvay Congress there was a general sense that the electron theory would have to be radically reformed in

---

[22] H. A. Lorentz, *Problems of Modern Physics*, ed. H. Bateman (Boston: Ginn, 1927), p. 221. Here Lorentz identifies himself as a physicist "of the old school" who says, "'I prefer the time that is measured by a clock that is stationary in the ether, and I consider this as the true time, though I admit that I cannot make out which of the two times is the right one, that of A or that of B'" (Ibid.).
[23] A. Einstein, "Die formale Grundlage der allgemeinen Relativitätstheorie," *Physikalische Zeitschrift* 14 (1913): 1249-1262.
[24] H. A. Lorentz to A. Einstein, January, 1915, Boerhaave Museum, cited in Illy, "Einstein Teaches Lorentz," p. 274. The text reads, "Ein 'Weltgeist,' der ohne an einem bestimmten Ort gebunden zu sein, das ganze betrachte System durchdränge, oder 'in dem' dieses System bestände, und der unmittelbar alle Ereignisse 'fühlen' könnte, würde natürlich sofort eins der Systeme U, U', u.s.w. vor den anderen auszeichnen." See comment by A. J. Knox, "Hendrik Antoon Lorentz, the Ether, and the General Theory of Relativity," *Archive for History of Exact Sciences* 38 (1988): 75.
[25] Lorentz to Einstein.
[26] Recall our discussion in chap. 1.

light of the advent of quantum physics.[27] Nonetheless, Lorentz continued to adhere to an approach to relativity theory which preserved the classical notions of space and time. A theory may be classified as Lorentzian just in case it affirms (i) physical objects are n-dimensional spatial objects which endure through time, (ii) the round trip vacuum propagation of light is isotropic in a preferred (absolute) reference frame R_0 (with speed $c=1$) and independent of the velocity of the source, and (iii) lengths contract and time rates dilate in the customary special relativistic way only for systems in motion with respect to R_0.[28] In light of our foregoing argument, any metaphysically adequate theory of space and time must be Lorentzian in this sense. In order to avoid confusion with Lorentz's original electron theory, we may call such theories neo-Lorentzian.

EINSTEINIAN VS. NEO-LORENTZIAN RELATIVITY

Although in semi-popular expositions of SR scorn is usually poured upon a neo-Lorentzian interpretation, particularly the Lorentz-FitzGerald contraction, the fact is that a small minority of physicists, including such notable figures as H. E. Ives, Geoffery Builder, and S. J. Prokhovnik, have continued the Lorentzian research programme down to the present day, so that the correct physical interpretation of relativity theory remains a matter of debate.[29] In fact, that debate has intensified in recent years, due to what one participant has called a "sea change" in the foundations of physics community concerning the viability of a Bohmian interpretation of quantum mechanics, which plausibly requires absolute simultaneity and, hence, absolute time.[30] It is acknowledged even by its detractors that the neo-Lorentzian version is empirically equivalent to the received, Einstein-Minkowskian version of SR, so that the decision between them must be made on the basis of non-empirical considerations.

It is often asserted that the received version of the theory is simpler and therefore to be preferred. However, as is well-known, one must be very cautious about the connection between the simplicity of a theory and its *truth*, as opposed to its *utility*.[31]

---

[27] See Russell McCormmach, "H. A. Lorentz and the Electromagnetic View of Nature," *Isis* 61 (1970): 486-488.
[28] See A. K. A. Maciel and J. Tiomno, "Analysis of Absolute Spacetime Lorentz Theories," *Foundations of Physics* 19 (1989): 507-508.
[29] See the proceedings of the biennial conferences sponsored by the British Society for the Philosophy of Science, "Physical Interpretations of Relativity Theory," Imperial College of Science and Technology, London. It is noteworthy that in Zahar's analysis, SR did not by itself empirically supersede Lorentz's programme. Zahar can claim that Einstein's research programme superseded Lorentz's only by taking GR to be continuous with SR and by then appealing to the empirical confirmation of GR. (Elie Zahar, "Why Did Einstein's Programme Supersede Lorentz's? (I & II)," *British Journal for the Philosophy of Science* 24 [1973]: 95-123; 223-262). But as we shall see below, GR is definitely not continuous with SR; it is a theory of gravitation, not a theory of relativity. It therefore follows that Einstein's programme never in fact did supersede Lorentz's, in the sense of demonstrating its superiority over Lorentz's.
[30] John Kennedy, in a paper presented at the Central Division Meeting of the American Philosophical Association, April 23-26, 1997, Pittsburgh, Pennsylvania.
[31] Cushing further observes, "Generally agreed upon, objective criteria of simplicity do not seem to exist. Hence, simplicity as a guide to truth (as opposed merely to economy of effort) in formulating and developing a theory or hypothesis can be quite subjective" (James T. Cushing, *Philosophical Concepts in Physics* [Cambridge: Cambridge University Press, 1998], p. 161).

What right do we have to infer that if the received version is simpler, it is therefore true, especially in light of the extraordinary metaphysical commitments it involves (no absolute space, no absolute time, but, plausibly, absolute spacetime)? Moreover, that simplicity is achieved by means of conventional definitions, which makes it difficult to regard the theory as being true rather than merely useful or expedient. One could also argue that counterbalancing the simplicity of the received version is the heuristic superiority of the neo-Lorentzian version.[32] Besides all this, the fact remains that we have good reasons for believing that a neo-Lorentzian theory is correct, namely, the existence of God in A-theoretic time implies it, so that concerns about which version is simpler become of little moment.

In any case, the claim that the received theory is simpler is simply wrong. Although Lorentz's own theory was more complicated than Einstein's, H. E. Ives was able to derive the Lorentz transformation equations from (i) the laws of conservation of energy and momentum and (ii) the laws of transmission of radiant energy. He showed that there is an apparent discrepancy in the equations for a particle governed by these laws which demands that the particle's mass vary with velocity. He then derived from these variations of dimensions the Lorentz transformations. "The space and time concepts of Newton and Maxwell are retained without alteration," he wrote, "It is the dimensions of the material instruments for measuring space and time that change, not space and time that are distorted."[33] On Ives's achievement, Martin Ruderfer comments that he succeeded in elevating Lorentz's *ad hoc* theory to an equal status with SR and did so with the same number of basic assumptions as Einstein, so that his theory has the same "beauty." "The Ives and Einstein interpretations represent two different, but equally valid, views of the same set of observations."[34] Hence, assertions that Einstein's version is simpler than a neo-Lorentzian theory are incorrect.

Neo-Lorentzian theorists, on the other hand, have complained that the received theory leaves us with physical effects without any causal explanation for them. Thus, for example, Builder asserts,

> The relative retardation of clocks, predicted by the restricted theory of relativity, demands our recognition of the causal significance of absolute velocities.... The observable effects of absolute accelerations and of absolute velocities must be ascribed to the interaction of bodies and physical systems with some absolute inertial system. We have no alternative but to identify this absolute system with the universe.[35]

Builder argues that the universe as a whole constitutes a unique, absolute inertial system, which he calls "the ether," that affects the masses and dimensions of bodies,

---

[32] See J. S. Bell, "How to Teach Special Relativity," in *Speakable and Unspeakable in Quantum Mechanics* (Cambridge: Cambridge University Press, 1987), pp. 66-80; see also S. J. Prokhovnik, "An Introduction to the Neo-Lorentzian Relativity of Builder," *Speculations in Science and Technology* 2 (1979): 226; G. Builder, "Ether and Relativity," *Australian Journal of Physics* 11 (1958): 279-297, reprinted in *Speculations in Science and Technology* 2 (1979): 239-241.
[33] Herbert E. Ives, "Derivation of the Lorentz Transformations," *Philosophical Magazine* 36 (1945): 392-401, reprinted in *Speculations in Science and Technology* 2 (1979): 247, 255.
[34] Martin Ruderfer, "Introduction to Ives' 'Derivation of the Lorentz Transformations'," *Speculations in Science and Technology* 2 (1979): 243.
[32] Builder, "Ether and Relativity," p. 230.

as well as the running of clocks, in accordance with their absolute speed. He contrasts this interpretation with the received view:

> The conceptual difficulties associated with the restricted theory all arise out of the denial that these absolute concepts are permissible, and out of the consequent attempts to avoid them in the presentation of the theory. It is frequently maintained that the theory has forced us to discard entirely the old-fashioned commonsense notions of time and space, but nothing comprehensible or definable has been offered in their place. Moreover, any questions as to what *causes* the relativity of simultaneity, the measured constancy of the velocity of light in all inertial reference systems, or the reciprocity of relativistic variations of length, of mass, and of clock rates, are avoided by vague references to the principle of relativity, to the four-dimensional character of spacetime, and so on.
>
> On the other hand, the presentation of the restricted theory in terms of the absolute concepts (following generally the lines of its development by Poincaré and Lorentz) involves no conceptual difficulties. The relativity of simultaneity, the reciprocity of relativistic variations, and the constancy of the measured velocity of light, then all appear simply as comprehensible effects of the motions, relative to the ether, of the bodies of observers and of the measuring instruments used.[36]

Developing Builder's approach, Prokhovnik interprets length contraction and time dilation as retarded potential effects. He explains that Builder's notion of a fundamental inertial frame I(F) implies that energy propagation is strictly isotropic with respect to this frame. A body moving with respect to this frame will drag its gravitational field with it. The theory of retarded potentials applies to the gravitational or electromagnetic field of a body moving relative to I(F). For such a body energy propagation is not isotropic and its co-moving fields are no longer symmetric, which leads to a whole chain of interacting anisotropy effects. If the field of a moving electron is compressed in the direction of motion, for example, then its surface must be similarly compressed to maintain an equipotential equilibrium state, and the moving electron takes the shape of an ellipsoid. A system of particles would contract in the direction of motion in order to sustain the interparticle equilibrium of the system. No substantial aether is required for the production of such effects; simply the existence of I(F) itself suffices, as Prokhovnik explains,

> Fitzgerald and Lorentz considered the contraction as an absolute effect due to movement relative to a unique reference frame associated with an aether. It is clear that the contraction can be even more satisfactorily considered as a secondary effect resulting from movement relative to a unique cosmologically-based fundamental reference frame, I(F), defined in terms of our greatly expanded view of the universe since 1930. The retarded potential field effect is a primary consequence of movement relative to I(F) and, as seen in the basis and derivation of this effect, it emerges essentially in consequence of the limiting velocity restriction on the transmission of energy; no aether or any other property of 'physical space' is required for the emergence of this effect.[37]

Length contraction can be considered as the reaction of a moving system to maintain its stationary-in-I(F) equilibrium state in the circumstance of an asymmetric gravitational field resulting from its motion in I(F). This phenomenon is thus as universal and general as the gravitational property of matter.

---

[36] Ibid., p. 240.
[37] S. J. Prokhovnik, *Light in Einstein's Universe* (Dordrecht, Holland: D. Reidel, 1985), pp. 84-85.

Following Builder, Prokhovnik sees time dilation as the consequence of the fact that a moving light clock has a longer unit of time and hence runs more slowly than a similar, stationary clock in I(F) (Figure 9.1).

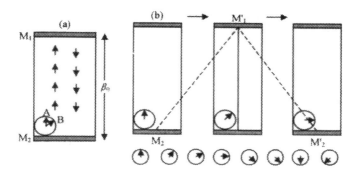

Figure 9.1. The light-pulse clock. (a) When at rest a light pulse is reflected back and forth from the mirrors M_1 and M_2; each return of the pulse to one end counts as a unit of time. The dial hand moves from A to B in that time. (b) When the clock is in motion the light travels the distance $M_2M'_1M'_2$ between ticks, so causing it to go slow. The dials underneath the moving clock are those stationary clocks distributed in space and used for comparison with the moving clock times.

This effect can also be generalized, according to Prokhovnik: "The effect will apply to all phenomena involving electromagnetic impulses and energy-exchanges; so it can be considered to apply to all clocks and physical phenomena and, by extension, also to biological phenomena."[38] Thus, clock retardation, the Twin Paradox, and asymmetrical aging become comprehensible phenomena.

The adequacy of such causal explanations is not the primary point here. Rather the salient point is that a neo-Lorentzian theory involves appeal to some sort of causal explanation for the physical effects predicted by SR. By contrast, in Einstein's theory, these effects simply follow, as we have seen, from Einstein's denial of the existence of a fundamental frame and his two postulates. Given his postulates, a new kinematics follows according to which reciprocal length contraction and clock retardation occur, but without any dynamical explanation.

As a theory of principle rather than a constructive theory, Einstein's SR is based on postulates which are characterized by their very non-empirical character. As Holton and Goldberg have emphasized, although Einstein insisted early on that his SR was empirically grounded, in fact the two postulates on which the theory is based "were postulates for which there was and can be no direct empirical confirmation"; their status is that of "a non-verifiable and non-falsifiable

---

[38] Ibid., p. 89. See also S. J. Prokhovnik, "The Logic of the Clock Paradox," paper presented at the International Conference of the British Society for the Philosophy of Science, "Physical Interpretations of Relativity Theory," Imperial College of Science and Technology, London, 16-19 September, 1988.

presupposition."[39] The only version of SR which is experimentally verifiable, according to Builder, "is the theory that the spatial and temporal coordinates of events, measured in any one inertial reference system, are related to the spatial and temporal coordinates of the same events, as measured in any other inertial reference system, by the Lorentz transformations."[40] But this verifiable statement only concerns measurements made in inertial reference systems and is neutral with regard to the Lorentzian and Einsteinian interpretations. As a constructive theory, the neo-Lorentzian approach promises to enrich our understanding of the causal structure of the world in a way that Einstein's cannot.

Now, as we have seen, the spacetime interpretation of SR does furnish explanations of time dilation and length contraction in terms of the spacetime distances involved. One will recall Figure 3.3, illustrating the Twin Paradox. In four-dimensional spacetime, the path of the travelling sibling is actually *shorter* than that of the earthbound twin.[41] Therefore, it is not surprising that his clock records less time. A similar account could be given of length contraction. Therefore, on a spacetime interpretation Einstein's SR does explain why these relativistic phenomena occur.

But here it needs to be seriously called into question whether any such metaphysical reality as spacetime actually exists. In the first place, while a spacetime interpretation of SR is arguably superior in some respects to a relativity interpretation of SR[42], there are no compelling reasons to prefer a spacetime ontology over a classical ontology of space and time as the latter comes to expression in a neo-Lorentzian theory of relativity. Spacetime realists are self-confessed, post-positivistic metaphysicians, and so cannot complain that the classical concepts of space and time are obnoxiously metaphysical. As Earman and Friedman write, "Newton's theory, and especially his conceptions of space and time, have often been criticized on operational and positivistic grounds; but it is rarely pointed out that relativity theory, which is often taken as vindicating Newton's philosophical critics, is not less subject to the same criticisms...."[43] When spacetime realists champion a spacetime ontology over a space and time ontology, virtually all their arguments are aimed at demonstrating the superiority of the spacetime interpretation of SR over the relativity interpretation of that theory. The

---

[39] Gerald Holton, "Where Is Reality? The Answers of Einstein," in *Science and Synthesis*, ed. UNESCO (Berlin: Springer Verlag, 1971), pp. 52-64; idem, "Mach, Einstein, and the Search for Reality," in *Ernst Mach: Physicist and Philosopher*, Boston Studies in the Philosophy of Science 6 (Dordrecht: D. Reidel, 1970), p. 178; idem, "Poincaré and Relativity," in *Thematic Origins of Scientific Thought: Kepler to Einstein* (Cambridge, Mass.: Harvard University Press, 1973), p. 190; so also Stanley Goldberg, *Understanding Relativity: Origin and Impact of a Scientific Revolution* (Boston: Birkhäuser, 1984), p. 105; idem, "Putting New Wine in Old Bottles: The Assimilation of Relativity in America," in *The Comparative Reception of Relativity*, ed. Thomas F. Glick, Boston Studies in the Philosophy of Science 103 (Dordrecht: D. Reidel, 1987), pp. 15-21.
[40] G. Builder, "The Constancy of the Velocity of Light," *Australian Journal of Physics* 11 (1958): 457-480; reprinted in *Speculations in Science and Technology* 2 (1971): 422.
[41] See Edwin F. Taylor and John Archibald Wheeler, *Spacetime Physics* (San Francisco: W. H. Freeman, 1966), p. 34; cf. Hermann Bondi, *Relativity and Common Sense* (New York: Dover, 1964), p. 67; L. Marder, *Time and the Space-Traveller* (London: George Allen & Unwin, 1971), p. 78.
[42] Recall our discussion in chap. 5.
[43] J. Earman and M. Friedman, "The Meaning and Status of Newton's Law of Inertia and the Nature of Gravitational Forces," *Philosophy of Science* 40 (1973): 341.

Lorentz-Poincaré approach is simply passed over in benign neglect, under the assumption that any such theory is no longer a viable option. Spacetime realists are all of them Einsteinians and so take no notice of contemporary neo-Lorentzians. The irony of this situation is that the self-restriction of spacetime realists to relativistic interpretations of time and space is rooted, as we have seen, in the very positivism and verificationism which would be inimical to their own approach to questions of time and space. Having freed themselves from the metaphysical straitjacket of positivism, spacetime realists now need to ask why they should continue to philosophize only within the familiar walls of Einsteinian relativity and should not rather begin to explore alternatives in previously proscribed realms.

Occasionally arguments are offered on behalf of a spacetime interpretation of relativity theory which are relevant to a neo-Lorentzian approach, and these merit our attention. For example, oftimes space and time theorists have accused spacetime realists of a gratuitous hypostatization of what is only a geometrical representation of space and time. There is no reason to think that because a thing's spatio-temporal location can be plotted on a Minkowski diagram, therefore an entity *spacetime* exists anymore than because temperature and pressure can be similarly related in a diagrammatic way, therefore an entity *temperaturepressure* exists. In response to this charge d'Abro points to the four-dimensional metric of Minkowski spacetime as its distinguishing feature, in virtue of which it is not like the artificial temperaturepressure continuum.[44] The temperaturepressure continuum can be separated into two distinct elements, but Minkowski spacetime cannot.

> In short, the reality of spacetime arises from Minkowski's discovery that it was possible to discover an invariant distance between two points in spacetime, holding for all observers; and that it was impossible to define any such invariant distance in space alone or in time alone, showing that space and time by themselves were phantoms. These last two concepts must henceforth be considered jointly, and no longer as separate entities.[45]

One can, of course, adopt a spacetime approach to Newtonian theory, as Friedman has done. But d'Abro's point is that there is no special four-dimensional metric for such a continuum; it reduces to a mere geometric representation of the Galilean transformations.[46] Like the temperaturepressure continuum, no real unification takes place.

This is an odd argument on behalf of realism, since it treats Minkowski spacetime preferentially over Newtonian spacetime. The would-be Newtonian spacetime realist is dismissed along with the temperaturepressure realist, even though he holds to a spacetime realism, too. His spacetime is a mathematical fiction, we are told, whereas Minkowski spacetime is not, since the latter alone has a four-dimensional metric. But the reason why it has such a four-dimensional metric is precisely because of its preclusion of relations of absolute simultaneity. In the limit case where $c \to \infty$, the elsewhere region in spacetime is squeezed out and the

---

[44] A. d'Abro, *The Evolution of Scientific Thought*, 2d rev. ed. (1927; rep. ed.: n. p.: Dover Publications, 1950), p. 347.
[45] Ibid., p. 447.
[46] Ibid., pp. 472-473.

Newtonian dichotomy between a universal, absolute past and future is recovered.[47] Thus the very existence of a spacetime metric and a lightcone structure in spacetime depends upon there not existing such a global, absolute separation of past and future. The Minkowskian spacetime realist can hardly justify relativistic spacetime over the Newtonian spacetime realist's classical spacetime on the grounds that only the former's has a spacetime metric, since it is a necessary condition of relations of absolute simultaneity (which are the distinguishing mark of classical spacetime) that no such metric exist in Newtonian spacetime. But if the presence of such a metric serves merely to differentiate Minkowskian from Newtonian spacetime, rather than to justify realism about the former over realism about the latter, then neither can it justify Minkowskian spacetime realism over space and time realism, since the latter does not differ metrically from Newtonian spacetime realism.

Probably at the root of many physicists' rejection of a neo-Lorentzian approach to relativity theory is the deep-seated conviction that comes to expression in Einstein's aphorism: "Subtle is the Lord, but malicious He is not."[48] That is to say, if there exists a fundamental asymmetry in nature, then nature will not conspire to conceal it from us by precisely countervailing effects. D'Abro expresses this sentiment clearly:

> If Nature was blind, by what marvelous coincidence had all things been so adjusted as to conceal a velocity through the ether? And if Nature was wise, she had surely other subjects to attend to, more worthy of her consideration, and would scarcely be interested in hampering our feeble attempts to philosophize. In Lorentz's theory, Nature, when we read into her system all these extraordinary adjustments *ad hoc*, is made to appear mischievous; it was exceedingly difficult to reconcile one's self to finding such human traits in the universal plan.[49]

One difficulty with this objection is that it seems to be guilty of greatly over-exaggerating the extent of the alleged conspiracy. After all, SR is a restricted theory of relativity: it is only uniform motion relative to the privileged frame that fails to manifest itself. But in all other cases of motion, the absolute character of that motion is disclosed. This is not to say that acceleration or rotation proves the reality of privileged space, but it is to say that, given the classical concepts of time and space, nature does not at all conspire to conceal either absolute motion or the privileged space from us. Moreover, as we shall see in the next chapter, there are modern equivalents of the classical aether which serve to disclose a privileged frame. Indeed, when Einsteinians complain that no evidence of a privileged space and time exist, one wonders what it would take to convince them of the contrary. If no empirical evidence of a fundamental frame is incapable of being explained away, then the supposed failure of nature to disclose such a frame to us becomes trivial.

---

[47] Olivier Costa de Beauregard, *Time, the Physical Magnitude*, Boston Studies in the Philosophy of Science 99 (Dordrecht: D. Reidel, 1987), pp. 59-60. Cf. Lucas and Hodgson, *Spacetime and Electromagnetism*, p. 228-238.

[48] "*Raffiniert ist der Herr Gott, aber boshaft ist er nicht,*" remark of Albert Einstein during a visit to Princeton, upon being informed that D. C. Miller, a former colleague of Michelson, had claimed to have detected the ether wind (cited in Abraham Pais, *'Subtle is the Lord...': The Science and Life of Albert Einstein* [Oxford: Oxford University Press, 1982], pp. 113-114).

[49] D'Abro, *Evolution of Scientific Thought*, p. 138.

The more difficult it is for nature to provide evidence of the existence of a privileged frame, the less compelling the charge that she is conspiring against us to conceal it.

But even considered in abstraction from these wider considerations, why accept the assumption that fundamental asymmetries in nature must disclose themselves to us? This assumption is by no means obvious, as Martin Carrier explains:

> Science would be an easy matter if the fundamental states of nature expressed themselves candidly and frankly in experience. In that case we could simply collect the truths lying ready before our eyes. In fact, however, nature is more reserved and shy, and its fundamental states often appear in masquerade. Put less metaphorically, there is no straightforward one-to-one correspondence between a theoretical and an empirical state. One of the reasons for the lack of such a tight connection is that distortions may enter into the relation between theory and evidence, and these distortions may alter the empirical manifestation of a theoretical state. As a result, it is in general a nontrivial task to excavate the underlying state from distorted evidence.[50]

On a neo-Lorentzian theory, Carrier's general remarks on distortions of a theoretical state in its empirical manifestation are quite literally true, for the result of uniform motion relative to privileged space is length contraction and clock retardation, phenomena which it will be recalled, are every bit as real under the Einsteinian theory.[51] If it is in general difficult to excavate the underlying state of nature from distorted evidence, if nature's fundamental states often appear in masquerade, then why are the relativistic phenomena which mask the privileged frame improbable on a classical ontology?

As for d'Abro's concern with finding "human traits in the universal plan," the neo-Lorentzian might plausibly appeal to the Anthropic Principle in response.[52] According to that principle, features of the universe can only be judged in their correct perspective when due allowance has been made for the fact that certain features of the universe are necessary if it is to contain observers like ourselves.[53] Since our existence depends on the maintenance of equilibrium states within us, the Lorentz-FitzGerald contraction and clock retardation are necessary pre-requisites of our existence as observers. Thus, nature's alleged conspiracy, when seen in anthropic perspective, seems much less mischievous. Unless it can be shown that length contraction and clock retardation are extremely improbable (and thus evidence of the universe's fine-tuning for intelligent life), then we should hardly be

---

[50] Martin Carrier, "Physical Force or Geometrical Curvature?" in *Philosophical Problems of the Internal and External Worlds*, ed. John Earman, Allen I. Janis, Gerald J. Massey, and Nicholas Rescher (Pittsburgh: University of Pittsburgh Press, 1993), p. 3. Tim Maudlin has emphasized and illustrated Carrier's point in his *Quantum Non-Locality and Relativity*, Aristotelian Society Series 13 (Oxford: Blackwell, 1994). He surveys what he characterizes as the "teratological collection" of theories attempting to explain the Bell Inequalities and integrate the EPR results with relativity theory and concludes, "One way or another God has played us a nasty trick" (Ibid., p. 241). One cannot dismiss neo-Lorentzian relativity on the grounds that it would be deceptive, since partisans of each theory could say the same of rival positions. "...the real challenge falls to the theologians of physics, who must justify the ways of a Deity who is, if not evil, at least extremely mischievous" (Ibid., p. 242).
[51] It is perhaps noteworthy that d'Abro holds that "The Fitzgerald contraction is no longer a real physical contraction, as it was assumed to be in Lorentz's theory" (d'Abro, *Evolution of Scientific Thought*, p. 151).
[52] I owe this point to Robin Collins.
[53] John D. Barrow and Frank J. Tipler, *The Anthropic Cosmological Principle* (Oxford: Clarendon Press, 1986), p. 15.

surprised at nature's "conspiracy." In any case, given the theistic perspective from which we approach these issues, even fine-tuning should not surprise us. Given that God designed the universe to contain intelligent beings like us, we should expect that He will have chosen laws of nature that serve to maintain the equilibrium essential to our existence. Even if, as d'Abro says, Nature is blind, God is not; and if Nature is not wise, God is. It is not Nature, then, who is concerned with our feeble selves, who deems us worthy subjects to attend to, but the Creator and Sustainer of the universe who is mindful of man. *Raffiniert ist der Herr Gott, barmherzig ist er auch.*

Earman and Friedman, in defense of spacetime realism over the classical ontology, have asserted that "neglect of spacetime structure in the classical context has led to a number of philosophical errors and oversights."[54] Presumably, they have reference, for example, to classical theorists' overlooking Galilean spacetime, which preserves absolute time without absolute space. But such examples serve to show only the heuristic utility of a spacetime methodological approach to questions of space and time. In the same way that possible worlds semantics is an illuminating and perhaps indispensable tool for exploring questions of modality, but does not commit one *ipso facto* to modal realism, so also spacetime methodology can enlighten without implying a spacetime ontology. The same point may be made concerning the avoidance of "philosophical errors"—though here one must add that a spacetime approach might also lead one to commit philosophical errors which a space and time approach would have avoided. For example, attempts to relativize acceleration and rotation by appeal to four-dimensional structures come to mind, representations which may be of heuristic value but have also led some thinkers to conclude erroneously that absolute motions such as Newton pointed to have been eliminated. Or again, due to the isomorphism of the space and time dimensions in spacetime, which differ only in sign, erroneous philosophical conclusions about the interchangeability of time and space, as in the Schwarzschild metric or the Hartle-Hawking cosmology, might be drawn, which a space and time approach would perhaps prevent. To return to "oversights", the neo-Lorentzian will charge that a perspectival, four-dimensional approach to relativistic phenomena has resulted in the neglect of the search for the dynamic causes of time dilation and length contraction, a major oversight. The obvious difficulty here is that one man's insights are another man's oversights, and it is difficult to tell which one is correct without independent justification of one point of view.

Earman elsewhere argues that even if spacetime can be ignored in the context of classical physics, one cannot neglect it in the context of relativistic physics, since spacetime is not uniquely separable into space and time.[55] In fact, he points out, there are relativistic spacetimes which resist every effort to separate them into a three-dimensional space enduring through a one-dimensional time. But Earman is speaking here of GR spacetimes, not SR spacetime; we shall have more to say about time in the context of GR in the next chapter. So far as SR is concerned, Einstein's

---

[54] Earman and Friedman, "Meaning and Status," p. 329; cf. John Earman, "Spacetime, or How to Solve Philosophical Problems and Dissolve Philosophical Muddles without Really Trying," *Journal of Philosophy* 67 (1970): 259-277.
[55] Earman, "Spacetime," pp. 259-260; cf. Earman and Friedman, "Meaning and Status," p. 352.

original formulation of the theory shows that a space and time approach is necessarily equivalent to a spacetime approach. Moreover, one cannot automatically assume that just because a particular spacetime can be modeled that such a spacetime represents a realistic possibility (think, for example, of imaginary spacetimes). One might well question whether models such as Gödel's, alluded to by Earman, which permit pathological results like closed timelike loops, are anything more than mathematical curiosities. In any case, we must not forget that relativity theory concerns physical time only, not metaphysical time. In a universe such as Gödel's the impossibility of delineating a one-parameter family of spacelike hypersurfaces covering spacetime would only show that there exists no physical measure of God's metaphysical time in such a universe. There would be an absolute present and absolute simultaneity, even if it were physically impossible to gather present events into a single hypersurface. Hence, the possibility of such relativistic spacetimes does not suffice to establish the ontological reality of spacetime.

It has been suggested that a neo-Lorentzian interpretation of SR is disfavored because of the excessive spacetime structure it posits with respect to the laws of motion.[56] Recognizing the failure of verificationist critiques of absolute space on the basis of its unobservability, Earman, for example, suggests that "this objection of unobservability is more accurately stated as an objection based on Occam's razor."[57] What Earman has in mind are criteria which would serve to establish what sort of structure spacetime is endowed with. He proposes two "symmetry principles" which he presents as conditions of adequacy on theories of motion:

SP1: Any dynamical symmetry of a theory T is a space-time symmetry of T.

SP2: Any space-time symmetry of a theory T is a dynamical symmetry of T.

What justification is there for these criteria? Earman explains, "Behind both principles lies the realization that laws of motion cannot be written on thin air alone, but require the support of various space-time structures. The symmetry principles then provide standards for judging when the laws and the space-time structure are appropriate to one another."[58] With respect to (SP1), which will be crucial in the case at hand, Earman writes,

> The motivation for (SP1) derives from combining a particular conception of the main function of laws of motion with an argument that makes use of Occam's razor. Laws of motion, at least in so far as they relate to particles, serve to pick out a class of allowable or dynamically possible trajectories. If (SP1) fails, the same set of trajectories can be picked out by the laws working in the setting of a weaker space-time structure. The theory that fails (SP1) is thus using more space-time structure than is needed to support the laws, and slicing away this superfluous structure serves to restore (SP1).[59]

A prime example of (SP1) at work is the demonstration that Newtonian mechanics does not, in fact, require Newtonian spacetime, but only Galilean spacetime. While

---

[56] This has been suggested to me by Yuri Balashov.
[57] John Earman, *World Enough and Space-Time* (Cambridge, Mass.: MIT Press, 1989), p. 48.
[58] Ibid., p. 486.
[59] Ibid., pp. 46–47.

188   CHAPTER 9

the spacetime symmetries of the theory are Newtonian, the dynamic symmetries are Galilean, a "clear violation" of (SP1).[60] By excising absolute space, one produces a theory whose spacetime symmetries and dynamic symmetries are both Galilean, in compliance with (SP1). Now it might similarly be argued—though Earman does not do so—that a neo-Lorentzian theory also violates (SP1) in that it posits more structure to spacetime, such as hyperplanes of absolute simultaneity, than is necessary to explain the symmetries required by relativistic laws of motion, that is, the symmetries of the Lorentz-Poincaré group. Therefore, such structures should be excised by Ockham's razor.

Now Earman is undoubtedly on to something important here. One should like to have some sort of constraint upon the postulation of gratuitous spacetime structures. But the difficulty with (SP1) and (SP2) is that they are too restrictive, or to put the point another way, they can be overridden by considerations broader than the laws of motion. Indeed, one of the central contentions of our present study is that questions of time and space are metaphysical in character, so that such considerations cannot be ignored. As we have seen, the foundations of Newtonian time and space were laid in Newton's theism, as Earman acknowledges, but Earman simply ignores such theological considerations, choosing instead to put a "modern gloss" on Newton's views.[61] Newton's postulation of absolute space was not gratuitous but based, rightly or wrongly, in his conception of divine omnipresence. I think that Newton's conception of God's immensity was flawed; but it was not irrelevant, and I have elsewhere argued at length that his position on divine temporality and, hence, absolute simultaneity, is eminently defensible. Such metaphysical considerations qualify any force which Earman's symmetry principles might possess.

In any case, it needs to be understood that Earman's principles *presuppose* a spacetime ontology and therefore cannot be employed to justify a spacetime interpretation of SR over a neo-Lorentzian space and time interpretation. Earman employs a spacetime approach to all the theories he considers, speaking not only of Newtonian spacetime but even of Machian spacetime. This involves him in frequent anachronisms: he speaks of Newton's "insistence that the structure of space-time is immutable;" he says that "If we take Newton at his word in the Scholium, the space-time setting for the theory of motion and gravitation of the *Principia* is supposed to be a full Newtonian space-time...;" he explains that for Newton "the space-time is given once and for all as an emanative effect of God...."[62] Of course, Newton himself insisted on and believed none of this, since his was a 3+1 ontology involving physical objects enduring through time. His laws of motion are not supported by various spacetime structures, as Earman assumes, for there are on Newton's view no such structures. Time and space have intrinsic structure, and motion is to be explained, not in terms of the geometrical structure of spacetime, but in terms of forces acting on bodies in space enduring through time. Earman's principles, then, take for granted a spacetime realism. At best, therefore, with respect to SR, Earman's symmetry principles could only serve to justify a

---

[60] Ibid., p. 48.
[61] Ibid., p. 10.
[62] Ibid., pp. 35-36, 45, 48.

Minkowskian spacetime realism over a Lorentzian spacetime realism. But if only a single slice of spacetime actually exists, then the question becomes whether the relativity interpretation is preferable to the neo-Lorentzian interpretation. And neo-Lorentzians and spacetime realists concur that the relativity interpretation is deficient. In short, one's ontology may warrant postulating more structure to space and time than symmetry principles alone would allow.

GR raises a further ostensible advantage of spacetime realism: in GR gravitation is understood in terms of the geometry of spacetime. As Nerlich urges, "...not only is spacetime curvature the fundamental explanatory concept of the theory, but the idea of spacetime geometry is actually used to reduce causal explanation by gravitational force in space during time."[63] Nerlich gives the example of orbital planetary motion, which on the space and time ontology would have to be regarded as motion of an object under a gravitational force. But on a spacetime ontology we can still apply Newton's first law to it in spacetime, seeing its trajectory as the geodesic path of a force-free body, thus reducing gravitation to spacetime curvature.

The question raised by the geometrization of gravitation in GR is whether this is to be understood instrumentally or realistically. For what it is worth, most physicists are apparently content to take the theory instrumentally, interpreting spacetime curvature as a geometrical representation of gravitation. According to Arthur Fine, few working, knowledgeable scientists give credence to the realist existence claims for entities like the four-dimensional manifold and associated tensor fields in GR; rather GR is seen as "a magnificent organizing tool" for dealing with certain gravitational problems in astrophysics and cosmology: "most who actually use it think of the theory as a powerful instrument, rather than as expressing a 'big truth'."[64] I think we can safely say that no disadvantage arises from treating the geometrization of gravitation as a heuristic device only.

On the contrary, it can be argued that a realist interpretation of spacetime actually obscures our physical understanding of nature by substituting geometry for a physical force, thereby impeding progress in connecting gravitational theory to particle physics. In the preface to his text *Gravitation and Cosmology* Steven Weinberg reflects,

---

[63] Graham Nerlich, *What Spacetime Explains* (Cambridge: Cambridge University Press, 1994), p. 181.
[64] Arthur Fine, *The Shaky Game: Einstein, Realism, and the Quantum Theory* (Chicago: University of Chicago Press, 1986), p. 123. See the statement of Steven Weinberg, *Gravitation and Cosmology: Principles and Explications of the General Theory of Relativity* (New York: John Wiley & Sons, 1972), p. 147:

> "...the nonvanishing of the tensor $R_{\lambda\mu\nu\kappa}$ is the true expression of the presence of a gravitational field.... Riemann...introduced the curvature tensor $R_{\lambda\mu\nu\kappa}$ to generalize the concept of curvature to three or more dimensions. It is therefore not surprising that Einstein and his successors have regarded the effects of a gravitational field as producing a change in the geometry of space and time. At one time it was even hoped that the rest of physics could be brought into a geometric formulation, but this hope has met with disappointment, and the geometric interpretation of the theory of gravitation has dwindled to a mere analogy, which lingers in our language in terms like 'metric,' 'affine connection,' and 'curvature,' but is not otherwise very useful. The important thing is to be able to make predictions about the images on the astronomers' photographic plates, frequencies of spectral lines, and so on; and it simply doesn't matter whether we ascribe these predictions to the physical effect of gravitational fields on the motions of planets and photons or to a curvature of space and time."

190   CHAPTER 9

> There was another, more personal reason for my writing this book. In learning general relativity, and then in teaching it to classes at Berkeley and M.I.T., I became dissatisfied with what seemed to be the usual approach to the subject. I found that in most textbooks geometric ideas were given a starring role, so that a student who asked why the gravitational field is represented by a metric tensor, or why freely falling particles move on geodesics, or why the field equations are generally covariant would come away with an impression that this had something to do with the fact that spacetime is a Riemannian manifold.
>
> Of course, this *was* Einstein's point of view, and his preeminent genius necessarily shapes our understanding of the theory he created. However, I believe that the geometrical approach has driven a wedge between general relativity and the theory of elementary particles. As long as it could be hoped, as Einstein did hope, that matter would eventually be understood in geometrical terms, it made sense to give Riemannian geometry a primary role in describing the theory of gravitation. But now the passage of time has taught us not to expect that the strong, weak, and electromagnetic interactions can be understood in geometrical terms, and too great an emphasis on geometry can only obscure the deep connections between gravitation and the rest of physics.[65]

Weinberg believes that the theory of gravitational radiation provides "a crucial link" between GR and the microscopic frontier of physics, since radiative solutions of Einstein's equations lead to the notion of a particle of gravitational radiation, the so-called graviton.[66] The geometrical approach of spacetime realism only impedes our gaining a more integral understanding of physics. Riemannian geometry, in Weinberg's view, should be understood "only as a mathematical tool" and "not as a fundamental basis for the theory of gravitation."[67]

In sum, while we may agree that a spacetime interpretation of SR is in some respects, at least, superior to a relativity interpretation of SR, there do not seem to be comparably good reasons to prefer a spacetime interpretation of SR over a neo-Lorentzian conception of space and time.

On the other hand, I think we do have good reasons for rejecting spacetime realism and, therefore, a spacetime interpretation of SR. Inherent to the concept of

---

[65] Weinberg, *Gravitation and Cosmology*, p. vii.
[66] Ibid., p. 251. Cf. Rovelli's judgment:
> "Einstein's identification between gravitational field and geometry can be read in two alternative ways:
> i. as the discovery that the gravitational field is nothing but a local distortion of spacetime geometry; or
> ii. as the discovery that *spacetime geometry is nothing but a manifestation of a particular physical field*, the gravitational field.
>
> The choice between these two points of view is a matter of taste, at least as long as we remain within the realm of nonquantistic and nonthermal general relativity. I believe, however, that the first view, which is perhaps more traditional, tends to obscure, rather than enlighten, the profound shift in the view of spacetime produced by general relativity.*

> * ...The recent efforts to construct a quantum theory of gravity, or a unified theory of all interactions (for instance, via string theory), or a thermodynamics of the gravitational field, and other efforts, suggest that the second point of view, namely viewing the metric/gravitational field as a field like any other, is the fruitful one" (Carlo Rovelli, "Halfway through the Woods: Contemporary Research on Space and Time," in *The Cosmos of Science*, ed. J. Norton and J. Earman [Pittsburgh: University of Pittsburgh Press, 1998], pp. 193-194, 219).

[67] Weinberg, *Gravitation and Cosmology*, p. viii.

spacetime is the indissoluble unification of space and time into a four-dimensional continuum. Hence, Minkowski's pronouncement that "Henceforth, space by itself, and time by itself, are doomed to fade away into mere shadows, and only a kind of union of the two will preserve an independent reality."[68] It seems to me, however, that we have powerful metaphysical grounds for believing that time can exist independently of space. In making his pronouncement, Minkowski forgot Kant's insight that time is a form applicable not only to the external world, but to consciousness as well.[69] Concerning Minkowski spacetime, Wenzl cautions:

> From the standpoint of the physicist, this is a thoroughly consistent solution. But the physicist will (doubtless) understand the objection, raised by philosophy, that time is by no means a merely physical matter. Time is, as Kant put it, the form not merely of our outer sense but also of our inner sense.... Should our experiences of successiveness and of memory be mere illusion...?[70]

A series of mental events alone is sufficient to set up a temporal sequence.[71] Thus, if we imagine God's counting down to the moment of creation: "..., 3, 2, 1, *Fiat lux*!" then the beginning of spacetime would be preceded by a metaphysical time associated with the mental events of counting which would be wholly independent of space. Whether such a count-down can be beginningless or whether metaphysical time must itself have also had a beginning need not concern us here; the point is that physical events having spatial co-ordinates are clearly not a necessary condition of temporal events. This is especially evident in the case of God, since He is an unembodied Mind with whose mental states it is impossible to associate any brain states. If He experiences a succession of contents of consciousness, He is clearly in metaphysical time even if space does not exist. Pointing out that there are two percepts which give birth to the concept of time, one in the physical realm and one in the mental, E. A. Ramige contends,

> The spacetime concept is found to be ever more of a necessity in the theories of modern physics. However, necessary as it is in certain realms, that does not thereby rule out the separate concepts of space and time, any more than the concept of the area of a figure rules out the separate concepts of length and breadth. In the non-physical realm which we shall go into in taking up the concept of God it is time rather than space that is important. Descartes had stressed that thought is a separate attribute of reality from

---

[68] H. Minkowski, "Space and Time," in *The Principle of Relativity*, by A. Einstein, *et al.*, with Notes by A. Sommerfeld, trans. W. Perrett and G. B. Jeffery (1923; rep. ed.: New York: Dover Publications, 1952), p. 75.
[69] Immanuel Kant, *Immanuel Kant's "Critique of Pure Reason,"* trans. Norman Kemp Smith (London: Macmillan, 1970), A34; B51 (p. 77).
[70] A. Wenzl, "Einstein's Theory of Relativity, Viewed from the Standpoint of Critical Realism, and its Significance for Philosophy," in *Albert Einstein: Philosopher-Scientist*, ed. P. A. Schilpp, Library of Living Philosophers 7 (LaSalle, Ill.: Open Court, 1949), pp. 587-588.
[71] A point emphasized by Reichenbach:
> "An act of thought is an event and therefore defines a position in time. If my experiences are always produced within the framework of a 'now,' that means that each act of thought defines a point of reference. We cannot escape the 'now' because the attempt to escape signifies an act of thought and therefore defines a 'now'" (Hans Reichenbach, "Les fondements logiques de la mécanique des quanta," *Annales de l'Institut Henri Poincaré* 13 [1952]: 157).

extension. 'What is conscious is not spatial and what is spatial is not conscious.' It is still held that in the mental realm space plays a less important role than time.[72]

Indeed, time is more fundamental than space, as Lucas is wont to emphasize.[73] For while a series of mental events is a sufficient condition of time, it is not sufficient to constitute space. Since God is incorporeal, He would not be spatial prior to the creation of the universe even though He were temporal. Thus, on the orthodox view of God, time is more fundamental than space, since God's mere thinking discursively is sufficient for the existence of time, but not of space.

Reinforcing one's scepticism about spacetime realism is that fact that in Minkowski spacetime the spacetime interval between timelike separated events takes imaginary values. Christensen comments,

> This is disturbing in the first place because imaginary numbers seem like a paradigm of a convenient conceptual fiction. They were originally invented to make it the case that all quadratic equations, not just some, would have two roots. The serious question, however, is not whether there are imaginary numbers—numbers being a peculiar lot of entities to begin with—but whether there could really be such a thing as mathematically imaginary *physical quantities*.... No one doubts the convenience of computing with such quantities, even in an eminently practical field such as electrical engineering; but in the end the imaginary parts of the result are discarded. Could it be otherwise in SR, rationally?[74]

On a relativity interpretation of SR, it is easy to dismiss this feature of Minkowski spacetime as a mathematical feature which does not correspond to any physical property in the world. But such a recourse is not open to the spacetime realist, who prides himself precisely on the interval as a luminous indication of space and time's unification. The spacetime interval is as real a magnitude—nay, more real, if Minkowski is to be believed—than the distance from New York to Los Angeles. But what sense is there in affirming that the spacetime separation of, say, my having lunch and my having dinner is imaginary?

In addition to these difficulties, spacetime realism raises a host of problems due to its entailment of the doctrine of perdurance, or the existence of four-dimensional objects comprised of temporal parts. Since these also constitute independent objections to a B-Theory of time, I have chosen to discuss such objections elsewhere.[75] In fact, commitment to an A-Theory of time, and thus to the objectivity of tense and temporal becoming, is itself sufficient for the rejection of spacetime realism. Thus, with little to commend spacetime realism over a neo-Lorentzian conception of space and time and with powerful objections lodged against it, we may conclude that there is no reason to adopt a spacetime interpretation of SR rather than a neo-Lorentzian approach to the problems of relativity theory. I can only agree with David Bohm when he complains that a Minkowski diagram is

---

[72] Eldon Albert Ramige, *Contemporary Concepts of Time and the Idea of God* (Boston: Stratford, 1935), pp. iii-iv.
[73] J. R. Lucas, *A Treatise on Time and Space* (London: Methuen, 1973), p. 3.
[74] F. M. Christensen, *Spacelike Time* (Toronto: University of Toronto Press, 1993), pp. 260-261.
[75] See my *Tenseless Theory of Time*, Pt. 2, Sec. 1.

"misleading" in showing the past and future as existent[76] and with Max Black when he writes,

> ...this picture of a 'block universe,' composed of a timeless web of 'world-lines' in a four-dimensional space, however strongly suggested by the theory of relativity, is a piece of gratuitous metaphysics.... Here, as so often in the philosophy of science, a useful limitation in the form of representation is mistaken for a deficiency in the universe.[77]

Minkowski's four-dimensional, mathematical space serves as a convenient calculational and diagrammatical aid, a mathematical *Hilfsmittel*, but says absolutely nothing about ontology. As Arzeliès states, "The four-dimensional continuum should therefore be regarded as a useful tool, and not as a physical 'reality'."[78] Since time can exist independently of space, Minkowski's spacetime is at best a representation of physical time and space as described by the equations of SR and cannot pretend to imply a four-dimensional ontology. Minkowski spacetime is a mathematical space in which the Lorentz transformation of coordinates can be diagrammatically displayed. Since it is analytically connected with the propositions of the theory it displays, it cannot *explain* anything in the theory but only exhibit the theory's content graphically. Diagrams are useful in mathematics, as Plantinga says, because they enable us to see connections, entertain propositions, and resolve questions that would otherwise be seen, entertained, or resolved only with the greatest difficulty; but they do not serve as explanations.[79] In the same way, spacetime diagrams are indispensable aids in providing a visualization of the propositional content of the theory without serving as explanations of why time dilation and length contraction occur.[80] Sellars, noting that Minkowski spacetime is simply a metrical picture of the location of events, not a depiction of things which come to be and cease to be, concludes, "It has too often been supposed by philosophers (and physicists) who discuss the Special Theory of Relativity that the Minkowski mathematical apparatus automatically carries with it a commitment to an ontology of 'events.' This...is simply a mistake."[81]

---

[76] David Bohm, *The Special Theory of Relativity*, Lecture Notes and Supplements in Physics (New York: W. A. Benjamin, 1965), pp. 174-175.

[77] Max Black, review of *The Natural Philosophy of Time*, by G. J. Whitrow, *Scientific American* 206 (April 1962): 181-182.

[78] Henri Arzeliès, *Relativistic Kinematics*, rev. ed. (Oxford: Pergamon Press, 1966), p. 258; cf. Mary F. Cleugh, *Time and Its Importance in Modern Thought*, with a Forward by L. Susan Stebbing (London: Methuen, 1937), pp. 66-67; Herbert Dingle, *The Special Theory of Relativity* (London: Methuen, 1940), pp. 43-44.

[79] Alvin Plantinga, "Reply to Robert Adams," in *Alvin Plantinga*, ed. James Tomberlin and Peter Van Inwagen, Profiles 5 (Dordrecht, Holland: D. Reidel, 1985), p. 378.

[80] For a graphic explanation of how non-explanatory such accounts are, see C. Misner, K. S. Thorne, and J. A. Wheeler, *Gravitation* (San Francisco: W. H. Freeman, 1973), pp. 32-33. See also J. D. North, "The Time Coordinate in Einstein's Restricted Theory of Relativity," *Studium Generale* 23 (1970): 222.

[81] Wilfrid Sellars, "Time and the World Order," in *Scientific Explanation, Space, and Time*, ed. Herbert Feigl and Grover Maxwell, Minnesota Studies in the Philosophy of Science 3 (Minneapolis: University of Minnesota Press, 1962), p. 578. Robert Geroch says, "If you like, 'four dimensions' is just a convenient way of describing the world and thinking about the world, nothing more....these pictures are merely a convenient way of looking at (not a necessary ingredient of) spacetime, while spacetime is merely a convenient way of looking at (not a necessary ingredient of) nature" (Robert Geroch, *General Relativity from A to B* [Chicago: University of Chicago Press, 1978], pp. 12-13). Dugas concludes,

CHAPTER 9

CONCLUSION

It seems to me, therefore, that despite the widespread aversion to a neo-Lorentzian interpretation of relativity theory, such antipathy is really quite unjustified. Admitted on all sides to be empirically equivalent to the Einsteinian interpretation, the neo-Lorentzian interpretation is neither *ad hoc* nor more complicated than its rival. The physical effects it posits are no less real in the received version, only there they appear as axiomatic deductions lacking causal explanations. Indeed, its fecundity in opening the question about physical causes is an important advantage of the neo-Lorentzian interpretation. Therefore, our philosophically grounded preference for the neo-Lorentzian interpretation, however distasteful to the majority of physicists, cannot count against the cogency of our position concerning the relation of absolute to relativistic time.

Of course, one could go on working within the theoretical framework of Einstein's theory, being accustomed to so working and thereby retaining the advantage of using a received view of the scientific community, and yet consistently affirm what I have said concerning God, metaphysical time, and measured time simply by eschewing a realist understanding of SR. So long as one accords to his theory a purely instrumentalist interpretation, one can employ it on pragmatic grounds without regarding it as even approximately true. But if one is interested in being a scientific realist on matters of time and space, then, in view of God's existence in metaphysical time and its implication of a set of absolutely simultaneous events being created by Him in the "now" of metaphysical time, one ought to affirm a neo-Lorentzian interpretation of the SR. Moreover, a moment's reflection reveals that the real work in our argument for a neo-Lorentzian interpretation of the formalism of SR is done, not by Newton's theistic hypothesis, by the metaphysic of an A-Theory of time. For if an A-Theory of time is correct, then Minkowski's spacetime interpretation of the SR formalism cannot be correct. Thus, given an A-Theory of time, our choice is between Einstein's relativity interpretation and a neo-Lorentzian interpretation. But we have already seen how fantastic and explanatorily impoverished the former is. Therefore, the neo-Lorentzian approach is to be preferred. Therefore, even thinkers who do not share our commitment to Newton's temporal theism but who are convinced of the objective reality of tense and/or temporal becoming ought to adopt a neo-Lorentzian approach to special relativity.

"Therefore it seems that Minkowski, carried away by a very natural enthusiasm for the remarkable geometrical synthesis that he had discovered, had to some extent gone beyond the relativistic doctrine, which does not in any way forbid that an observer should reason and calculate, as in daily life, in terms of space and time" (René Dugas, *A History of Mechanics*, with a Foreword by Louis de Broglie, trans. J. R. Maddox [New York: Central Book Company, 1955], p. 501).

CHAPTER 10

GOD'S TIME AND GENERAL RELATIVITY

A final issue now needs to be engaged: Given our commitment to theism, do we have some idea of what measured time coincides with God's metaphysical time, or in other words, what clock time is the true time? The answer to this question will take us from Special into General Relativity, as we seek to gain a cosmic perspective on time.

THE GENERAL THEORY OF RELATIVITY

The Special Theory is a restricted theory, since it concerns only reference frames which are in a state of uniform motion. Although the Special Theory can be adapted so as to analyze the notion of non-inertial (that is, accelerating or decelerating) frames of reference, it does not serve to relativize the motion of such frames by rendering them equivalent to inertial frames. As Einstein wrote, when we speak of the "special principle of relativity"—namely, that "If a system of co-ordinates K is chosen so that, in relation to it, physical laws hold good in their simplest form, the *same* laws also hold good in relation to any other system of co-ordinates K' moving in uniform translation relative to K"—, then "The word 'special' is meant to intimate, that the principle is restricted to the case when K' has a motion of uniform translation relatively to K, and that the equivalence of K' and K does not extend to the case of non-uniform motion of K' relatively to K."[1]

Troubled by the non-equivalence of inertial and non-inertial frames, Einstein labored for the decade subsequent to the publication of his SR on a General Theory of Relativity, which had as its aim the enunciation of a General Principle of Relativity which would serve, in turn, to render physically equivalent all inertial and non-inertial frames alike. He completed his General Theory by 1915 and in the following year published his definitive statement of it in his article "The Foundations of General Relativity Theory," which appeared in *Annalen der Physik*. He boasted that his theory "takes away from space and time the last remnant of physical objectivity."[2] It was supposed to be, in effect, the final destruction of Newton's absolute space and time.

In fact, however, Einstein was only partially successful in achieving his aims. He did not succeed in enunciating a tenable General Principle of Relativity after the pattern of the Special Principle, nor was he able to show the physical equivalence of all reference frames. He did succeed in drafting a revolutionary and complex theory of gravitation, which has been widely hailed as his greatest intellectual achievement.

---

[1] A. Einstein, "The Foundations of General Relativity Theory," in *General Theory of Relativity*, ed. C. W. Kilmister, Selected Readings in Physics (Oxford: Pergamon Press, 1973), pp. 141-172. The original paper appeared in *Annalen der Physik* 49 (1916): 769. On the ambiguity of Einstein's statement of the Special Principle of Relativity, recall the discussion on pp. 25-27.
[2] Ibid., p. 148.

The so-called General Theory of Relativity is thus something of a misnomer: it is really a theory of gravitation and not an extension of the Special Theory of Relativity from inertial reference frames to all reference frames.[3] Since our interest is primarily in the conception of time in GR, we may forego an extensive exposition of the theory in favor of focusing on its implications for God's relationship to time.

In a conscious effort to cast GR as parallel to SR, Einstein opens his paper with an epistemological analysis of the Postulate of Relativity and the need for an extension thereof to non-inertial frames. He begins, "In classical mechanics, and no less in the special theory of relativity, there is an inherent epistemological defect which was, perhaps for the first time, clearly pointed out by Ernst Mach."[4] He illustrates this defect by means of the following *Gedankenexperiment* (Figure 10.1).

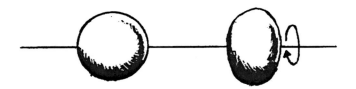

Figure 10.1. The sphere which becomes ellipsoidal is identified as the rotating sphere.

Two fluid spheres, S_1 and S_2, of the same size and nature hover freely in space in a gravitation-free environment. Let either mass, as judged by an observer at rest relative to the other mass, freely rotate with a constant angular velocity about a common axis. Let us now suppose that by means of measurements taken by instruments at rest relative to S_1 and S_2 respectively it is determined that the surface of S_1 is a sphere and the surface of S_2 an ellipsoid of revolution. "What is the reason for this difference in the two bodies?" queries Einstein. "No answer can be admitted as epistemologically satisfactory, unless the reason given is an *observable fact of experience*. The law of causality has not the significance of a statement as to the world of experience, except when *observable facts* ultimately appear as causes and effects."[5]

This *Gedankenexperiment* is obviously inspired by Newton's own illustration of the two globes connected by a cord, revolving about their common center of gravity (Figure 10.2).

---

[3] See the very frank discussion by Hermann Bondi, "Is 'General Relativity' Necessary for Einstein's Theory of Gravitation?" in *Relativity, Quanta, and Cosmology in the Development of the Scientific Thought of Albert Einstein*, ed. Francesco De Finis, 2 vols. (New York: Johnson Reprint Corp., 1979), pp. 179-186. According to Bondi, any notion of equivalence between inertial and accelerated observers is "physically meaningless," which goes to show "how void of significance any general principle of relativity must be." But because "a physically sound formulation of Einstein's theory of gravitation exists not involving the physically empty concept of general relativity," one may admire and embrace Einstein's theory of gravitation while rejecting his route to it. "It is perhaps rather late to change the name of Einstein's theory of gravitation, but general relativity is a physically meaningless phrase that can only be viewed as a historical memento of a curious philosophical observation."
[4] Einstein, "Foundations of General Relativity," p. 143.
[5] Ibid.

Figure 10.2. Newton's twin sphere experiment. The tension of the cord enables one to measure the absolute rotation of the system.

Newton argues that even if this rotating system were placed "in any immense vacuum, where nothing external or sensible existed with which the balls could be compared," still we could by measuring the tension of the cord find "both the quantity and the direction of this circular motion."[6] Thus, the rotating globes, like the more celebrated illustration in the *Scholium* two paragraphs earlier of the rotating bucket of water, serve to disclose to us absolute motion and, hence, absolute place (absolute motion being the translation of a body from one absolute place to another) and, hence, absolute space (absolute place being a part of absolute space which a body takes up).[7]

But Einstein's Machian verificationism will not permit such an explanation of the difference between the spheres. He writes,

> Newtonian mechanics does not give a satisfactory answer to this question. It pronounces as follows:—The laws of mechanics apply to the space R_1, in respect to which the body S_1 is at rest, but not to the space R_2, in respect to which the body S_2 is at rest. But the privileged space R_1 of Galileo, thus introduced, is a merely *factitious* cause, and not a thing that can be observed. It is therefore clear that Newton's mechanics does not really satisfy the requirement of causality in the case under consideration, but only apparently does so, since it makes the factitious cause R_1 responsible for the observable difference in the bodies S_1 and S_2.[8]

On the positivistic assumption that a causal statement is meaningless unless the cause as well as the effect be an observable fact of experience, the Newtonian appeal to absolute space is ruled out of court, since absolute space is not a thing that can be observed. I shall not belabor the point how utterly untenable and outmoded such a positivistic outlook is. Let us rather look at Einstein's alternative explanation.

---

[6] Isaac Newton, *The Principia*, trans. I. Bernard Cohen and Anne Whitman, with a Guide by I. Bernard Cohen (Berkeley: University of California Press, 1999), p. 414.
[7] Ibid., pp. 6-7.
[8] Einstein, "Foundations of General Relativity," pp. 143-144.

As Friedman explains,[9] Einstein was confronted with two possible routes of explaining non-symmetrical, differential effects, such as those evinced by the spheres, without recourse to absolute motion: (1) Extend the Principle of Relativity for inertial motions to the class of all motions, so that no differential effects would appear. The Special Principle requires that no differential effects arise between two bodies in relative inertial motion; an extension of this principle would require that S_1 and S_2 exhibit no such effects either. But (1) is untenable because such differential effects would in fact exist between S_1 and S_2. This alternative would therefore fall into that class of answers which, as Einstein puts it, "may be satisfactory from the point of view of epistemology, and yet be unsound physically, if it is in conflict with other experiences."[10] (2) Adopt Mach's Principle. Instead of regarding S_2 as in absolute motion with respect to R_1, we could regard S_2 as in merely relative motion with respect to some third class of objects, such as the fixed stars. According to Mach's Principle, Newtonian absolute motion is determined wholly by these distant masses. Were S_1 and S_2 to exist in empty space, there would be no differential effects in them. This is the alternative which Einstein adopts, but in so doing he misconstrues it as an extension of the Principle of Relativity. He writes,

> The satisfactory answer can only be that the physical system consisting of S_1 and S_2 reveals within itself no imaginable cause to which the differing behavior of S_1 and S_2 can be referred. The cause must therefore lie *outside* this system. We have to take it that the general laws of motion, which in particular determine the shapes of S_1 and S_2, must be such that the mechanical behaviour of S_1 and S_2 is partially conditioned, in quite essential respects, by distant masses which we have not included in the system under consideration. These distant masses and their motions relative to S_1 and S_2 must then be regarded as the seat of the causes (which must be susceptible to observation) of the different behaviour of our two bodies S_1 and S_2. They take over the rôle of the factitious cause R_1. Of all imaginable spaces R_1, R_2, etc., in any kind of motion relatively to one another, there is none which we may look upon as privileged *a priori* without reviving the above-mentioned epistemological objection. *The laws of physics must be of such a nature that they apply to systems of reference in any kind of motion.* In this way we arrive at an extension of the postulate of relativity.[11]

In this paragraph, Einstein makes a number of mistakes. First, Mach's Principle requires not just that the mechanical behavior of the spheres be partially conditioned by the distant masses, but that it be entirely determined by those masses insofar as effects relating to absolute acceleration and rotation are concerned. Second, even if such a strategy were successful, it would not yield an extension of the Principle of Relativity, for the differential effects remain real physical effects. Whether we take the mean matter of the universe to be at rest and S_2 rotating relatively to it or S_2 to be at rest and the average matter of the universe to be circling around it, the fact remains that there are produced in S_2 differential effects as a result of the gravitation of the distant masses of the universe. Finally, Mach's strategy simply fails to eliminate absolute motion. For according to Mach's Principle, S_2's rotation is to be explained as a rotation relative to the mean matter distribution of the universe. But what if the average matter of the universe is not itself at rest? Nothing prevents the

---

[9] See his lucid commentary in Michael Friedman, *Foundations of Spacetime Theories* (Princeton: Princeton University Press, 1983), pp. 204-215.
[10] Einstein, "Foundations of General Relativity," p. 143.
[11] Ibid., p. 144.

mean matter distribution of the universe from undergoing itself an absolute rotation. But then that rotation obviously cannot itself be explained as a rotation relative to the average matter of the universe. In the end Mach's Principle fails to eliminate absolute motion.

These confusions are compounded by Einstein's confirmatory argument for a General Principle of Relativity, namely, the alleged physical equivalence of arbitrarily accelerated reference frames and frames under the influence of gravitational fields. He invites us to envision a reference frame K' which is moving in uniformly accelerated motion relative to the Galilean reference frame K. A mass sufficiently distant from other masses and moving uniformly relative to K will have an accelerated motion relative to K'. Einstein asks,

> Does this permit an observer at rest relatively to K' to infer that he is on a 'really' accelerated system of reference? The answer is in the negative; for the above mentioned relation of freely movable masses to K' may be interpreted equally well in the following way. The system of reference K' is unaccelerated, but the spacetime region in question is in a gravitational field, which generates the accelerated motion of the bodies relatively to K'.[12]

This interpretation is possible, Einstein explains, because the gravitational field possesses the remarkable property of imparting the same acceleration to all bodies. The mechanical behavior of accelerated bodies is the same as that of stationary bodies in a gravitational field. Therefore, either K or K' can be regarded from the physical standpoint as being stationary. This well-known physical fact favors an extension of the Principle of Relativity.

In his commentary on Einstein's use of the so-called Principle of Equivalence as an argument for a General Principle of Relativity, Friedman notes that Einstein is perfectly correct that in the Newtonian Theory of gravitation, one cannot distinguish K from K'.[13] One would be thus far justified in enunciating the Principle of Equivalence

> E. All Galilean reference frames are physically equivalent (or physically indistinguishable).

If we supplement (E) with the Special Principle of Relativity

> R_2. If two frames are indistinguishable according to a spacetime theory T, they should be theoretically identical according to T,

then we should appear to have a General Principle of Relativity that relativizes acceleration just as the Special Principle relativized velocity.

But, says Friedman, this argument fails for two reasons. First, even if (E) were true, it is still too restrictive to yield a General Principle of Relativity because Galilean reference frames do not include rotating frames of reference. Rotating frames of reference can be physically distinguished from non-rotating frames by the presence of Coriolis forces, which unlike gravitational or inertial forces, depend on the velocity of the particle acted upon. There is no Principle of Equivalence which governs rotating as well as arbitrarily accelerating frames. But secondly, (E) is in

---

[12] Ibid., pp. 144-145.
[13] See Friedman, *Spacetime Theories*, pp. 191-204 for the following critique.

any case false because according to GR arbitrarily accelerating frames are not physically equivalent or indistinguishable. Only a certain class of accelerating frames within SR become physically indistinguishable within GR. This is because in GR the inertial frames in flat Minkowski spacetime are replaced with inertial trajectories of freely falling particles in curved Riemannian spacetime. GR introduces a variable curvature into the flat Minkowski manifold, the degree of curvature of a given region of spacetime being dependent on the distribution of mass and energy. The metric for Minkowski spacetime must therefore be replaced with a new metric

$$ds^2 = \sum_{ij=0}^{3} g_{ij} dx_i dx_j \tag{1}$$

where the g_{ij}'s are not constant. Straight lines in this curved manifold are geodesics and are construed as the world lines of freely falling particles. There are no inertial frames within GR because bodies moving inertially are those which freely follow four-dimensional "straight lines" or geodesics. These inertial trajectories thus play the same role in GR as did inertial frames in SR. Freely falling frames are equivalent to inertial frames only at a single spacetime point or on a single trajectory. SR and GR are thus infinitesimally equivalent at every point of spacetime, but they differ in how those points are connected into a spacetime manifold. The difference between SR and GR is that in the latter inertially moving bodies follow geodesics of a non-flat spacetime connection instead of geodesics of a flat connection as in the former. The inertial trajectories serve as a standard which gives rise in the General Theory to an absolute distinction between inertial motion and non-inertial motion. Inertial motions are those whose trajectories in spacetime are geodesics of the postulated connection, and non-inertial motions are those whose trajectories deviate from the geodesics. Thus, the Principle of Equivalence does not eliminate privileged reference frames or states of motion within GR. Reference frames whose motions describe geodesics in spacetime are privileged in that their motion is natural, whereas motion which deviates from geodesic lines is accelerated. In particular GR permits an absolute, cosmic rotation of matter, as the Gödel and Ozsváth-Schücking models illustrate.[14] Hence, the Principle of Equivalence does not serve to render all reference frames indistinguishable.

Einstein's overstatement of his case seems to result from his conflation of physical equivalence and general covariance. But as Friedman points out, the inference that two reference frames are physically equivalent if the laws of nature take the same form in both is a *non sequitur*.[15] The notions of "sameness of form" and covariance correspond to the notions of physical equivalence and relativity only in flat spacetime theories, but in non-flat spacetime theories this correspondence breaks down because there is no standard formulation of the theory in the usual sense. The only standard formulation is the generally covariant formulation holding

---

[14] Kurt Gödel, "A Remark about the Relationship between Relativity Theory and Idealistic Philosophy," in *Albert Einstein: Philosopher-Scientist*, ed. P. A. Schilpp, Library of Living Philosophers 7 (LaSalle, Ill.: Open Court, 1949), pp. 557-562; I. Ozsváth and E. Schücking, "Finite Rotating Universe," *Nature* 193 (1962): 1168-1169.
[15] Friedman, *Spacetime Theories*, pp. 207-214.

in all coordinate systems, and this is too weak a condition to guarantee physical equivalence of all reference frames. Friedman concludes,

> So Einstein's vision of a Leibnizian, or a relationalist, theory rests on a double confusion: a confusion of the strategy of Machianization with the strategy of relativization and a confusion of the notions of physical equivalence and relativity with the notions of sameness of form and covariance. This double confusion is reinforced to the point of irresistibility by the principle of equivalence....
>
> Clearly, then, general relativity realizes neither of our two strategies for relativizing motion. It does not satisfy a generalized or extended relativity principle, since inertially moving frames are just as distinguishable from accelerating and rotating frames as in all previous theories; and it does not conform to the strategy of Machianization, since absolute rotation has essentially the same status—namely, independence from external masses—as in all previous theories. In the end, therefore, general relativity does not solve Einstein's problem of the rotating globes (or, equivalently, Newton's problem of the rotating bucket). General relativity, like all previous 'absolutist' theories, predicts that if S_1 and S_2 were alone in the universe, it would still be possible for one and only one of them to experience distorting differential effects.[16]

The failure of Einstein's epistemological and physical arguments aimed at relativizing acceleration and rotation does not, however, entail the existence of Newtonian absolute space and time. For within the context of GR these "absolute" motions are conceived to exist, not with respect to some absolute space and absolute time, but rather with respect to spacetime. All events are embedded in a four-dimensional spacetime manifold. Within this manifold a privileged class of inertial trajectories representing the world lines of freely falling particles exists. A particle whose world line deviates from one of these four-dimensional straight lines is moving non-inertially. "Thus, acceleration and rotation are essentially four-dimensional notions, whereas rest and velocity are essentially three-dimensional," explains Friedman. "In moving from a three- to a four-dimensional 'container' or embedding space, we eliminate the latter notions while retaining the former."[17] This can be seen by recalling Galilean spacetime, in which absolute time exists but not absolute space. This theory distinguishes inertial from non-inertial motion, but concepts of absolute rest and absolute velocity find no place in the theory because a body which occupies the same spatial point on two planes of simultaneity relative to one reference frame will occupy two different points at the relevant times relative to a different reference frame in motion with respect to the first. An observer associated with the first reference frame would claim that the body is at rest, while an observer associated with the second would hold that it is in motion. Thus, the distinction between absolute rest and velocity is vanquished. But Galilean spacetime, while lacking absolute space, still makes sense of "absolute" motions like acceleration and rotation by interpreting these with respect to spacetime, rather than space. An object is accelerated just in case its world line deviates from a four-dimensional geodesic trajectory; an object is rotating just in case its world lines are "twisted" relative to the four-dimensional geodesic trajectories (Figure 10.3).

---

[16] Ibid., pp. 208-209, 210-211.
[17] Ibid., p. 17.

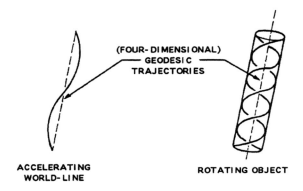

Figure 10.3. Acceleration and rotation as four-dimensional motions.

In moving from Galilean to relativistic spacetime, we drop absolute time as well as absolute space, but acceleration and rotation are still definable with respect to the four-dimensional spacetime manifold. Thus, although "absolute" motions—"absolute" in the sense of "non-relativized"—still exist within relativity theory, they do not guarantee the existence of absolute space and time, as Newton thought. Spacetime itself becomes, as Einstein held, a sort of relativistic ether which serves to define such motions, but is not itself a reference frame.[18] Of course, such considerations do not prove that absolute space and time do *not* exist, for one may reject, as we have done, the spacetime realism underlying the above analysis. Still this analysis does show that, *pace* Newton, non-relativistic motions do not of themselves necessitate the existence of absolute space and time.

COSMIC TIME

It might appear, therefore, that GR has nothing more to contribute to the understanding of time than SR. They differ simply over the presence of curvature in spacetime; if one adds a condition of flatness to GR, then SR results. But such a conclusion would be hasty, indeed, for GR serves to introduce into relativity theory a cosmic perspective, enabling one to draft cosmological models of the universe governed by the gravitational field equations of GR. Within the context of such cosmological models, the issue of time resurfaces dramatically.

Einstein himself proposed the first GR-based cosmological model in his paper, "Cosmological Considerations on the General Theory of Relativity" in 1917.[19] The model describes a spatially finite universe which possesses at every time t the

---

[18] A. Einstein, "Ether and the Theory of Relativity," in *Sidelights on Relativity* (New York: Dover Publications, 1903), pp. 16-17.

[19] A. Einstein, "Cosmological Considerations on the General Theory of Relativity," in *The Principle of Relativity*, by Albert Einstein, *et al.*, with Notes by A. Sommerfeld, trans. W. Perrett and G. B. Jeffery (rep. ed.: New York: Dover Publications, 1952), pp. 177-188.

geometry of the surface of a sphere in three dimensions with a constant radius R. The model is characterized by the static metric

$$ds^2 = -dt^2 + R^2 [dr^2 + \sin^2 r \, (d\theta^2 + \sin^2\theta d\phi^2)] \qquad (2)$$

Time, which is decoupled from space, extends from $-\infty$ to $+\infty$. Thus, spacetime takes on the form of a sort of four-dimensional cylinder, temporal cross-sections of which are the 3-spheres. In order to bring such a model into conformity with his field equations, Einstein was compelled to adopt a term Λ, the so-called cosmological constant, which counteracted gravitation and so preserved the static three-dimensional space through time. By setting $\Lambda > 0$, one generates a weak repulsion between bodies, which keeps the universe equiposed between gravitational collapse and cosmological expansion. Kanitscheider draws our attention to the sort of time coordinate which shows up in the metric of Einstein's model:

> It represents in a certain sense the restoration of the universal time which was destroyed by SR. In the static world there is a global reference frame, relative to which the whole of cosmic matter finds itself at rest. All cosmological parameters are independent of time. In the rest frame of cosmic matter space and time are separated. For fundamental observers at rest, all clocks can be synchronized and a worldwide simultaneity can be defined in this cosmic frame.[20]

Thus, cosmological considerations prompted the conception of a cosmic time which measures the duration of the universe as a whole.

Nor was this cosmic time limited to Einstein's static model of the universe. Expansion models, which trace their origin to de Sitter's 1917 model of an empty universe,[21] may also involve a cosmic time. De Sitter showed that Einstein's field equations may be satisfied by a gravitational field in the absence of any matter whatsoever in a world characterized by the metric

$$ds^2 = -dt^2 + e^{2tH} [dr^2 + r^2 (d\theta^2 + \sin^2\theta d\phi^2)] \qquad (3)$$

The de Sitter universe also appears to be static, but that is only because it is empty. If particles of matter are introduced into it, one discovers that de Sitter space is in fact expanding because the distance between any two arbitrary points given by specific values of r, θ, and ϕ increases with time by a factor of e^{tH} due to the repulsive force of the cosmological constant Λ. De Sitter spacetime resembles a pseudo-sphere with three-dimensional cross-sections of positive curvature. Its geodesics were increasingly distant from each other in the past, approach each other to a minimal distance at $t = 0$, and then re-separate out toward infinity in the future. Thus, in 1917 two GR-based cosmological models had been drafted: one in which a material world exists, but is static, and another which is dynamic, but lacks a material world—a situation which inspired Eddington's crack: "Shall we put a little

---

[20] Bernulf Kanitscheider, *Kosmologie* (Stuttgart: Philipp Reclam., Jun., 1984), p. 155. See also G. J. Whitrow, *The Natural Philosophy of Time*, 2d ed. (Oxford: Clarendon Press, 1980), pp. 283-284.
[21] Willem de Sitter, "On the Relativity of Inertia," in *Koninglijke Nederlandse Akademie van Wetenschappen Amsterdam. Afdeling Wis-en Natuurkundige Wetenschappen. Proceedings of the Section of Science* 19 (1917): 1217-1225.

motion into Einstein's world of inert matter or shall we put a little matter into de Sitter's *primum mobile*?"[22]

It was the Russian physicist Aleksandr Friedmann who put together these elements in 1922 to produce a cosmological model of an expanding, material universe.[23] His model universe was characterized by ideal homogeneity and isotropy and set $\Lambda = 0$ (which was equivalent to omitting it, thus preserving the form of Einstein's original equation). Friedmann's solution to Einstein's field equation was

$$\dot{R}^2 = \frac{8\pi G \rho R^2 - k}{3} \qquad (4)$$

where $\dot{R}^2$ is the rate of increase of the scale factor R with time, G is the gravitational constant, ρ is the matter density, and k is the curvature constant which may take the values of 0, 1, or −1. The most startling implication of the Friedmann model is that as one traces the expansion back in time the universe becomes increasingly dense until one arrives at a state of infinite density before which the universe did not exist. Indeed, this state represents a spacetime singularity at which all spatial and temporal dimensions become zero, so that it marks the boundary to physical time itself. Cosmic time could not exist at or prior to the singularity, so that cosmic time must be finite in the past, thereby implying a definite, finite age of the universe. It is difficult to exaggerate how amazing such a prediction was, for it revealed that, in the ideal case at least, the original GR equations implied the finitude of the past and *creatio ex nihilo*.

In 1930, one year after Edwin Hubble's explanation of the observed red-shift of galactic light as a Doppler effect due to a universal, isotropic galactic recession, Eddington demonstrated that the static Einstein model was radically unstable.[24] Even small changes in the density would upset the balance between gravitation and the cosmological constant, so that a cosmic expansion or collapse would result; moreover, so much as the mere transportation of matter from one part of the universe to another would cause the former region to expand and the latter to collapse. The following year Einstein recommended that the cosmological constant be dropped from the equations, later recalling it as "the greatest blunder of my life."[25] Thus, the Big Bang model of the universe—despite a later, temporary challenge from the Steady State model—came to be the controlling paradigm of GR-based cosmological models.

The nature of the cosmic time which measures the duration of the universe in such models deserves our further scrutiny. In constructing a cosmological model, one overlooks the inhomogeneity of matter on the galactic and sub-galactic scales and takes instead an extremely large-scale viewpoint of the universe, on which these

---

[22] Arthur S. Eddington, *The Expanding Universe* (Cambridge: Cambridge University Press, 1952), p. 46.
[23] A. Friedmann, "Über die Krümmung des Raumes," *Zeitschrift für Physik* 10 (1922): 377-386.
[24] A. S. Eddington, "On the Instability of Einstein's Spherical World," *Monthly Notices of the Royal Astronomical Society* 19 (1930): 668-678.
[25] Albert Einstein, quoted in George Gamow, *My World Line* (New York: Viking Press, 1970), p. 44.

inhomogeneities become negligible. On this very large scale, galaxies and galactic clusters can be viewed as particles of a homogeneous and isotropic dust or gas filling the universe. We may even choose to ignore the particulate structure of this gas and treat it instead as an idealized, perfect fluid, characterized by a four-dimensional velocity u, a mass-energy density ρ, and a pressure p. The 4-velocity u has reference to a hypothetical observer who is at rest relative to the material substratum in his region and who therefore observes the immediately proximate galaxies to have no mean motion. Such an observer is frequently referred to, after the manner of E. A. Milne, as a "fundamental observer" and his associated galactic particle as a "fundamental particle." The mass-energy density ρ also has reference to such a fundamental observer, being the material density and radiation density of the cosmological fluid as observed by a fundamental observer in his reference frame. The pressure p is the kinetic pressure of the galaxies determined by both matter and radiation.

In the Friedmann model, the cosmological fluid is homogeneous and isotropic. But how are these notions to be understood? Intuitively, something is homogeneous if it is everywhere the same at a given moment of time. But when we take a cosmological perspective, this notion necessitates the existence of a cosmic time. Such a universal time can be constructed by assigning a time parameter to spacelike hypersurfaces which are distinguished by natural symmetries in the spacetime. A spacelike hypersurface is a three-dimensional spatial slice of spacetime, the prefix "hyper-" serving to alert us to the fact that the surfaces which dissect spacetime are not the two-dimensional planes which appear in our diagrams, but three-dimensional spaces. By foliating spacetime into such slices we can construct a cosmic time by ordering these slices serially according to a time parameter. The cosmic time so constructed will bear a special relationship to the fundamental observers, whose local planes of simultaneity, calculated by the standard SR clock synchronization procedure, will fit together to coincide with the cosmic hypersurface. Misner, Thorne, and Wheeler explain,

> In Newtonian theory there is no ambiguity about the concept 'a given moment of time.' In special relativity there is some ambiguity because of the nonuniversality of simultaneity, but once an inertial frame has been specified, the concept becomes precise. In general relativity there are no global inertial frames (unless spacetime is flat); so the concept of 'a given moment of time' is completely ambiguous. However, another, more general concept replaces it: the concept of a three-dimensional spacelike hypersurface. This hypersurface may impose itself on one's attention by reason of natural symmetries in the spacetime. Or it may be selected at the whim or convenience of the investigator.... At each event on a spacelike hypersurface, there is a local Lorentz frame whose surface of simultaneity coincides locally with the hypersurface. Of course, this Lorentz frame is the one whose 4-velocity is orthogonal to the hypersurface. These Lorentz frames at the various events on the hypersurface do not mesh to form a global inertial frame, but their surfaces of simultaneity do mesh to form the spacelike hypersurface itself.
>
> The intuitive phrase 'at a given moment of time' translates, in general relativity, into the precise phrase 'on a given spacelike hypersurface.' The investigator can go further. He can 'slice up' the entire spacetime geometry by means of a one-parameter family of such spacelike surfaces. He can give the parameter that distinguishes one such slice from the next the name of 'time'.... The successive slices of 'moments of time' may shine with simplicity or may only do a tortured legalistic bookkeeping for the dynamics [of the geometry of the universe]. Which is the case depends on whether the

typical spacelike hypersurface is distinguished by natural symmetries or, instead, is drawn arbitrarily.[26]

Several features of this explanation deserve comment. First, although one may slice spacetime into various hypersurfaces wholly arbitrarily, certain spacetimes have natural symmetries that guide the construction of cosmic time.[27] GR does not itself lay down any formula for dissecting the spacetime manifold of points; it has no inherent "layering." Theoretically, then, one may slice it up at one's whim. Nevertheless, certain models of spacetime, like the Friedmann model, have a dynamical, evolving physical geometry, a geometry that is tied to the boundary conditions of homogeneity and isotropy of the cosmological fluid, and in order to ensure a smooth development of this geometry, it will be necessary to construct a time parameter based on a preferred foliation of spacetime. For example, in 1935 H. P. Robertson and A. G. Walker independently showed that a homogeneous and isotropic universe requires that space be possessed of a constant curvature and be characterized by the metric:

$$\frac{dr^2 + r^2(d\theta^2 + \sin^2\theta d\phi^2)}{(1 + kr^2/4)^2} \qquad (5)$$

In the metric for spacetime, the spatial geometry is dynamic over time:

$$ds^2 = -dt^2 + R(t)\frac{dr^2 + r^2(d\theta^2 + \sin^2\theta d\phi^2)}{(1 + kr^2/4)^2} \qquad (6)$$

In this equation, called the Robertson-Walker line element, t represents cosmic time, the proper time of a fundamental observer. It is detached from space and serves to render space dynamic. The geometry of space is thus time-dependent. The factor $R(t)$ determines that all spatial structures of cosmic proportions, for example, a triangle demarcated by three galactic clusters or fundamental particles, will either shrink or stretch through the contraction or expansion of space, in this case into a similar smaller or larger triangle. The boundary condition of homogeneity precludes other geometrical changes such as shear, which would preserve the area but not the shape of the triangle. The condition of isotropy further precludes that the triangle should be altered in such a way as to preserve both its area and shape while nonetheless undergoing a rotational change of direction. Thus, in a Friedmann universe there are certain natural symmetries related to the dynamic geometry which serve as markers for the foliation of spacetime and the assigning of a cosmic time parameter. Of course, there are other cosmological models which do not involve homogeneity and isotropy and so may lack a cosmic time altogether.[28] Cosmic time

---

[26] Charles W. Misner, Kip S. Thorne, and John Archibald Wheeler, *Gravitation* (San Francisco: W. H. Freeman, 1973), pp. 713-714.
[27] See Kanitscheider, *Kosmologie*, pp. 182-197.
[28] Kanitscheider comments,
"On the other hand it should also be emphasized concerning the geometric side of this world model that the simple, form-preserving (dynamic), physical geometry can be traced back to the boundary conditions which have been laid down and by no means possess either a logically *a priori* or physically necessary character. If one eases the boundary conditions, one obtains world models with shear and rotation, and they, too, ...can be brought into harmony with the Einsteinian gravitational theory" (Ibid., p. 188).

is thus not nomologically necessary, and its actual existence is an empirical question.

Secondly, cosmic time is fundamentally parameter time and only secondarily co-ordinate time.[29] Physical time can be related in two quite different ways to the manifold in which motion is represented. If it is part of that manifold, then it functions as a coordinate. If it is external to that manifold, then it functions as a parameter. In Newton's physics time functioned only as a parameter. Motion takes place in absolute space and is parameterized by absolute time. Similarly, in Einstein's original formulation of SR, relativistic time functions only as a parameter. Einstein rejected the existence of absolute space and a fundamental rest frame in favor of a plurality of relatively moving inertial spaces, each of which was characterized by a time parameter which registered the proper time for that inertial frame. There was no absolute parameter time, only separate parameter times assigned to their respective inertial frames. The familiar spacetime formulation of SR used in virtually all contemporary expositions of the theory, according to which time is a co-ordinate (along with the three spatial co-ordinates) of an event in spacetime, derives from Minkowski. The spacetime formalism of Minkowski, in which time is part of the manifold in which motion is represented and so functions as a co-ordinate, was a wonderful aid to the visualization and comprehension of relativity theory; but, as we have seen, even Newton's theory can be cast in terms of a spacetime formalism in which time functions as a co-ordinate of events in spacetime.[30] Both theories admit of either a spacetime formulation (in which spacetime is the manifold) or a space and time formulation (in which the manifold is space(s), and time is a parameter). In the spacetime formulation, time functions as both a co-ordinate (locating events in the manifold) and as a parameter (recording the lapse of proper time along an observer's inertial trajectory), the chief difference between the two theories being that in SR parameter time loses the universality it possesses in Newtonian spacetime (that is, simultaneity becomes relative). When it comes to GR, it is unclear, according to Kroes, whether the theory could be formulated in terms of space and time rather than spacetime. He observes that differences in coordinate time values generally have no direct physical significance in GR (this is because of the variable spacetime geometry or gravitational fields which distort the co-ordinate grids laid on them). But insofar as time functions as a parameter in GR,[31] it is a more fundamental notion of time because it does possess

---

[29] See Peter Kroes, *Time: Its Structure and Role in Physical Theories*, Synthèse Library 179 (Dordrecht: D. Reidel, 1985), pp. 60-96.

[30] Such is the treatment of Friedman, *Spacetime Theories*, for all the theories he discusses, including Newtonian spacetime (pp. 71-124).

[31] There are three choices of time parameter available in GR, according to Misner, Thorne, and Wheeler: (i) the original time variable t. "This quantity gives directly proper time elapsed since the start of the expansion. It is the time available for the formation of galaxies. It is also the time during which radioactive decay and other physical processes have been taken place" (*Gravitation*, p. 730). (ii) the expansion factor $R(t)$. Since this factor grows with time, it serves to distinguish one phase of the expansion from another, thus serving as a parametric measure of time in its own right. The ratio of $R(t)$ at two different times gives the ratio of the dimensions of the universe at those two times. (iii) the arc-parameter measure of time $\eta(t)$. During the time interval dt, a photon traveling on hypersphere of radius $R(t)$ covers an arc in radians equal to $d\eta = dt/R(t)$. (In a model where the curvature constant $k=0$ or -1, the

direct physical significance. Parameter time can serve as a direct measure of the time elapsed between two events. Moreover, parameter time is well-suited, according to Kroes, for accommodating the A-theoretic notion of temporal becoming. While there is no intrinsic difference between past and future in co-ordinate time, there exists such a distinction in parameter time. Thus, the "flow" of time could relate to parameter time. Kroes writes,

> In the space and time formulation of Newtonian physics, the increase of parameter time represents the objective flow of absolute time; for increasing values of parameter time, the distribution of the particles in space will be different, and therefore there is change and becoming with regard to parameter time. However, the same kind of reasoning, applied to parameter time in the spacetime formalism of relativity theory, leads to the conclusion that parameter time has an objective flow (but with the proviso that the flow of parameter time is not universal).[32]

Because parameter time in the Special Theory is the proper time of each inertial observer and because no inertial frame is preferred, the "flow" of parameter time is not universal. But insofar as cosmic time plays a role in GR-based cosmological models, that universality is restored. It is highly significant, then, that cosmic time appears fundamentally as a parameter time in GR, though it can be used to generate co-ordinate time as well, as we shall see. As a parameter, it is not part of the spacetime manifold, and it thus measures the duration of the universe in an observer-independent way, that is to say, the lapse of cosmic time is the same for all observers. Moreover, cosmic time supplies a physical time which is well-suited to accommodate the philosophical notion of temporal becoming.

Thirdly, cosmic time is intimately related to a class of fundamental observers whose individual hyper-planes of simultaneity mutually combine to align with the hypersurface demarcating the cosmic time.[33] These hypothetical observers are conceived to be moving along with the cosmological fluid so that, although space is expanding and they are therefore mutually receding from each other, each is in fact at rest with respect to space itself. As time goes on and the expansion of space proceeds, each fundamental observer remains in the same place—his spatial co-ordinates do not change—though his spatial separation from fellow fundamental observers increases. Because of this mutual recession, the class of fundamental observers do not serve to define a global inertial frame, technically speaking, though all of them are at rest. When each of them utilizes the light signal method of synchronization, interesting relativistic effects arise due to their relative recessional motion. Since the spatial distance between them grows with time, a light signal from fundamental observer F to another such observer F' will have farther to travel on its return leg of its journey than on its out-going leg (Figure 10.4).

---

words "hypersphere" and "arc" should be replaced with their appropriate analogues.) Small values of the arc parameter indicate early times in the universe, large values later times.

[32] Kroes, *Time*, p. 96. Dorato also maintains that "cosmic time would be an ideal candidate with respect to which to define a world-wide, mind-independent becoming" (Mauro Dorato, *Time and Reality: Spacetime Physics and the Objectivity of Temporal Becoming*, Collana di Studi Epistemoligici II [Bologna: CLUEB, 1995], p. 189).

[33] See S. J. Prokhovnik, *Light in Einstein's Universe* (Dordrecht: D. Reidel, 1985), chaps. 4, 5, 6.

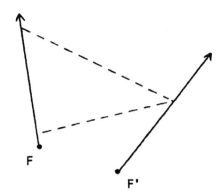

Figure 10.4. Because of the distance between fundamental observers F and F' increases with time, a light signal will have to travel a longer distance on its return trip than on its outbound trip in an exchange between F and F'.

F will therefore calculate that F''s clock is running slow; but, of course, the converse is also the case. Here we do have "pure relativity" without absolute effects. It is purely a reciprocal measurement phenomenon. F and F''s reciprocal observations of time dilation and length contraction have no absolute significance. They result merely from the manner of observation and are in fact similar to the familiar analogy of the mutual observation of diminishing size when two observers retreat from each other. Because F and F' will draw their planes of simultaneity orthogonal to their world lines, their surfaces of simultaneity will not be aligned. But since each is at rest with respect to space, his hyper-plane of simultaneity will coincide locally with the hypersurface of cosmic time. Were he in motion with respect to the cosmological fluid, then his plane of simultaneity would be at an angle with the local hypersurface. But in virtue of being at rest, he can be guaranteed that locally events which he judges to be simultaneous will lie on the hypersurface. Thus, the local regions of the planes of simultaneity of fundamental observers all blend together and coincide with the hypersurface, much as a circle is formed by the points of intersection of all the straight lines which are tangents of its circumference. This has two important implications: first, that the proper time of each fundamental observer coincides with cosmic time and, second, that all the fundamental observers will agree as to what time it is.

By employing the notions explained above, one is now prepared to define "homogeneity" and "isotropy":

> Homogeneity of the universe means, then, that through each event in the universe there passes a spacelike 'hypersurface of homogeneity' (physical conditions identical at every event on this hypersurface). At each event on such a hypersurface, the density, ρ, and pressure, p, must be the same; and the curvature of spacetime must be the same....

210   CHAPTER 10

> *Isotropy of the universe means that, at any event, an observer who is 'moving with the cosmological fluid' cannot distinguish one of his space directions from the others by any local physical measurement.*[34]

With these definitions in hand, one is now prepared to construct a cosmic co-ordinate time.[35] Choose any hypersurface of homogeneity S_1 and to all the events on it assign the co-ordinate time t_1. Next lay out a grid of spatial co-ordinates (x_1, x_2, x_3) on S_1 and then, using the world lines of the fundamental particles project this grid through successive hypersurfaces, thus co-ordinating all of spacetime. The system of co-ordinates is thus co-moving with the fluid, so that the fluid is always at rest relative to the space co-ordinates. For the time co-ordinate of any event P, one uses t_1 plus the lapse of proper time, as measured along the world line of a fundamental particle that passes through P, from S_1 to P. Letting ΔT represent the lapse of proper time, the time t of P will be

$$t(P) = t_1 + \Delta T \tag{7}$$

Milton K. Munitz explains,

> If we imagine observers attached to various particles of the homogeneous and isotropic cosmological fluid, then all such observers will be at rest with respect to any given hypersurface, yet they will move through a sequence of such hypersurfaces as the fluid itself moves through spacetime. The paths of timelike geodesics of free material particles in the cosmological fluid are orthogonal to the three-dimensional spatial hypersurface. Each observer measures proper time by his own clock along his own timelike geodesic. The intervals of each timelike geodesic of a material particle, orthogonal to a given hypersurface, will measure proper time along its own individual world line. Let various observers set their individual clocks to a certain time when they observe the density of matter in their respective neighborhoods to have the same given value. Thus, if the galaxies are uniformly distributed throughout space, then at a given moment of cosmic time the same density of galaxies will hold at any region of space. The homogeneity of the model assures the retention of the synchronized clocks (and so of the measures of cosmic time) with change in density as the universe expands or contracts. (Any departures from the measurement of cosmic time—that is, any irregularities in time-keeping among different timekeepers—will be assigned to purely local or peculiar variations in velocity, and to local variations in gravitational fields associated with particular astronomical systems....) ...All observers situated on particles of the cosmological fluid can thus employ a common set of synchronous temporal coordinates. By adding the same interval of elapsed proper time to the coordinate time t_0 of our present, initial, arbitrary hypersurface, we arrive at different hypersurfaces, each of which is identified by its distinctive value of the cosmic time t. The cosmic time coordinate will be constant for each hypersurface, but will change with changing hypersurfaces. Therefore, clocks attached to co-moving particles may be said to measure cosmic time t.[36]

One thus supplements the cosmic parameter time with a cosmic co-ordinate time.

It is noteworthy that deviations from this time are purely local effects to be explained due to velocity (SR) or to gravitation (GR). Thus, on a cosmic scale, we have that universality of time and absolute simultaneity of events which the Special Theory had denied. Whitrow asserts, "...in a universe that is characterized by the

---

[34] Misner, Thorne, and Wheeler, *Gravitation*, p. 714.
[35] Ibid., pp. 715-716.
[36] Milton K. Munitz, *Cosmic Understanding* (Princeton: Princeton University Press, 1986), pp. 97-98.

existence of a cosmic time, relativity is reduced to a local phenomenon, since this time is world-wide and independent of the observer."[37] Kanitscheider explains,

> ...with the parameter t we can so order all slices through spacetime (the homogeneous hypersurfaces) that an unequivocal earlier/later relation can be set up worldwide. Within such a slice $t = t_0$ (in a three-space) the material quantities p and ρ, as well as the physical geometry, are everywhere the same. Isotropy implies, moreover, that a particle of the cosmological fluid traces a worldline that orthogonally intersects the hypersurface of homogeneity. One recognizes that there is here again a privileged reference frame; to an observer at rest relative to the substratum, who swims along with the fluid, the universe has a simple form in material structure and spacetime geometry....
>
> The particular form of the motion of matter in this class of models suggests the utilization of a co-moving co-ordinate system, in which a worldwide, absolute simultaneity is defined. This is, however, no contradiction to the SR, since here the universe itself, with its limited possibility of movement, serves as an instrument of synchronization. The special relativistic time dilation, which we are acquainted with through local experiments, still holds as before for clocks moving relatively to the substratum. Nevertheless, the proper times of all observers who are at rest with respect to the flowing (expanding or contracting) substratum can be harmoniously fitted into a cosmic time.[38]

Based on a cosmological, rather than a local, perspective, cosmic time restores to us our intuitive notions of universal time and absolute simultaneity which SR denied. The question, then, becomes an empirical one: does cosmic time exist? The answer to that question comes from the evidence for large scale (scales of $\sim 10^8$ light years or larger) homogeneity and isotropy in the universe. In models like Gödel's and Ozsváth and Schücking's, there is posited a worldwide rotation of the homogeneous substratum, so that the isotropy condition of the Friedmann model is violated and the proper times of fundamental observers cannot be fitted together into a universal time. However, the observational evidence for cosmic isotropy, particularly for the isotropy of the cosmic microwave background radiation, which has been measured by the COBE satellite to an accuracy of one part in 100,000, makes it very likely that our actual universe does approximate a Friedmann universe. Martin Rees focuses on two important pieces of evidence that our universe is kinematically described by a Robertson-Walker metric: (i) *The Hubble Constant: the Present Scale of the Observable Universe.* The tight relation between the red shift and apparent magnitude of the brightest elliptical galaxies in clusters tells us that their motions accord with those expected in a Robertson-Walker spacetime for a set of co-moving objects whose distances with respect to each other are smaller than the overall radius of curvature of the hypersurfaces of homogeneity. The constant of proportionality between recession velocity and distance, that is, the Hubble Constant, is thus the key parameter determining the scale of the observable universe and the time scale over which its properties can change. (ii) *The Isotropic Microwave Background.* According to the hot Big Bang model, the material in the early universe would constitute a hot plasma, strongly coupled to a radiation field. When adiabatic expansion had cooled the material to approximately 4,000°K, the plasma would recombine and the radiation would no longer undergo scattering or absorption. The photons emitted then have been red-shifted to the microwave end

---

[37] Whitrow, *Natural Philosophy of Time*, p. 371.
[38] Kanitscheider, *Kosmologie*, pp. 186-187.

of the spectrum and are now observed as a remarkably isotropic radiation background filling all of space. This radiation background establishes that the gross kinematics of the universe are exceedingly symmetrical with a greater precision than the data on which the Hubble Law is based. The measured isotropy implies that the observable universe can be very precisely described by a Robertson-Walker metric. "Unless we adopt an anti-Copernican viewpoint, this forces us to adopt a Robertson-Walker metric."[39] In addition to these kinematical considerations, Rees adds (iii) *Evidence for large-scale homogeneity from radio, optical, and x-ray observations.* The distribution of radio, optical, and x-ray sources on the sky supports large scale isotropy, the x-ray radiation background supplying a particularly sensitive probe for large scale matter distribution. Rees concludes, "The most remarkable outcome of fifty years of observational cosmology has been the realization that the universe is more isotropic and uniform than the pioneer theorists of the 1920's would ever have suspected."[40] "Consequently, we have strong evidence that the universe as a whole is predominantly homogeneous and isotropic," states Whitrow, "and this conclusion...is a strong argument for the existence of cosmic time."[41] In fact, Hawking has shown that existence of stable causality, that is, the absence of any null or time-like closed causal paths, is a necessary and sufficient condition for the existence of cosmic time.[42] Thus, far from "taking away from space and time the last remnant of physical objectivity," as Einstein thought at first, GR through its cosmological applications gives back what SR had purportedly removed. Whitrow concludes,

> The concept of the relativity of simultaneity on which, in 1905, Einstein based his Special Theory of Relativity at first appeared to eliminate from physics any idea of an objective world-wide lapse of time according to which physical reality could be regarded as a linear succession of temporal states or layers. Instead each observer was regarded as having his own sequence of temporal states and none of these could claim the prerogative of representing the objective lapse of time. Nevertheless, a quarter of a century later, theoretical cosmologists who made use of the physical ideas and mathematical techniques associated with relativity theory were led...to re-introduce the very concept which Einstein had begun by rejecting....
>
> From the cosmological point of view..., it would therefore seem that neither the equivalence of all observers in uniform relative motion (Special Relativity) nor of all observers in any form of relative motion (General Relativity) can be accepted without restriction....once the existence of a world-wide distribution of matter, albeit of extremely low mean density, becomes an essential feature of the problem under investigation, then certain frames of reference and observers must be specially distinguished, namely those which move with the mean velocity of matter in their neighborhood. In the cosmological models which we have discussed..., the local times of all these 'privileged' observers fit together, into one world-wide time called 'cosmic time.'[43]

---

[39] Martin J. Rees, "The Size and Shape of the Universe," in *Some Strangeness in the Proportion*, ed. Harry Woolf (Reading, Mass.: Addison-Wesley, 1980), p. 293.
[40] Ibid., p. 301.
[41] Whitrow, *Natural Philosophy of Time*, p. 307.
[42] S. W. Hawking, "The Existence of Cosmic Time Functions," *Proceedings of the Royal Society of London* A 308 (1968): 433-435.
[43] Whitrow, *Natural Philosophy of Time*, p. 302.

Relativity theory thus gives back with its right hand what it had taken away with its left.

GOD AND COSMIC TIME

Now it is my contention that since the inception of the universe and the beginning of physical time, this cosmic time plausibly coincides with God's metaphysical time, that is, with Newton's absolute time. It therefore provides the correct measure of God's time and thus registers the true time, in contrast to the multiplicity of local times registered by clocks in motion relative to the cosmological substratum. Already in 1920, on the basis of Einstein and de Sitter's cosmological models, Eddington hinted at a theological interpretation of cosmic time:

> In the first place, absolute space and time are restored for phenomena on a cosmical scale.... The world taken as a whole has one direction in which it is not curved; that direction gives a kind of absolute time distinct from space. Relativity is reduced to a local phenomenon; and although this is quite sufficient for the theory hitherto described, we are inclined to look on the limitation rather grudgingly. But we have already urged that the relativity theory is not concerned to deny the possibility of an absolute time, but to deny that it is concerned in any experimental knowledge yet found; and it need not perturb us if the conception of absolute time turns up in a new form in a theory of phenomena on a cosmical scale, as to which no experimental knowledge is yet available. Just as each limited observer has his own particular separation of space and time, so a being co-extensive with the world might well have a special separation of space and time natural to him. It is the time for this being that is here dignified by the title 'absolute.'[44]

A couple of items in this remarkable paragraph deserve comment. First, Eddington rather charitably interprets SR as positing merely an epistemic limitation on our temporal notions rather than an ontological limitation on time and space. But as friend and foe alike have emphasized,[45] Einsteinian SR requires metaphysical, not merely epistemological, commitments concerning the non-existence of absolute space and time. Otherwise, one winds up with the Lorentz-Poincaré interpretation of the theory, which is, in truth, the position which Eddington is describing. Second, Eddington is quite willing to call cosmic time "absolute" in view of its independence from space, that is to say, its status as a parameter. Relativistic time is, as Lorentz and Poincaré maintained, only a local time, whereas cosmic time, being non-local, gives the true time. Third, although in 1920 there was no empirical evidence for cosmic time, within a few short years astronomical evidence confirmed the prediction of the Friedmann model of a universal expansion and, hence, of cosmic

---

[44] Arthur Eddington, *Space, Time and Gravitation,* Cambridge Science Classics (Cambridge: Cambridge University Press, 1920; rep. ed.: 1987), p. 168.

[45] Adolf Grünbaum asserts, "In short, it is because no relations of absolute simultaneity *exist* to be measured that measurement cannot disclose them; it is not the mere failure of measurement to disclose them that *constitutes* their non-existence, much as that failure is *evidence* for their non-existence" (Adolf Grünbaum, *Philosophical Problems of Space and Time,* 2d ed., Boston Studies in the Philosophy of Science 12 [Dordrecht: D. Reidel, 1973], p. 368). Richard Swinburne plumps for a neo-Lorentzian interpretation of SR, commenting, "One can describe the Universe of Special Relativity perfectly intelligibly by supposing that its equations show a limit to our knowledge of absolute simultaneity, not a limit to its existence" (Richard Swinburne, *Space and Time,* 2d ed. [London: Macmillan, 1981], p. 201).

time. The veil of epistemic limitation had been torn away by empirical science. Finally, this cosmic time would be the time of an omnipresent being whose reference frame is the hypersurface of homogeneity itself. Is Eddington recalling here Poincaré's *"intelligence infinie"*, who classified everything according to his universal frame of reference, just as finite observers classify events according to their local frames? Cosmic time is not merely the "fusion" of all the proper times recorded by the separate fundamental observers, but, even *more* fundamentally, it is the time which measures the duration of the omnipresent being which co-exists with the universe. As the measure of the proper time of the universe, cosmic time also measures the duration of and lapse of time for a temporal being co-extensive with the world. For Eddington, it is the time of this being that deserves to be called "absolute."

The theological application is obvious. In Whitrow's interpretation, "Eddington endeavoured to explain away Einstein's re-introduction of cosmic time by regarding this concept as one of the prerogatives of Newton's ubiquitous Deity, and this is beyond the scope of experimental science."[46] But the theistic philosopher need see nothing disingenuous about such an identification. It makes perfectly good sense to interpret the lapse of cosmic time as measuring the lapse of God's time. Thus, Fitzgerald, after asserting, "I see no way to save the idea of a worldwide surge of Absolute Becoming that would be intimately associated with God's temporal way of experiencing the world,"[47] recalls that some GR-based models have a cosmic time and admits that the theologian might take this to be a privileged time. He expresses hesitancy because Gödel showed that not all models have this cosmic time; still, he admits, this solution may not be a blind alley and he recommends that the theologian should explore it further.[48]

Now God's metaphysical time cannot be said to be identical with cosmic time, since the former is capable of exceeding the latter, in that metaphysical time could precede physical time (recall God's counting down to creation). Nevertheless, since the inception of cosmic time, the moments of God's time would seem to coincide with the moments of cosmic time. When we reflect that God is causally related to the cosmos, sustaining it in being moment by moment, then it seems difficult to deny that the duration measured by cosmic time is also the duration of God's temporal being. If the duration of the universe measured in cosmic time is 15 billion years since the singularity, then is not the duration of God's creatorial activity in

---

[46] Whitrow, *Natural Philosophy of Time*, p. 284.
[47] Paul Fitzgerald, "The Truth about Tomorrow's Sea Fight," *Journal of Philosophy* 66 (1969): 325.
[48] Ibid., p. 326. Fitzgerald later rejected the identification of cosmic time with God's time because cosmic time, being a statistical matter based on mean matter density, allows a range of regions of the universe to be classed as simultaneous. "This means that strictly speaking, several mutually incompatible 'cosmic times' will be definable, each equally usable for the gross purposes of the astronomer, and none sufficiently preferable to the others to justify identifying it with 'God's time'" (Idem, "Relativity Physics and the God of Process Philosophy," *Process Studies* 2 [1972]: 256). The problem is that Fitzgerald is still operating with a reductionistic view of time which equates time with physical time. But if God's time is metaphysical and cosmic time a sensible measure thereof, then it does not matter, as Newton saw, whether this measure is more or less accurate. Cosmic time gives a rough measure of God's time. Moreover, as a result of measurements made since the launch of the COBE satellite in 1989, we now have very precise measurements of the isotropy of the cosmic microwave background radiation, thereby honing the measure of cosmic time.

metaphysical time also 15 billion years? In God's "now" the universe has (present tense) certain specific and unique properties, for example, a certain radius, a certain density, a certain background temperature, and so forth; but in the cosmic "now" it has all the identical properties, and so it is with every successive "now." Is it not obvious that these "now's" coincide and designate one and the same present?

Perhaps we can state this consideration a bit more formally by means of the following proposed principle:

> P: For any constantly and non-recurrently changing universe U and temporal intervals x, y large enough to permit change, if the physical description of U at x is the same as the physical description of U at y, then x and y coincide.

Given that in metaphysical time there is a temporal interval or duration during which a certain physical description of the universe is true and that in cosmic time a similar interval exists, then it follows from P that those intervals of metaphysical and cosmic time coincide. It seems to me, therefore, that God's time and cosmic time ought naturally to be regarded as coincident since the inception of cosmic time. I do not mean to say that there are in fact two times rather than one; rather I mean simply to reaffirm Newton's distinction between absolute (=metaphysical) time and relative (=physical) time. The latter is merely a sensible measure of the former, and my suggestion is that cosmic time is a sensible measure of God's time since creation.

Such an affirmation will be typically met with passionate disclamations. Such protestations strike me, however, as being for the most part misconceived. In our weighing this issue, two questions need to be kept distinct: (i) Does cosmic time provide a sensible measure of God's metaphysical time? (ii) Is cosmic time in some sense absolute?

COSMIC TIME AS A MEASURE OF GOD'S TIME

The first question is scarcely ever directly addressed. But very frequently assertions are made about the nature of cosmic time which would seem to imply that cosmic time enjoys no special status such that we should be warranted in identifying it as a sensible measure of God's time any more than we should be warranted in picking out some arbitrary inertial frame and identifying the time associated with it as privileged. The choice of cosmic time as a measure of God's time could be said to be arbitrary because nothing in GR requires that we slice up spacetime in one way rather than another. The theory allows one's three-dimensional hyper-surfaces to dissect spacetime in any arbitrary way, so that the selection of one foliation over another is a conventional choice. Kroes explains,

> If cosmic time functions are considered *in abstracto*, i.e., without relating them to the notion of the evolution of the universe, it is immediately clear that the existence of these functions does not contradict the basic principles of RT. If a spacetime M admits the definition of one cosmic time function f, then infinitely many other cosmic time functions can be defined; f is by no means unique. But this infinity of different cosmic time functions contains members which generate different total temporal orders of the events. RT does not, however, prescribe such a choice; it does not specify which is the 'real' cosmic time function with its corresponding 'real' total temporal order.

216   CHAPTER 10

The foregoing can be stated in a different way. The choice of a particular cosmic time function introduces a family of simultaneity planes in the spacetime manifold; each simultaneity plane consists of all those events m of M for which $f(m)$ has the same value. But the simultaneity of events defined by f in the neighbourhood of a point of such a simultaneity plane will coincide only for a special observer, in the appropriate state of motion, with the simultaneity of events defined by the Einstein convention for synchronizing clocks. Thus to each cosmic time function f, there corresponds a special group of observers for whom the standard definition of simultaneity gives locally the same result as simultaneity defined by the cosmic time function f. Therefore, if it were possible to single out on the basis of objective physical principles, a unique cosmic time function as the 'real' one, then it would also be possible to single out a preferred group of observers. In that case, however, there would clearly be a clash with GR, according to which all observers, in whatever state of motion, are equivalent from a physical point of view.

Since it is impossible to single out, on physical grounds, a class of preferred observers for whom the definition of simultaneity has objective physical significance, it is impossible to determine which cosmic time function is the 'real' one.[49]

Eddington graphically expressed this point in drawing our attention to the difference between a pile of sheets of paper and a solid paper block. "The solid block is the true analogy for the four-dimensional combination of spacetime; it does not separate naturally into a particular set of three-dimensional spaces piled in time order. It can be redivided into such a pile; *but it can be redivided in any direction we please.*"[50] The implication would seem to be that the identification of our cosmic time as the sensible measure of God's time is wholly conventional.

But the shortcoming of this answer to our question is that such a response is not based on the whole story. It is true that the General Theory itself does not mandate a specific foliation of spacetime; but this is only to consider the theory, as Kroes puts it, *in abstracto*, apart from any *de facto* boundary conditions arising from the nature of material reality. The answer explicitly ignores "the notion of the evolution of the universe" and considers just a manifold of points. Once we introduce, however, considerations concerning the *de facto* distribution of matter and energy in the universe, then certain natural symmetries emerge which disclose to us the preferential foliation of spacetime and the real cosmic time in distinction from artificial foliations and contrived times. Michael Shallis, reflecting on the relativity of simultaneity in SR, writes,

It is also possible, however, to take a single clock as standard, taking it to define a universal time coordinate, and to relativize everything to it.... Of course, the choice of a coordinate time is, to a certain considerable extent, arbitrary—in principle one could take any clock as one's standard. But in a cosmological context, it is natural to take as standard a clock whose motion is *typical* or *representative* of the motion of matter in

---

[49] Kroes, *Time*, pp. 15-16.
[50] Eddington, *Space, Time and Gravitation*, p. 34. Cf. Graves's comment that it makes no difference to the validity of the (tensor) initial value equations how we define a hypersurface or what sort of coordinates we use on it. The choice of an initial hypersurface is "wholly arbitrary" (John Cowperthwaite Graves, *The Conceptual Foundations of Contemporary Relativity Theory*, with a Foreword by John Archibald Wheeler [Cambridge, Mass.: MIT Press, 1971], pp. 250-252). So also Vesslin Petkov, "Simultaneity, Conventionality, and Existence," *British Journal for the Philosophy of Science* 40 (1989): 75, who agrees with Putnam that all events are therefore equally real.

general—one which simply 'rides along', so to speak, with the overall expansion of the universe.[51]

Kroes himself admits that when we turn from the theory considered *in abstracto* to describing the actual evolution of the universe, "certain 'natural' cosmic time functions force themselves upon us."[52] In universes with fundamental observers one may introduce a special co-ordinate system which "distinguishes itself from all the other coordinate systems by the fact that the spacelike hyperplanes of constant t coincide with the hyperplanes of homogeneity;" relative to this group of observers, we can "speak properly of 'the' evolution of the universe."[53] To return to Eddington's analogy of the paper block, suppose that only by foliating the block into a stack of sheets do we discover that on each sheet is a drawing of a cartoon figure and that by flipping through the sheets successively, we can see this figure, thus animated, proceed to pursue some action. Any other slicing of the block would result merely in a scrambled series of ink marks. In such a case, it would be silly to insist that any arbitrary foliation is just as good as the foliation which regards the block as a stack of sheets. But analogously, the Robertson-Walker metric discloses to us the natural foliation of spacetime for our universe. It would, indeed, be disingenuous to insist that the universe is not *really* expanding homogeneously and isotropically in approximation of the Friedmann model, that it does not *really* have a certain spacetime curvature, density, and pressure, that it has not *really* been about 15 billion years since the singularity, but that any arbitrary foliation and contrived time will yield equally appropriate descriptions of the way the universe actually is and evolves.[54] Eddington realized this, of course, and we have seen that he recognized the privileged status of our cosmic time, though he emphasized that no experimental knowledge of it was as yet available. Today, however, the situation has considerably changed, as P. C. W. Davies explains:

> At any given place in the universe, there is only one reference frame in which the universe expands isotropically. This privileged reference frame defines a privileged time scale (the time as told by a clock at rest in that frame). Two separated places have their privileged reference frames in mutual motion, because of the expansion of the universe. Nevertheless, the time measured by the entire collection of imaginary standard clocks are obviously correlated such that the global condition (e.g. average separation of two galaxies) of the universe appears the same at equal times as registered by every privileged clock (assuming they are all properly synchronized). Happily, the earth is moving very slowly relatively to the local privileged frame in our vicinity of the universe, so that Earth time is a fairly accurate measure of cosmic time.[55]

---

[51] Michael Shallis, "Time and Cosmology," in *The Nature of Time*, ed. Raymond Flood and Michael Lockwood (Oxford: Basil Blackwell, 1986), pp. 68-69. Cf. Asghar Qadir and John Archibald Wheeler, "York's Cosmic Time Versus Proper Time as Relevant to Changes in Dimensionless 'Constants,' K-Meson Decay, and the Unity of the Black Hole and Big Crunch," in *From SU(3) to Gravity*, ed. Errol Gotsman and Gerald Tauber (Cambridge: Cambridge University Press, 1985), pp. 383-394.
[52] Kroes, *Time*, p. 16.
[53] Ibid., p. 17.
[54] Of course, as Dorato points out, the existence of a plurality of cosmic times which is merely the result of change of origin or a change of scale within the global time function does nothing to undermine the objectivity of cosmic time (Dorato, *Time and Reality*, p. 201; cf. p 192, n. 12).
[55] P. C. W. Davies, "Spacetime Singularities in Cosmology and Black Hole Evaporations," in *The Study of Time III*, ed. J. T. Fraser, N. Lawrence, and D. Park (Berlin: Springer Verlag, 1978), p. 76. I have corrected the spelling errors in the quotation.

Thus, not only do we know that a privileged cosmic time in which the universe evolves exists, but because the earth is approximately at rest with respect to our galactic fundamental particle, we also have a fair idea of what time it is!

Thus, it is possible to single out on physical grounds—not from the theory alone, but from *de facto* material conditions in the universe—a special group of preferred observers, the fundamental observers, that serve to define a privileged cosmic time, which deserves to be called the real cosmic time in counterdistinction to other mathematically possible functions. Nor does this conclusion in any way clash with GR. That theory does not, *pace* Kroes, succeed in establishing the equivalence of all observers from a physical point of view, as we have seen, and there simply is no General Principle of Relativity that requires that no privileged time exists from the cosmological point of view.

In view of the importance of this answer to our first question, it seems worthwhile to underscore it by means of two supporting arguments. First, *a fundamental frame for light propagation is plausible in view of modern cosmology.* As Prokhovnik explains,[56] the Robertson-Walker metric is expressed in such a way that every fundamental particle has a fixed set of co-ordinates which do not vary with time. The particles are not conceived to be moving through space, but to be each at the same place in space as space itself expands with time. In view of the equivalence of all fundamental observers with respect to their view of the universe and the laws of nature, the interval $d\tau$, signifying proper time, is invariant for all such observers. The invariance of the interval is restricted to the set of reference frames associated with the fundamental observers, and each such observer can be considered at the origin of such a frame. Therefore, the ensemble of fundamental observers serves to define a unique cosmological frame which may properly be referred to as the *fundamental reference frame.* The members of the ensemble of equivalent frames are merely representations of the fundamental frame in terms of different points of origin of the coordinate system. Hubble's Law $w = r/t$ (velocity equals radius over time) manifests itself isotropically only to fundamental observers and thus provides an observable criterion for distinguishing any representation of the fundamental frame. The constant c represents the velocity of light with respect to the reference system defined by the Robertson-Walker metric, namely, that associated with the set of fundamental observers. "Hence, the formulation of the Robertson-Walker metric implies that the propagation of light takes place relatively to a unique cosmological reference frame associated with the large scale distribution of matter in the universe."[57]

The existence of a unique fundamental reference frame also has important implications for our interpretation of SR as well. For the existence of a fundamental reference frame for light propagation implies that the speed of light is isotropic only with respect to observers which are stationary relative to the fundamental frame (the

---

[56] Prokhovnik, *Light in Einstein's Universe*; idem, "The Logic of the Clock Paradox," paper presented at the International Conference of the British Society for the Philosophy of Science, "Physical Interpretations of Relativity Theory," Imperial College of Science and Technology, London, 16-19 September, 1988; idem, "The Twin Paradoxes of Special Relativity—Their Resolution and Implications," *Foundations of Physics* (Preprint). Cf. Swinburne, *Space and Time*, chaps. 11, 12.

[57] Prokhovnik, *Light in Einstein's Universe*, p. 53.

fundamental observers). With respect to an observer which is moving with a velocity v relative to the fundamental frame, light will approach such an observer with the speed of $c + v$ in the line of v and overtake it with a speed of $c - v$ in the line of v. These departures from constant c will not, however, be detectable to such observers due to length contraction and time dilation. Length contraction results from the endeavor of a moving system of particles to maintain its internal equilibrium. Time dilation follows as a result of the anisotropy of the velocity of light and length contraction. Einstein's Light Postulate is now seen to concern the *measurement* of light's velocity in various inertial frames, and the relativity of simultaneity which results from his synchronization procedure becomes intelligible in view of observers who are in motion relative to the fundamental frame's regarding their inertial systems as being at rest for the purpose of clock synchronization. Prokhovnik concludes,

> The anisotropy consequences..., which affect moving bodies and the observations of moving observers, provide a complete physical interpretation, free of any ambiguity, of Special Relativity. They show how and why the Light Principle operates in respect to all inertial frames, they explain why any local experiment designed to detect an absolute velocity is bound to yield a null-effect. The existence of a fundamental reference frame provides a physical basis for the absolute anisotropy effects, and their interaction produces the (local) observational equivalence of all inertial frames in respect to the laws of nature.... Only by cosmological astronomical observations can we discern the existence of the fundamental frame and our movement relative to it.[58]

Given the existence of this unique fundamental reference frame, one thus arrives at a perspicuous neo-Lorentzian interpretation of Special Relativity Theory. The unpopularity of such an interpretation no doubt accounts for the widespread resistance to the notion of a unique fundamental reference frame; nonetheless, the existence of such a frame is implied by the existence of the ensemble of fundamental observers and is disclosed in observational astronomy.

Secondly, *modern equivalents of the aether serve to establish a preferred reference frame.* In a remarkable about-face of science, the old, despised aether of classical physics has been replaced by modern ethers, which serve a similar function in distinguishing a privileged rest frame co-extensive with space itself. According to Duffy, "...in recent years there are many signs that the aether concept is playing an ever increasing role in physics."[59] There are several plausible candidates, each of which will serve nicely as an *ersatz*-ether for the nineteenth century aether:

(i) *The cosmological fluid.* The "gas" of fundamental particles is itself a sort of ether in that it is co-extensive with and at rest with respect to space. It therefore serves to single out a universal frame of reference and thereby assumes the function of the classical aether. Heller, Klimek, and Rudnicki remark,

---

[58] Prokhovnik, "Twin Paradoxes," pp. 8-9.
[59] Michael Ciaran Duffy, "The Modified Vortex Sponge: a Classical Analogue for General Relativity," paper presented at the International Conference of the British Society for the Philosophy of Science, "Physical Interpretations of Relativity Theory," Imperial College of Science and Technology, London, 16-19 September, 1988. According to the eminent Italian physicist Franco Selleri, "the absence of the notion of an ether has important, negative consequences for the possibility of our rationally understanding the world of physics and notably the nature of time" (Franco Selleri, "Le principe de la relativité et la nature du temps," *Fusion* 66 [1997]: 54). Selleri's article is strikingly confirmatory of the understanding of SR defended in this book.

> We may talk of symmetries in the energy-mass distribution...only after distinguishing a certain universal frame of reference in which these symmetries appear in a natural way. The existence of such a particular frame of reference resembles the concept of the aether in classical electrodynamics.[60]

These three authors therefore refer to the universal reference frame as the "neo-ether of Relativistic Cosmology." Prokhovnik points out that this universal reference frame can be regarded as a sort of inertial frame (though it is expanding), in the sense that a fundamental particle stationary with respect to this frame remains so.[61] A material body will maintain its motion relative to the expanding cosmological substratum in the absence of any extraneous causes, thus satisfying Newton's first law. So although the cosmological fluid is radically different from the classical aether in that it exists in a systematically expanding frame, still it has a universal connotation and is able to fulfill a number of important roles for which the classical aether was invented. Specifically, this modern ether allows in a natural way for the existence of (1) universal cosmic time, (2) 3-spaces of constant curvature orthogonal to the time lines, and (3) a frame of reference co-moving with the substratum.[62] Therefore, just as the classical aether was regarded as the physical realization of the fundamental reference frame, so the gas of fundamental particles serves to distinguish physically an equally fundamental frame.

(ii) *The microwave background radiation.* The cosmic microwave background radiation fills all of space and is remarkably isotropic for any observer at rest with respect to the expansion of space. The radiation background will be anisotropic for any observer in motion with respect to an observer whose spatial coordinates remain fixed. It is therefore a sort of ether, serving to distinguish physically a fundamental universal reference frame.[63] What is especially astounding is that recent tests have actually been able to detect the earth's motion relative to the radiation background, thus fulfilling nineteenth century physics' dream of measuring the aether wind! Smoot, Gorenstein, and Muller discovered that the earth is moving relative to the radiation background with a velocity of 390±60 km/sec in the direction of the constellation Leo. They comment,

> Thus, *except for a component that varies as* $\cos(\theta, \eta)$ the cosmic blackbody radiation is isotropic to 1 part in 3,000.
> The cosine anisotropy is most readily interpreted as being due to the motion of the earth relative to the rest frame of the cosmic blackbody radiation—what Peebles calls the 'new aether drift'.[64]

Commenting on this result, Kanitscheider remarks, "The cosmic background radiation thereby furnishes a reference frame, relative to which it is meaningful to

---

[60] Michael Heller, Zbigniew Klimek, and Konrad Rudnicki, "Observational Foundations for Assumptions in Cosmology," in *Confrontation of Cosmological Theories with Observational Data*, ed. M. S. Longair (Dordrecht: D. Reidel, 1974), p. 3.
[61] Prokhovnik, *Light in Einstein's Universe*, pp. 76-78, 126; so also Geoffery Builder, "Ether and Relativity," *Australian Journal of Physics* 11 (1958): 279-297, reprinted in *Speculations in Science and Technology* 2 (1979): 230-242.
[62] Heller, *et al.*, "Foundations for Assumptions in Cosmology," pp. 4-5.
[63] Ibid., p. 4.
[64] G. F. Smoot, M. Y. Gorenstein, and R. A. Muller, "Detection of Anisotropy in the Cosmic Blackbody Radiation," *Physical Review Letters* 39 (1977): 899.

speak of an *absolute* motion."[65] What Michelson and Morley failed to detect using visible light radiation, twentieth century physicists discovered using microwave radiation. Mansouri and Sexl comment, "The discovery of cosmic background radiation has shown that cosmologically a preferred system of reference does exist. This system is defined and singled out much more unambiguously to be a candidate for a possible 'ether frame' than was the solar rest frame of Einstein's days."[66] By means of this empirically distinguished frame of reference one can, in Stapp's words, "define an absolute order of coming into existence."[67] One can only speculate whether, had this microwave background radiation and the measure of our motion relative to it been known to Einstein prior to 1905, he would have claimed that no fundamental frame exists relative to which all local inertial frames are in motion.[68]

(iii) *The quantum mechanical vacuum*: Underlying all of physical reality is the quantum mechanical vacuum, which is not "nothing," but rather a sea of evanescent particles forming by fluctuations of the energy field and returning to it almost at once; as Hönl and Dehnen explain:

---

[65] Kanitscheider, *Kosmologie*, p. 256.
[66] Reza Mansouri and Roman U. Sexl, "A Test of the Theory of Special Relativity: I. Simultaneity and Clock Synchronization," *General Relativity and Gravitation* 8 (1977): 497-498.
[67] Henry Pierce Stapp, "Quantum Mechanics, Local Causality, and Process Philosophy," *Process Studies* 7 (1977): 176. In order to rescue the relativity of simultaneity, Stapp is driven to decouple temporal order from the order of coming into existence. Not only is this expedient counter-intuitive; it is ultimately futile, since in some frame temporal order will coincide with the absolute order of becoming, and thereby that frame will be distinguished.
[68] Consider Arthur I. Miller's paper, "On Some Other Approaches to Electrodynamics in 1905," in *Strangeness in the Proportion*, pp. 66-91, in connection with P. A. M. Dirac's remark in discussion that
"In one respect Einstein went far beyond Lorentz and Poincaré and the others, and that was in asserting that the Lorentz transformation would apply to the whole of physics and not merely to the phenomena based on electrodynamics..., which is going far beyond what the people who were working with electrodynamics were thinking about. And, of course, in a way Einstein was wrong, because the Lorentz transformation does not apply to everything. There is the microwave radiation, which does provide an absolute velocity. It provides an ether, but the real importance of Einstein's work was to show how Lorentz transformations dominate physics" (P. A. M. Dirac, "Discussion," *Strangeness in the Proportion*, pp. 110-111).
Cf. Elie Zahar, "Why Did Einstein's Programme Supersede Lorentz's? (II)," *British Journal for the Philosophy of Science* 24 (1973): 243-244.
Cf. Nathan Rosen's comment:
"In view of the existence of the Hubble effect, it appears that the universe is expanding. It also appears that there exists a frame of reference—nearly coinciding with that of the solar system—in which the universe presents isotropic appearance. This holds for the Hubble effect and also the microwave background radiation. In other words there exists a fundamental frame in the universe. From the equations of the general relativity theory one can also show that, in such an expanding universe, an observer carrying out mechanical and optical experiments in his laboratory in principle can determine the motion of the laboratory with respect to this fundamental frame of reference.
One cannot help wondering what would have happened had Einstein been aware of the existence of this fundamental reference frame at the time he was looking for a generalization of the special relativity theory that would describe gravitation. Would he have developed the same general relativity theory that he actually did?" (Nathan Rosen, "Bimetric General Relativity and Cosmology," *General Relativity and Gravitation* 12 [1980]: 494.)

222   CHAPTER 10

> The vacuum of modern physics is not identical with empty space—with Democritus' μη ov; it is to be taken for the 'ground state of the universe' and it possesses structure. That means that it contains all elementary particles virtually, in consequence of which it is polarizable and able to produce real particles and antiparticles after sufficient supply of energy. Therefore the vacuum is to be considered as its own substratum besides the classical matter and the metrical field.[69]

All physical reality is ultimately a manifestation of this quantum mechanical substratum. The quantum realm supplies the grounds for a modern equivalent of the ether in various ways, among which we may mention two:

(a) *Quantum electrodynamics*. In the classical electrodynamics of Maxwell, radiation was thought of as waves in the all-pervasive aether. But with the triumph of Special Relativity Theory, electromagnetic energy was regarded as traveling through the empty vacuum, and the aether medium was regarded as non-existent. As quantum theory penetrated electrodynamics, however, the vacuum was discovered to be anything but empty and to be the seat of fluctuating electromagnetic fields. P. A. M. Dirac, who pioneered the development of relativistic, quantum electrodynamics, regarded the idealized state of the quantum vacuum as a suitable candidate for an ether, commenting,

> Physical knowledge has advanced very much since 1905, notably by the arrival of quantum mechanics, and the situation has again changed. If one re-examines the question in the light of present day knowledge, one finds that the aether is no longer ruled out by relativity, and good reasons can now be advanced for postulating an aether.[70]

Dirac's ether is the quantum vacuum considered in abstraction from matter and field but which serves as the dynamical arena in which virtual events occur. It differs from the classical aether in that it is not amenable to a mechanical description; but it nonetheless plays a role similar to that of its classical predecessor, as Rompe and Treder explain:

> Dirac's quantum vacuum…is a physical entity for which one cannot define a magnitude of motion and even not the velocity zero, but which defines an inertial system within special relativity. And, furthermore, all particles, the whole substantial world of atomism and the real physical fields connected according to Einstein and de Broglie with atomism, represent excitations of the vacuum.[71]

As a physically real, universal, all-pervasive, inertial system, Dirac's quantum vacuum deserves, according to Sir Edmund Whittaker in his monumental *History of the Theories of Aether and Electricity*, to be regarded as the modern equivalent of the classical aether, which relativity theory had cast away:

---

[69] H. Hönl and H. Dehnen, "The Aporias of Cosmology and the Attempts at Overcoming Them by Nonstandard Models," in *Old and New Questions in Physics, Cosmology, Philosophy, and Theoretical Biology*, ed. Alwyn van der Merwe (New York: Plenum Press, 1983), p. 150.

[70] P. A. M. Dirac, "Is There an Aether?" *Nature* 168 (29 November 1951): 906-907; cf. P. A. M. Dirac and L. Infeld, "Is There an Aether?" *Nature* 169 (26 April 1952): 772.

[71] R. Rompe and H.-J. Treder, "Is Physics at the Threshold of a New Stage of Evolution?" in *Quantum, Space and Time—The Quest Continues*, ed. Asim O. Barut, Alwyn van der Merwe, and Jean-Pierre Vigier, Cambridge Monographs on Physics (Cambridge: Cambridge University Press, 1984), pp. 603-604. See also Alexander W. Stern, "Space, Field, and Ether in Contemporary Physics," *Science* 116 (November 1952): 493-496.

> As everyone knows, the aether played a great part in the physics of the nineteenth century; but in the first decade of the twentieth, chiefly as a result of the failure of attempts to observe the earth's motion relative to the aether, and the acceptance of the principle that such attempts must always fail, the word 'aether' fell out of favour, and it became customary to refer to the interplanetary spaces as 'vacuous'; the vacuum being conceived as mere emptiness, having no properties except that of propagating electromagnetic waves. But with the development of quantum electrodynamics, the vacuum has come to be regarded as the seat of the 'zero-point' oscillations of the electromagnetic field, of the 'zero-point' fluctuations of electric charge and current, and of a 'polarization' corresponding to a dielectric constant different from unity. It seems absurd to retain the name 'vacuum' for an entity so rich in physical properties, and the historical word 'aether' may fitly be retained.[72]

This modern quantum mechanical ether thus serves to delineate a fundamental reference frame with respect to which, like the classical aether, privileged velocities occur and, hence, privileged spatial and temporal relations may be established. Ironically, due to an ether compensatory mechanism, no 'ether wind' can be detected due to an observer's motion through the vacuum.[73] In this respect it resembles even more closely the classical aether. This new ether, in the words of Igor Novikov, "can restore the concepts of absolute rest and absolute motion. Indeed, the motion relative to this medium would be the motion...with respect to absolute space."[74]

(b) *The EPR experiment and Bell's Inequalities*: A second way in which quantum physics serves to disclose a privileged reference frame and absolute simultaneity concerns the startling experimental results obtained on what was originally a thought experiment aimed at exposing the deficiencies of Bohr's Copenhagen Interpretation of quantum measurement situations. According to one recent commentator, the experimental verification of violations of Bell's Inequalities "constitutes the most significant event of the last half-century" for those interested in the fundamental structure of the physical world.[75] Let us first look at the original thought experiment before examining recent findings and their implications.

As is well-known, Einstein staunchly opposed the indeterminist interpretation of quantum physics in favor of a "hidden variables" interpretation, according to which elementary particles actually do possess complementary properties like position and momentum, even if these are incapable of being simultaneously measured by us; and

---

[72] Edmund Whittaker, *A History of the Theories of Aether and Electricity*, 2 vols. (rep. ed.: New York: Humanities Press, 1973), I:v.
[73] I. D. Novikov explains,
> "When an observer starts moving in this medium, the oncoming energy flow does meet him and it may seem that the observer could measure this flow (this would be the 'wind'). However, another oncoming energy flow, due to negative pressure, will also be there. This flow has negative sign, its magnitude equal to that of the former flow and exactly canceling it. As a result, no 'wind' is produced. Whatever the motion of the observer by inertia, he always measures the same energy density of the vacuum (if it is nonzero) and the same negative pressure, so no 'wind' will be created by the motion. The vacuum is the same for all observers moving by inertia with respect to one another..." (Igor D. Norikov, *The River of Time*, trans. Vitaly Kisin [Cambridge: Cambridge University Press, 1998], p. 175).

[74] Ibid., p. 174.
[75] Tim Maudlin, *Quantum Non-Locality and Relativity*, Aristotelian Society Series 13 (Oxford: Blackwell, 1994), p. 4.

between 1927 and 1935, in a prolonged dispute with Niels Bohr, he proposed a number of thought experiments, each aimed at demonstrating that the quantum physical description of a particle is really incomplete.[76] The most celebrated of these thought experiments was proposed in 1935 in collaboration with Boris Podolsky and Nathan Rosen and, hence, bears the initials of its concocters as the EPR experiment.

The EPR thought experiment is based on a reality criterion, which supplies a sufficient condition for some physical quantity's being real: "If, without in any way disturbing a system, we can predict with certainty (i.e., with probability equal to unity) the value of a physical quantity, then there exists an element of physical reality corresponding to this physical quantity."[77] Thus, if one could predict accurately, say, the position of a particle without in any way disturbing it via a measurement process, the particle must actually have a position. Accordingly, Einstein, Podolsky, and Rosen imagine the following scenario: two photons are fired in opposite directions along the line of motion. Complementarity forbids that both the momentum and position of one of the photons be measured simultaneously. But because the photons are identical, we know that if we determine the momentum of photon 1, then photon 2 must possess precisely the same momentum, even though no measurement is carried out on photon 2. Nor can our measurement of photon 1's momentum be said to disturb photon 2, for causal influences cannot propagate faster than the speed of light. Yet at the very instant that the momentum of photon 1 is established, photon 2 must have an identical momentum. Therefore, according to the EPR reality criterion, photon 2 must really possess a momentum. But, quite evidently, we could just as easily have chosen to measure photon 1's position instead of its momentum. But then by parity of reasoning, photon 2 would also have to possess a real position. So whether we choose to measure photon 1's momentum or position, photon 2, while remaining causally isolated, must also possess at the same time the relevant property, as really as does photon 1. But since the EPR reality criterion expresses a modal proposition, photon 2 must possess these properties not merely whenever measurements *are* carried out on photon 1, but whenever such measurements—and corresponding predictions—*can* be carried out. So long as we can predict photon 2's position and momentum without disturbing it, photon 2 must actually possess those properties. The quantum mechanical description of the system is therefore essentially incomplete: any description of the system will include specific values of only one of the complementary properties, even though the EPR experiment shows that the particles actually possess both. There must therefore exist hidden variables in any quantum mechanical situation which we are incapable of determining.

---

[76] The reader may note that Einstein's attitude toward quantum mechanical descriptions in quantum theory was thus the precise opposite of his positivistic attitude toward absolute simultaneity in relativity theory! This strange inconsistency was not lost on others, who queried Einstein about it, only to receive the reply, "A good joke should not be repeated!"

[77] A. Einstein, B. Podolsky, and N. Rosen, "Can Quantum Mechanical Description of Physical Reality Be Considered Complete?" *Physical Review* 47 (1935): 777, reprinted in *Quantum Theory and Measurement*, ed. John Archibald Wheeler and Wojciech Hubert Zurek, Princeton Series in Physics (Princeton: Princeton University Press, 1983), p. 138.

The EPR thought experiment remained just that until 1964, when J. S. Bell showed that any hidden variables theory which preserved locality—that is to say, prohibited action at a distance—must make statistical predictions which disagree with those made by quantum mechanics.[78] Suddenly, the EPR experiment had been seen to lead to testable results. In his paper, Bell notes that David Bohm's hidden variables interpretation of quantum mechanics escapes Von Neuman's alleged proof that no such theory is possible due to the fact that Bohm's interpretation has a grossly non-local structure. Bell's Theorem is the proof that any theory which reproduces exactly the quantum mechanical predictions must be non-local. In place of EPR's use of the complementary properties of position and momentum, Bell borrows Bohm's example of spin ½ particles moving freely in opposite directions and so correlated that if the measurement by use of magnets of the spin component in one particle yields the value +1, then measurement of the spin component in the other particle must yield the value −1, and *vice versa*. He then assumes "that if the two measurements are made at places remote from one another the orientation of one magnet does not influence the result obtained with the other"[79] and then attempts to elicit a contradiction with quantum mechanical predictions, thereby falsifying the locality assumption.

Most contemporary discussions of Bell's Theorem utilize measurements on the polarization of photons rather than their spin orientation. Light which is polarized possesses an electric field which is orthogonal to the direction of motion and which is oriented in a certain direction (for example, vertically or horizontally). Polarization can characterize not only wavelengths of light, but individual photons as well. In the EPR situation, the individual photons are so correlated that once one of them is measured and so acquires a polarization, both have the same polarization. The polarization detectors A and B are constructed to measure photon polarization either along or across the axis of the detector. If the two detectors are oriented at the same angle, their polarization correlation equals 1, a perfect match in the distribution of photons detected vertically or horizontally polarized with respect to the axis. If, on the other hand, B is set at a 90° angle to A's orientation, then their polarization correlation is 0, a perfect mismatch in the distribution of photons detected to be vertically or horizontally polarized. The rub comes when the angle between the settings of the detectors varies between zero and ninety degrees: here statistical correlations known as Bell's Inequalities emerge. For example, we may set A such that its directional setting differs from B's by some angle θ such that for every four photon pairs measured, A and B agree on the vertical/horizontal polarization of only three; for one photon pair out of four A will measure it as vertical and B as horizontal or *vice versa*. Recalling then the locality condition presupposed by Bell and beginning with the settings of A and B aligned, let us turn the setting of B such that it is at the angle θ with respect to A and then turn A in the opposite direction such that the angle between the settings of A and B is 2θ. Since the rate of mismatches at θ is one out of four, the rate at 2θ cannot be greater than

---

[78] J. S. Bell, "On the Einstein Podolsky Rosen Paradox," *Physics* 1 (1964): 195-200, reprinted in *Quantum Theory and Measurement*, pp. 403-408.
[79] Ibid., p. 403.

two out of four (though it could be less, since a pair of deviations from the detectors' aligned value could coincide to produce a match). This ratio is an example of a Bell Inequality. Quantum mechanics, however, predicts that at the angle 2θ, the mismatches would be *greater* than two out of four.

Since Bell's Theorem was explicated, a number of EPR-type experiments have been run, and the most precise of these, notably the experiments of Alain Aspect at the Institut d'Optique d'Orsay and the long distance tests of Tittel, Brendel, Gisin, and Zbinden of the University of Geneva, have fully vindicated the predictions of quantum mechanics.[80] The breaching of the Bell Inequalities therefore necessitates abandonment of the locality assumption which underlay the EPR thought experiment. According to Bell, "It is the requirement of locality, or more precisely that the result of a measurement on one system be unaffected by operations on a distant system with which it has interacted in the past, that creates the essential difficulty."[81] Therefore, according to Rae, "nearly everyone working in this area" has been convinced "that all local hidden-variable theories can now be discounted."[82] And it is worth emphasizing that abandonment of locality is not dependent upon a realist interpretation of quantum physics: the breaching of the Bell Inequalities can be demonstrated on the macro-level, so that even if quantum physics is someday superseded, we seem to be stuck with non-locality.[83]

The demonstration that reality is non-local seems to leave us in a dilemma, either horn of which has, in turn, important implications for the existence of an ether. Our first alternative is to hold that adjustments in the polarization at A have a causal effect on the polarization at B. Since the collapse of the wave function occurs instantaneously over arbitrarily large distances, "Quantum theory suggests that measurement at A, say, causes an instantaneous change at B, and this seems to be confirmed by experiment."[84] But such an instantaneous influence establishes absolute simultaneity and thus requires a re-interpretation of quantum theory along neo-Lorentzian lines. In his 1964 paper, Bell concluded, "…there must be a mechanism whereby the setting of one measuring device can influence the reading of another instrument, however remote. Moreover, the signal involved must propagate instantaneously, so that such a theory could not be Lorentz invariant."[85] Such instantaneous causal connections serve to establish an absolute reference frame in which the events at A and B are simultaneous. Hence, Bell, later pondering the implications of Aspect's experiments, comments,

> I think it's a deep dilemma, and the resolution of it will not be trivial; it will require a substantial change in the way we look at things. But I would say that the cheapest

---

[80] See Alain Aspect and Philippe Grangier, "Experiments on Einstein-Podolsky-Rosen-type correlations with pairs of visible photons," in *Quantum Concepts in Space and Time*, ed. R. Penrose and C. J. Isham (Oxford: Clarendon Press, 1986), pp. 1-15; W. Tittel, J. Brendel, N. Gisin, and H. Zbinden, "Long Distance Bell-type Tests Using Energy-Time Entangled Photons," *Physical Review* A 59/6 (1999): 4150-4163.
[81] Bell, "On the Einstein Podolsky Rosen Paradox," p. 403.
[82] Alastair I. M. Rae, *Quantum Physics: Illusion or Reality?* (Cambridge: Cambridge University Press, 1986), p. 45.
[83] See Nick Herbert, *Quantum Reality* (Garden City, N. Y.: Doubleday, 1985), pp. 234-236.
[84] Euan Squires, *The Mystery of the Quantum World* (Bristol: Adam Hilger, 1986), p. 100; cf. p. 102.
[85] Bell, "On the Einstein Podolsky Rosen Paradox," p. 107.

resolution is something like going back to relativity as it was before Einstein, when people like Lorentz and Poincaré thought that there was an aether—a preferred frame of reference—but that our measuring instruments were distorted by motion in such a way that we could not detect motion through the aether....that is certainly the cheapest solution. Behind the apparent Lorentz invariance of the phenomena, there is a deeper level which is not Lorentz invariant....what is not sufficiently emphasized in textbooks, in my opinion, is that the pre-Einstein position of Lorentz and Poincaré, Larmor and Fitzgerald was perfectly coherent, and is not inconsistent with relativity theory. The idea that there is an aether, and these Fitzgerald contractions and Larmor dilations occur, and that as a result the instruments do not detect motion through the aether—that is a perfectly coherent point of view.... The reason I want to go back to the idea of an aether here is because in these *EPR* experiments there is the suggestion that behind the scenes something is going faster than light. Now if all Lorentz frames are equivalent, that also means that things can go backward in time....[this] introduces great problems, paradoxes of causality, and so on. And so it is precisely to avoid these that I want to say there is a real causal sequence which is defined in the aether.[86]

If, then, we allow for causal connections between the events at A and B, the EPR experiment implies the existence of absolute simultaneity and an ether frame.

And, in fact, such a theory exists in the form of de Broglie-Bohm pilot wave model. Bohmian quantum mechanics is mathematically consistent and consonant with all experimental results but is a deterministic theory featuring super-luminal causal influences. According to Henry Stapp,

The simplest picture of nature compatible with quantum theory is the model of David Bohm. It explains all of the empirical facts of a relativistic quantum theory, including, in particular, the impossibility of transmitting 'signals' (i.e., controlled messages) faster than light. In spite of this complete agreement with relativistic quantum theory at the level of observed phenomena, and the strict prohibition of all observable faster-than-light effects, Bohm's model is based explicitly on the postulated existence of an advancing sequence of preferred global 'nows,' which single out a preferred reference frame for defining absolute simultaneity.[87]

Bohm's interpretation thus demotes SR from "the status of absolutely universal foundational theory and instead demands relativistic invariance only of the 'observational' content of a physical theory"—precisely Lorentz's approach.[88] Unfortunately, the same epistemological positivism which underlay Einstein's SR also served to torpedo Bohm's interpretation of the formalism of quantum mechanics. Bohm ascribed the indifference to his theory to "a general philosophical

---

[86] "John Bell," interview in P. C. W. Davies and J. R. Brown, *The Ghost in the Atom* (Cambridge: Cambridge University Press, 1986), p. 48-49. Cf. J. S. Bell, "The paradox of Einstein, Podolsky, and Rosen: action at a distance in quantum mechanics?" *Speculations in Science and Technology* 10 (1987): 279. According to Vigier,

"...far from destroying the Einstein-de Broglie causal point of view, it may very well turn out that quantum non-locality would act in their favor, provide evidence of the 'aether's' existence, strengthening (instead of weakening) their basic idea that our universe is a causal, deterministic machine—and show that Einstein was basically right in the Bohr-Einstein controversy" (Jean-Pierre Vigier, "Louis de Broglie—Physicist and Thinker," in *Quantum, Space and Time*, pp. 9-10).

See also J. S. Bell, "On the Impossible Pilot Wave," in *Quantum, Space and Time*, pp. 66-76.

[87] Henry Stapp to D. R. Griffin, April 16, 1992, cited in David Ray Griffin, "Hartshorne, God, and Relativity Physics," *Process Studies* 21 (1992): 109-110.

[88] James T. Cushing, *Philosophical Concepts in Physics* (Cambridge: Cambridge University Press, 1998), p. 337.

point of view containing various branches such as 'positivism,' 'operationalism,' 'empiricism,' and others and which began to attain widespread popularity among physicists during the twentieth century."[89] In fact, as we have seen, Einstein's verificationist-operational approach to problems of time and space actually served to engender a similar approach to quantum physics.[90] The strategy of the defenders of Copenhagen orthodoxy was simple:

> Simply postulate that what cannot be measured—does not exist. By defining concepts operationally through a procedure for their measurement, and then applying the quantum formalism to the analysis of the measurement procedure, we will obtain nothing less but deductions from the quantum formalism (such as, for example, the uncertainty relations). In this way an illusion is created that the features of the theory (such as uncertainty) belong to the very definitions of the concepts used, and that they inevitably follow from a logical analysis of conditions of experience.[91]

In fact the triumph of the Copenhagen Interpretation was secured only by the "dictatorial help" of operationalism.[92] With the collapse of positivism, a Bohmian approach to quantum physics is receiving renewed attention.[93] Callender and

---

[89] David Bohm, *Causality and Chance in Modern Physics* (London: Routledge, Kegan & Paul, 1957), p. 97.

[90] Cushing remarks, "The prevalence of an empiricist-operationalist philosophical tendency among Heisenberg, Pauli, and Bohr can be traced in part (somewhat ironically, given Einstein's later view) back to Einstein's 1905 relativity papers. This operationalist approach, one aspect of which was an eschewal of unobservable entities in a theory, seems to have made a great impression and to have exerted a profound influence upon young German physicists" (Cushing, *Philosophical Concepts in Physics*, p. 287). In fact, Heisenberg described his abandonment of the "Kantian category of causality" as the natural continuation of Einstein's overthrow of Kantian space and time as forms of intuition! (Mara Beller, "Bohm and the 'Inevitability' of Acausality," in *Bohmian Mechanics and Quantum Theory: An Appraisal*, ed. James T. Cushing, Arthur Fine, and Sheldon Goldstein, Boston Studies in the Philosophy of Science 184 [Dordrecht: Kluwer Academic Publishers, 1996], p. 214).

[91] Beller, "Bohm," p. 220.

[92] Ibid., p. 221.

[93] See, e.g., Antony Valentini, "Pilot-Wave Theory of Fields, Gravitation and Cosmology," in *Bohmian Mechanics and Quantum Theory*, pp. 45-66; Tim Maudlin, "Space-Time in the Quantum World," in *Bohmian Mechanics and Quantum Theory*, pp. 285-307. Valentini points out that the pilot wave model naturally singles out a preferred rest frame. "The nonlocality acts instantaneously across a true 3-space, defining an absolute simultaneity and a true time *t*" (Valentini, "Pilot-Wave Theory," p. 56). What emerges "is not Einstein's special relativity but Lorentz's earlier interpretation of the Lorentz transformations" (Ibid.). Such an absolute 3+1 approach to electrodynamics is simpler and therefore preferable to the relativistic approach. Valentini extends the absolute 3+1 approach into general relativity as well, postulating an absolute, curved 3-space which evolves in absolute time. Nonlocality distinguishes and maintains the absolute slicing of spacetime into 3-D hypersurfaces. Maudlin argues that if we are to avoid backward causation, we must either postulate a foliation of spacetime into spacelike hyperplanes which serve to define a preferred synchronization between widely separated events or else posit directly a synchronization parameter. The most straightforward route to integrating quantum theory with relativity theory, he states, is to add some structure to spacetime and, hence, to reject relativity as the complete story. He asserts,

> "Indeed, given that this is the most obvious way to frame Bohmian or collapse theory in a non-classical space-time, we ought to take it as a benchmark. What Bohm's theory and the collapse theories seem to need is something like the classical notion of simultaneity: a fundamental physical relation between events at space-like separation. In effect such a relation would induce a *foliation* of spacetime, a division of the space-time manifold into a stack of space-like hyperplanes. Putting a measure over those hyperplanes yields an absolute time function in terms of which the Bohmian dynamics or the collapse dynamics can be framed. If there is something objectionable about

Weingard have applied Bohmian quantum mechanics cosmologically and report that "when cosmological factors are considered, the de Broglie-Bohm interpretation remains the only satisfactory interpretation of quantum theory."[94] This model enables one to resolve the problem of time without having to split worlds, multiply minds, or even worry about observers collapsing wavefunctions, in contrast to the Many Worlds Interpretation, the Many Minds Interpretation, or the Copenhagen Interpretation. Time in the Bohmian model remains essentially the immutable, external, unobservable, unique time of Newtonian mechanics, in which the dynamic variables of cosmology, like the radius of the universe and the scalar field, evolve. This implies that Bohmian cosmology is not generally covariant: "The laws define a preferred time."[95] Callender and Weingard recognize that the admission of a preferred time may be upsetting to some, but they note that the goal of cosmology is to watch the evolution of physical quantities over time, which Bohmian cosmology, in contrast the infinite number of possible time parametrizations (most of which are pathological) compatible with SR, permits.

Earlier I alluded to a "sea-change" in the attitude toward Bohmian quantum mechanics which has occurred in the foundations of physics community largely as a result of EPR and the Bell Inequalities. Adoption of such an interpretation would require abandonment of an Einsteinian/Minkowskian approach to relativity theory in favor of a non-relativistic approach. But an increasing number of theorists today are willing to give a serious look at Bohm's interpretation. Even if the details of his theory undergo revision and development, the commitment to a preferred time will remain one of the defining features of such an approach.

Suppose, on the other hand, that we reject a causally deterministic interpretation in favor of some interpretation according to which wave function collapse does occur, the photons at A and B being somehow *correlated*, but not *causally connected*. On this interpretation, the composite state consisting of the two photons with their respective polarization detectors constitutes a single system, which is in a definite state.[96] To affect the behavior of one photon via measurement is to disturb the *whole* system. When the wave function of a photon at A collapses, there is an immediate and correlated collapse of the wave function of its counterpart at B. But clearly, even though this interpretation denies the superluminal causal influence from A to B posited by the hidden variables theory, it still just as effectively abrogates the relativity of simultaneity, since the collapse of the paired wave

---

adding a foliation to space-time, we should consider just how objectionable it is, since there is no point in doing something even *more* objectionable just to retain the relativistic account of space-time" (Maudlin, "Space-Time in the Quantum World," p. 295).
See Maudlin's ensuing, interesting discussion about what would be so terrible, after all, about positing such structure. He indicts contemporary theorists for an "obsessive attachment to Relativity" (Ibid., p. 305).
[94] Craig Callender and Robert Weingard, "The Bohmian Model of Quantum Cosmology," in *PSA 1994*, ed. David Hull, Micky Forbes, and Richard M. Burian (East Lansing, Mich.: Philosophy of Science Association, 1994), p. 218.
[95] Ibid., p. 224.
[96] See Niels Bohr, "Can Quantum-Mechanical Description of Physical Reality Be Considered Complete?" *Physical Review* 48 (1935): 696-702, reprinted in *Quantum Theory and Measurement*, pp. 145-151.

functions is simultaneous. Therefore, asseverations that such an interpretation would not run contrary to the received interpretation of SR because that theory prohibits only *signals* of superluminal velocities, not instantaneous correlations, are quite beside the point.[97] The point is that such correlations furnish the means of establishing relations of absolute simultaneity and a fundamental frame.[98] In his acclaimed *Quantum Non-Locality and Relativity*, Tim Maudlin repeatedly underscores this point. Although Maudlin interprets the counterfactual dependence relation which exists between the separated polarization events in EPR as a causal connection, his emphasis lies on the fact that relativity cannot make physical sense of instantaneous collapse of the wave function. He writes,

> The classical formulations of non-relativistic quantum theory also posit an instantaneous non-local change: wave collapse. When our two photons leave the source each is in an indefinite state of indefinite polarization (or each is not in any state of polarization). At the moment that the first photon is observed the wave function describing the pair undergoes a sudden change such that the unobserved partner assumes a definite state of polarization which matches its partner. In this way the perfect correlations of the EPR experiment are maintained (and the correlations which violate Bell's inequality generated).
>
> In Minkowski spacetime this theory of wave collapse no longer makes sense. The collapse can be instantaneous in at most one reference frame, leading to two possibilities: either some feature of the situation picks out a preferred reference frame, with respect to which the collapse is instantaneous, or the collapse is not instantaneous at all.[99]

The problem posed by instantaneous collapse of the wave function for relativity theory can be clarified by realizing that since, according to SR, simultaneity is relative to reference frames, the collapse of the wave function for spatially separated photons will itself become relative to a reference frame, as Figure 10.5 illustrates.

---

[97] Such assertions may not even be true. It is very difficult to see why the entanglement-based quantum cryptography suggested by the experiments of the Geneva Group (see note 80) would not involve the instantaneous transmission of information even in the absence of superluminal propagation of causal influences. James Franson of Johns-Hopkins, when asked how identical random-number sequences generated simultaneously by widely separated particles differs from information, could only say, "That's a difficult question, and I don't think anyone could give you a coherent answer. Quantum theory is confirmed by experiments, and so is relativity theory, which prevents us from sending messages faster than light. I don't know that there's any intuitive explanation of what that means" ("Far Apart, 2 Particles Respond Faster than Light," *New York Times* [22 July 1997], p. C1).

[98] See Henry P. Stapp, "Are Faster-Than-Light Influences Necessary?" in *Quantum Mechanics versus Local Realism*, ed. Franco Selleri, Physics of Atoms and Molecules (New York: Plenum Press, 1988), pp. 71-72, who points out that Bohr and Heisenberg themselves effectively reject the EPR locality assumption. See the very interesting statement by Werner Heisenberg, *The Physical Principles of the Quantum Theory* (New York: Dover, 1930), p. 39.

[99] Maudlin, *Quantum Non-Locality*, p. 196; cf. pp. 137-138, 144.

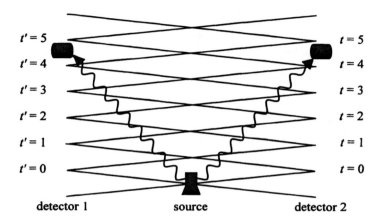

Figure 10.5. The EPR experiment in two reference frames. For the unprimed frame, the right photon strikes the detector first, and instantly the left photon acquires a definite polarization *before* it strikes detector 1; but for the primed frame the left photon is detected first and at the same time the right photon acquires a definite polarization *before* it strikes detector 2.

Whether a photon is definitely polarized before striking its detector depends upon one's reference frame. In some frames a photon is in an indefinite state until measured by a classical apparatus and its interaction with it will be stochastic; but relative to other frames that same photon is fully determinate before it reaches the detector and its interaction with it wholly deterministic. Thus, the determinacy or indeterminacy of a photon at the same spacetime point is hyperplane dependent. The problem, as Maudlin points out, is that

> polarization is not such an intrinsically relational matter. In a universe with only one photon, the photon could have a definite state of polarization. Indeed, one has the strong intuition that whether or not a photon is polarized should be a matter purely of the intrinsic state of it, independent of considerations about hyperplanes.[100]

The theory of the hyperplane-dependence of quantum states achieves Lorentz invariance only by exacting an exorbitant price:

> It does so via a fundamental rejection of the ontological foundations of both common sense and of classical physics. Indeed, the rejection is more radical than either Relativity or non-relativistic quantum theory suggest. Photons in non-relativistic quantum theory may be in strange states of indefinite polarization, but at least they are perfectly determinate strange states. Photons in [the hyperplane dependence] theory can be both polarized and not polarized. Measurement events can be both deterministic and stochastic. It all depends on the particular hyperplane to which one assigns the photon. And Nature assigns the photons even-handedly to all the hyperplanes at once.[101]

---

[100] Ibid., p. 210.
[101] Ibid., p. 212.

The problems attending hyperplane dependence in SR are magnified when we move to GR. Due to the curvature of spacetime, one must make quantum states dependent upon any spacelike hypersurface. Maudlin points out that the quantum state of a particle may differ relative to two hypersurfaces even though they coincide where they intersect the particle (Figure 10.6).

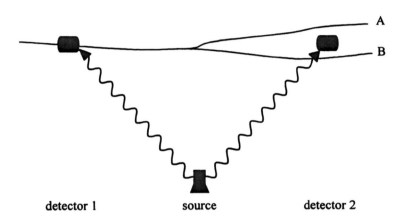

Figure 10.6. The left photon intersects hypersurfaces A and B at a place where they constitute a common surface. Yet relative to A the photon has a definite polarization, whereas relative to B it does not.

After surveying various attempts to integrate Bell's Inequalities with relativity theory, Maudlin concludes that these theories "entail such severe dislocations of our physical view that one must seriously consider whether our grounds for adhering to Relativity are really strong enough to justify such extreme measures."[102] "Indeed, the cost exacted by those theories which retain Lorentz invariance is so high that one might rationally prefer to reject Relativity as the ultimate account of spacetime structure."[103] The postulation of preferred hyperplanes of simultaneity in the structure of spacetime is, in fact, the only position discussed by Maudlin which does not face severe objections. The only objection to such an approach to quantum theory and EPR is that it requires one to reject Einsteinian Relativity. Maudlin writes,

> If a preferred family of hyperplanes is part of the intrinsic structure of spacetime then the fundamental postulate of the Theory of Relativity is false. Relativity postulates that the metrical structure is the *only* intrinsic spatio-temporal structure and that, more specifically, nothing intrinsic to spacetime picks out a preferred reference frame. A family of hyperplanes distinguishes one frame over all others, namely the frame in which the hyperplanes are simultaneity slices. This would not demand the elimination

---

[102] Ibid., p. 239. Cf. P. H. Eberhard, "Bell's Theorem and the Different Concepts of Locality," *Il Nuovo Cimento* 46B (1978): 392-419.
[103] Maudlin, *Quantum Non-Locality*, p. 220.

of any relativistic structure, but would undercut the relativistic democracy of frames. In a relativistic milieu, absolute simultaneity implies a unique privileged frame, allowing the definition of Absolute Motion and Rest. So option 1 can be adopted only by those who are willing to throw out the relativistic world-view.[104]

In light of the above, it is little wonder that Karl Popper regards the Aspect experiments as the first crucial experiments between Lorentz's and Einstein's interpretations of the formalism of Special Relativity. He remarks,

> The reason for this assertion is that the mere existence of an infinite velocity entails that of an absolute simultaneity and thereby of an absolute space. Whether or not an infinite velocity can be attained *in the transmission of signals* is irrelevant for this argument: the one inertial system for which Einsteinian simultaneity coincides with absolute simultaneity...would be the system at absolute rest—whether or not this system at absolute rest can be experimentally identified.[105]

If there is action at a distance, advises Popper, "it would mean that we have to give up Einstein's interpretation of special relativity and return to Lorentz's interpretation and with it to Newton's absolute space and time."[106] Popper goes on to observe that none of the formalism of SR need be given up, but only Einstein's interpretation of it. "If we now have theoretical reasons from quantum theory for introducing absolute simultaneity, then we would have to go back to Lorentz's interpretation."[107]

The above three physical realities—the cosmological fluid, the microwave background radiation, and the quantum mechanical vacuum—all serve to revitalize in a new guise the concept of the ether. James T. Cushing, after arguing on the basis of Bohm's quantum mechanics for the existence of a preferred frame, remarks, "As a curious point, there *does* today seem to be a universal preferred frame defined by the asymmetry of the cosmic background radiation.... Today...the aether has re-emerged through quantum phenomena!"[108] Bohm himself, maintaining that there is "a unique spacetime frame, in terms of which 'simultaneous contact' would be

---

[104] Ibid., p. 202.
[105] Karl Popper, "A Critical Note on the Greatest Days of Quantum Theory," in *Quantum, Space and Time*, p. 54; cf. idem, *Quantum Theory and the Schism in Physics* (Totowa, N. J.: Rowman & Littlefield, 1982), pp. xviii, 20. Even Popper's use of the expression "infinite velocity" is misleading, since the salient point is the simultaneous collapse of the correlated two wave functions, *as if* they were joined by an influence of infinite velocity. See Abner Shimony, "Metaphysical Problems in the Foundations of Quantum Mechanics," *International Philosophical Quarterly* 18 (1978): 13.
[106] Popper, *Quantum Theory*, p. 29. Actually, as Maudlin points out, one may either return to a Newtonian spacetime and show how electromagnetic effects in rods and clocks conceal the fundamental Newtonian structure or else one can retain the relativistic metric at the fundamental level and add some spacetime structure, such as preferred foliation, to it. Maudlin himself prefers to abandon relativity by means of the latter route because it is more straightforward. (Maudlin, "Space-Time in the Quantum World," p. 297; cf. pp. 295, 306). On either account, as Callender notes, temporal becoming "could occur either with respect to this extra structure or with respect to the underlying neo-Newtonian structure" (Callender, "Shedding Light on Time," p. 8).
[107] Ibid., p. 30.
[108] James T. Cushing, "What Measurement Problem?" in *Perspectives on Quantum Reality*, ed. Rob Clifton, University of Western Ontario Series in Philosophy of Science 57 (Dordrecht: Kluwer Academic Publishers, 1996), p. 175; cf. James T. Cushing, *Quantum Mechanics: Historical Contingency and the Copenhagen Hegemony*, Science and Its Conceptual Foundations (Chicago: University of Chicago Press, 1994), pp. 188-192. Popper also points out that there are "independent arguments" for a return to Lorentz's approach, as required by EPR, "especially since the discovery of the microwave background radiation" (Popper, *Quantum Theory*, p. 30).

specified," comments, "Empirically, this should be close to the frame in which the mean velocity of the 3°K radiation background in space is zero."[109] Since the microwave background radiation is isotropic only with respect to the frame in which the fundamental particles of the expanding universe are at rest, the fundamental frame of the cosmic expansion fills the role of the preferred frame required by quantum theory.[110] According to Prokhovnik, "The notion of non-local causality, discussed by Bell, requires a criterion of simultaneity which has some absolute significance; it is seen that a cosmological basis for a universal measure of (cosmic) time resolves this problem....the existence of an observationally-based fundamental frame is invaluable not only for the understanding of our universe and of Special Relativity, but also to make sense of quantum theory along the lines proposed, for example, by Dirac...and John Bell."[111]

While differing markedly from the classical aether, the new ethers play the same essential role in marking off a fundamental reference frame, the frame of expanding physical space itself, in which privileged relations of simultaneity may be established. Lorentz's prediction has been fully vindicated: "In my opinion it is not impossible that in the future this road, indeed abandoned at present, will once more be followed with good results, if only because it can lead to the thinking out of new experimental results."[112]

It seems to me, therefore, that we have quite convincing grounds for holding that cosmic time is, indeed, a privileged time. Its privileged status is implied by the existence of a fundamental reference frame for light propagation in line with modern cosmology and by the existence of modern ethers which serve to demarcate a preferential reference frame. Cosmic time is *the* time of the duration of the universe. Therefore, according to the argument presented above based on principle *P*, cosmic time and God's metaphysical time now coincide. The lapse of cosmic time measures the lapse of God's time, and the "now" of God's time coincides with the "now" of cosmic time.

About the only way to avoid this conclusion would be to claim that our physical, expanding space does not coincide with absolute space, but is in motion relative to it, so that God's metaphysical time does not coincide with cosmic time either. The question is not whether our 3-space is embedded in some higher dimensional space, but rather whether our physical 3-space coincides with metaphysical 3-space. But

---

[109] David Bohm to D. R. Griffin, May 17, 1992, cited in Griffin, "God and Relativity Physics," p. 110. When Callender and Weingard contrast the cosmic time of Bohmian cosmology with "the arbitrary parameter found in general relativity," the contrast concerns the arbitrariness permitted by GR taken *in abstracto* (Callender and Weingard, "Bohmian Model," p. 227). GR cosmic time and Bohmian cosmic time may well be extensionally equivalent, even though intensionally diverse.

[110] According to Valentini,
> "...the absolute 3-space along which subquantum nonlocality acts...will necessarily coincide with the observed rest-frame defined by the uniform microwave background. We therefore predict that, if the nonlocality ever becomes directly observable, it will be found to propagate along the hypersurface defined by the observed cosmological rest frame. (Note that York slicing coincides with cosmological rest for a Friedmann model.)" (Valentini, "Pilot-Wave Theory," p. 63).

[111] S. J. Prokhovnik, "A Cosmological Basis for Bell's Views on Quantum and Relativistic Physics," preprint.

[112] H. A. Lorentz, *The Einstein Theory of Relativity* (New York: Orentano's, 1920), pp. 61-62.

the problem with such an objection is that while God's existence (if He is changing) entails the existence of metaphysical time, His existence in no wise implies the existence of metaphysical space. While Newton correctly inferred the existence of metaphysical time from God's eternity, his deduction of metaphysical space from God's omnipresence is a *non sequitur*. If we thus reject Newton's curious doctrine that absolute space is God's *sensorium*, then we simply have no grounds for positing some metaphysical space which is non-coincident with physical space. After all, the objection under consideration does not spring from relativity theory; it could have been pressed with equal force against classical aether theories: maybe the aether itself was moving with respect to absolute space! It was simply assumed that the aether was at rest with respect to space itself and so served as its physical embodiment. In the context of modern cosmology, it is not clear that it even makes sense to speak of our expanding 3-space moving relative to a metaphysical 3-space.[113] The proper question is not whether physical space is moving in some metaphysical space, but rather whether any physically definable reference frame should be thought of as privileged and thus coincident with metaphysical space. On this score it must be said that the same Robertson-Walker metric that serves to distinguish a privileged time serves equally to demarcate a privileged reference frame that may be taken to coincide with metaphysical space as squarely as did the aether, except that this privileged frame is now seen to be expanding.[114] Maudlin observes that it is difficult to imagine what evidence for the existence of Newton's absolute space and time could be stronger than the occurrence of velocity terms in fundamental laws of nature, velocity terms which do not seem to refer to velocities relative to any material.[115] Hence, it seems to me that the objection is quite groundless and that there is no good reason to think that the isotropically expanding space of modern cosmology is not coincident with space itself. Therefore, I conclude, in answer to our first question, that God's metaphysical time does coincide with cosmic time.

THE ABSOLUTENESS OF COSMIC TIME

That leads us on to our second question: is cosmic time in some sense absolute? We have already had occasion to say something about this in our answer to the previous question. Since cosmic time coincides with God's metaphysical time since the moment of creation, it records, in effect, Newton's absolute time. We have seen

---

[113] In Milne's kinematic theory of relativity, the material universe is viewed as expanding into a static, empty 3-space. But even here, it is not physical space which is expanding in metaphysical space. And Milne's theory preserves a cosmic time. In any case, such a move on our objector's part would mean his abandoning Einsteinian relativity theory, which was supposed to be the basis for the whole objection to absolute becoming which is at issue here. For Milne's theory see E. A. Milne, *Kinematic Relativity* (Oxford: Clarendon Press, 1948); idem, *Modern Cosmology and the Christian Idea of God* (Oxford: Clarendon Press, 1952). For discussion, see Grünbaum, *Space and Time*, chap. 13.

[114] See Fitzgerald's critique of Swinburne's failure to posit a privileged space as well as a privileged time on the basis of the Robertson-Walker metric. Swinburne cannot have it both ways, he asserts. "Either the Robertson-Walker frame...gives us privileged cosmic instants and also privileged places lasting through time, or it gives us neither" (Paul Fitzgerald, Critical notice of *Space and Time*, by R. Swinburne, *Philosophy of Science* 43 [1976]: 631).

[115] Maudlin, *Quantum Non-Locality*, p. 192.

that several thinkers are therefore quite willing to speak of cosmic time as absolute. John Barrow, for example, asserts,

> ...whereas Einstein taught us that in the world of special relativity (where gravity is ignored) there is no absolute standard of time that is better than any other, and by use of which everyone in the universe can agree on the time unambiguously, this is no longer true in general relativity. In relativistic cosmology there exist absolute times. For example, in an expanding Universe of uniform density, anyone in the Universe could use the local measure of that density to determine the absolute time since the beginning of the expansion.[116]

Park therefore concludes, "Special Relativity questions absolute simultaneity, but relativistic cosmology reestablishes an absolute frame of reference (that of the universe as a whole) relative to which Aristotle's statement about a universal present makes sense."[117] Agazzi also asserts that "*cosmic time...is essentially an equivalent of the old absolute time, and in particular it allows one to speak of the past, present, and future of the universe.*"[118] Frank Tipler concurs: "this unique foliation of spacetime by constant mean extrinsic curvature hypersurfaces defines absolute space and time in general relativity: the hypersurfaces are absolute space, and the timelike trajectories which are everywhere normal to the hypersurfaces are absolute time."[119] In answer to the question, "Why should we exalt this foliation of spacetime above all others?" Tipler gives three reasons: (1) It is a natural foliation, in the sense that it is defined by the global distribution of matter and gravitational waves. (2) We adopt the constant mean extrinsic curvature foliation as the standard of space and time for the same reason we require clocks to measure universal time in Newtonian spacetime, namely, we adopt the time and space standard so that motion looks simple. (3) In a universe which is very close to the Friedmann case of homogeneity and isotropy (as our universe is) the rest frames of the foliation will coincide with the rest frames of the Cosmic Background Radiation. Tipler concludes, "The development of physics...has proven that absolute space and time do in fact exist, and we have succeeded in measuring our absolute motion. Newton's magnificent intuition, as shown in the *Principia*, has been fully vindicated."[120]

---

[116] John Barrow, *The World within the World* (Oxford: Oxford University Press, 1988) p. 234. Barrow's further discussion of which is the fundamental cosmic time has to do more with cosmic timekeeping and, despite his disclaimers, treats cosmic time on the pattern of Zeno's paradoxes, as is pointed out by Andreas Bartels, *Kausalitätsverletzungen in allgemeinrelativistischen Raumzeiten*, Erfahrung und Denken 68 (Berlin: Duncker & Humboldt, 1986), p. 112.

[117] David Park, "What is Time?" in *Time, Creation, and World Order*, Acta Jutlandica 74:1, Humanistic Series 72 (Aarhus, Denmark: Aarhus University Press, 1999), p. 22.

[118] Evandro Agazzi, "The Universe as a Scientific and Philosophical Problem," in *Philosophy and the Origin and Evolution of the Universe*, ed. Evandro Agazzi and Alberto Cordero, Synthèse Library 217 (Dordrecht: Kluwer Academic Publishers, 1991), p. 29.

[119] Frank J. Tipler, "The Sensorium of God: Newton and Absolute Space," in *Newton and the New Direction of Science*, ed. G. V. Coyne, M. Heller, and J. Zyncinski (Vatican City: Specola Vaticana, 1988), p. 222.

[120] Ibid., p. 224. Prokhovnik offers a more qualified endorsement:
> "Of course, cosmic time has no absolute connotation—it relates directly to the apparent nature of the cosmos—but it does provide a universal measure of time which 'flows' uniformly and independently of any local phenomena, and so fulfills Newton's desideratum for the nature of the ultimate variable associated with all physical changes" (Prokhovnik, *Light in Einstein's Universe*, p. 127).

But the suggestion that cosmic time is comparable to Newton's absolute time is likely to be met with heated resistance. Unfortunately, much of the disagreement seems due simply to the failure to keep clearly in mind Newton's distinction between absolute and relative time. Virtually all the objections are based on showing that cosmic time is in some way relative time and then concluding that it cannot therefore be absolute time—as if Newton had not himself distinguished the two! The pertinent question is whether the relative time kept by the ideal clock of a fundamental observer provides a measure of absolute time and so in that derivative sense can be said to share its absoluteness. Most critics have failed to keep clear the sense in which cosmic time can be said to be "absolute," and have compounded that failure by failing to appreciate the notion of the coincidence of cosmic time with metaphysical time.

For example, it is frequently said that cosmic time does not represent a return to Newtonian absolute time because cosmic time would not exist independently of all physical events as would absolute time.[121] But such an objection only reminds us that cosmic time is physical time rather than metaphysical time. One can reject cosmic time's absoluteness in the sense of "non-relational" without repudiating the absoluteness of cosmic time in the sense that cosmic time represents the true time (in the Lorentz-Poincaré sense) as opposed to the merely local time. The objection is impotent against the claim that cosmic time and metaphysical time are presently *coincident*, though not *identical*. Because these times are not identical, cosmic time need not share all the properties of absolute time, such as its allegedly non-relational character, and yet its moments still coincide with the moments of metaphysical time. In virtue of that coincidence, cosmic time may be quite properly said to be absolute in the sense that it gives the true time.

Or again, it is frequently objected that cosmic time is contingent and therefore cannot be regarded as absolute. We saw, for example, that Fitzgerald is worried about the existence of GR-based models which lack a cosmic time.[122] Since cosmic time does not exist in such models, it cannot be said to represent the true time. But all that follows from the existence of models lacking a cosmic time is that cosmic time *contingently* coincides with metaphysical time.[123] In virtue of that coincidence, it records the true time in this world. Our world is characterized by cosmic time, and its absence in other cosmological models is wholly irrelevant to whether it coincides with God's time in the actual world. The contingency of cosmic time thus says nothing against its privileged status in this world; in fact, a relationalist can consistently maintain that even metaphysical time exists contingently, for if God had chosen to exist absolutely changelessly permanently and never created a world, there would be no events at all and, hence, not even metaphysical time. The existence of both metaphysical and physical, cosmic time is thus a contingent fact dependent upon God's will.

On the basis of cosmic time's universality, Michael Shallis is also willing to say that it is like Newton's absolute time (Shallis, "Time and Cosmology," p. 71).
[121] Whitrow, *Natural Philosophy of Time*, pp. 34-36, 283-302.
[122] Fitzgerald, "Truth about Tomorrow's Sea-Fight," p. 326; see also Alan Padgett, *God, Eternity and the Nature of Time*, (New York: St. Martin's, 1992), pp. 128-129.
[123] Dorato also makes this point (Dorato, *Time and Reality*, p. 204).

Again, it is sometimes objected that cosmic time is not a vindication of Newton's absolute time because it is itself the by-product of relativity and is bound up with space. Munitz, for example, explains,

> One should not think, therefore, of what we are here calling 'cosmic time' as in any way equivalent to Newton's concept of absolute time. For relativity theory, time is not an independent continuum whose properties can be determined apart from space. Therefore, the assignment of a temporal structure to the universe as a physical system is intertwined with a determination of its spatial structure. Cosmic time is one dimension of the spacetime structure of the universe as a whole.[124]

But the fact that cosmic time is a formulation within a spacetime theory says nothing in and of itself against its absoluteness, since even Newtonian theory can be given such a formulation, as we have seen. Moreover, Munitz is not sufficiently discriminatory between cosmic parameter time and cosmic coordinate time. Only the latter is a dimension of the spacetime structure; but as a parameter cosmic time is independent of space. In the Robertson-Walker line element, the striking thing about the cosmic time parameter is precisely its independence from space. Of course, the parameter is assigned to certain hypersurfaces in spacetime, but then so is Newtonian time in a spacetime formulation of Newtonian theory. The point in saying that the cosmic time parameter is independent of space is not that it is unrelated to space, but that a mere change in spatial coordinates is not a sufficient condition for a change in the cosmic time. In any case, the objection becomes irrelevant once it is recalled that we are claiming, not the identity of Newtonian metaphysical time and cosmic time, but their coincidence since creation. For even if cosmic time were bound up with space, that fact says nothing against its coinciding with God's time and so being the privileged physical time.

A related objection is that cosmic time cannot be the basis for the construction of an absolute Newtonian time because cosmic time is observer-dependent and, hence, relativistic. Thus, Kroes admits that in universes in which fundamental observers exist one may introduce a special co-ordinate system which is distinguished by the fact that the spacelike hypersurfaces of constant t coincide with the hypersurfaces of homogeneity, so that relative to the group of fundamental observers, one may speak properly of "the" evolution of the universe. But, he insists, this temporal order is still observer-dependent: "it is based upon a preferred group of observers, viz., the group of fundamental observers."[125] Without this group of observers, the total temporal order has no physical meaning whatsoever. Therefore, one cannot construct an absolute Newtonian time on the basis of this cosmic time.

But Kroes's objection merely trades on the ambiguity of the expression "observer-dependent." Co-ordinate cosmic time does rely on the use of certain fundamental observers to establish origins of the co-moving coordinate system and in that sense can be said to be observer-dependent. But the time so constructed is the same for all observers, in whatever inertial frame they find themselves, and so is observer-independent. Every observer, whatever his state of motion, will measure events as occurring at the same values of cosmic time. Therefore, as Kroes says, the

---

[124] Munitz, *Cosmic Understanding*, p. 96.
[125] Kroes, *Time*, pp. 17-18.

fundamental observers are a *preferred* group of observers. Cosmic time is privileged, and relativity is thereby overcome.

The fact that cosmic time is relative to a reference frame, namely, the fundamental frame, is not incompatible with its coincidence with absolute time. Otherwise one might as well charge that the time of nineteenth century physics was not absolute either, since it was relative to the aether rest frame! The point is that in both cases, the reference frame in question is preferred and therefore the time kept with respect to it measures absolute time.

One of the most intriguing indications that cosmic time is the physical equivalent of Newtonian absolute time is the surprising demonstration by E. A. Milne and W. H. McCrea that all the results of GR-based, Friedmann cosmology can be recovered by Newtonian physics and in a way that is simpler than Einstein's cumbersome tensor calculus! Milne and McCrea were able to reproduce all the results of Big Bang cosmology by means of a material universe expanding in empty, classical space through classical time.[126] Comparing relativistic and Newtonian cosmology, Kerszberg observes, "as far as the prediction of the overall history of the universe is concerned, the equivalence seems to be total."[127] This implies, in Bondi's words, that GR "cannot be expected to explain any major features in any different or better way than Newtonian theory."[128] In particular the concept of cosmic time in GR-based models corresponds to absolute time in the Newtonian model. Schücking points out that the main asset of the Milne-McCrea formulation was that it gave exactly the same equations for the time development of the universe as the Friedmann theory and yet allowed a much simpler derivation.[129] The history of the universe described by the variation of the scale factor $R(t)$ in the Robertson-Walker line element is identical in the two theories, even though in the one the scale factor $R(t)$ is determined by Einstein's gravitational field equations, while in the other only Newtonian absolute time and Euclidian geometry come into play.[130] All this is not to suggest that Newtonian theory is correct after all; we have already seen how Lorentz was forced to modify Newtonian physics on the local level. But the equivalence of Milne-McCrea Newtonian cosmology with GR-based, Friedmann cosmology is a convincing demonstration that cosmic time is, indeed, the physical equivalent of Newtonian absolute time. Thus, Bondi compares cosmic time with Newton's uniform, omnipresent, and even-flowing time, which enables all observers to synchronize their clocks to a single time.[131] Kerszberg concludes, "On the whole, the equivalence between Newtonian and relativistic cosmology only reinforces the conviction that cosmic time is indeed a necessary ingredient in the formalisation of a

---

[126] E. A. Milne, *Relativity, Gravitation and World Structure* (Oxford: Clarendon Press, 1935); idem, "A Newtonian Expanding Universe," *Quarterly Journal of Mathematics* 5 (1934): 64-72; W. H. McCrea, "On the Significance of Newtonian Cosmology," *Astronomical Journal* 60 (1955): 271-274. For discussion see Peter T. Landsberg and David A. Evans, *Mathematical Cosmology: An Introduction* (Oxford: Clarendon Press, 1977).
[127] Pierre Kerszberg, "On the Alleged Equivalence between Newtonian and Relativistic Cosmology," *British Journal for the Philosophy of Science* 38 (1987): 349.
[128] H. Bondi, *Cosmology*, 2d ed. (Cambridge: Cambridge University Press, 1960), p. 89.
[129] E. L. Schücking, "Newtonian Cosmology," *Texas Quarterly* 10 (1967): 274.
[130] Bondi, *Cosmology*, p. 105; Kerszberg, "Equivalence," p. 349.
[131] Bondi, *Cosmology*, pp. 70-71.

relativistic cosmology, however alien to general relativity and congenial to Newton's theory the notion of universal synchronisation might seem."[132]

CONCLUSION

In conclusion, then, we have seen that when one moves from SR into GR, the application of the latter theory to cosmology yields a cosmic time, which is plausibly regarded as being the physical time which measures God's time and therefore registers to a good degree of approximation the true time. Although the theological significance of cosmic time is rarely discussed, sometimes contemporary thinkers come close. Consider, for example, Milne's notion of a *world map*. Munitz comments,

> Let us imagine a superhuman observer—a god—who is not bound by the limitations of the maximum velocity of light. Such an observer could survey in a single instant the entire domain of galaxies that have already come into existence. His survey would not have to depend on the finite velocity of light. It would not betray any restriction in information of the kind that results from the delayed time it takes to bring information about the domain of galaxies to an ordinary human observer situated in the universe, and who is therefore bound by the mechanisms and processes of signal transmission. The entire domain of galaxies would be seen instantaneously by this privileged superhuman observer. His observational survey of all galaxies would yield what Milne calls a 'world map.'[133]

A world map gives precisely the content of what God knows to be happening at any moment of cosmic time. It lays out, in effect, His perspective on the world. Kanitscheider proceeds,

> The theorist...would like to draw up, as Milne put it, a *world map*. On it the state of the world at a specific moment of cosmic time is indicated. All points of space on a hypersurface of spacetime are at once grasped and physically described. Such a slice through the happening of events corresponds, of course, to no datum of observation; rather it concerns a theoretical construction. Only a hypothetical, spiritual observer, who could visit all points on the hypersurface without any delay, would be able to achieve such an overview and could confirm the statement that $H(t)$ does in fact possess the same value at every place.[134]

Kanitscheider explains that we earthbound observers have only a *world picture*, in which distant parts of the world actually belong to earlier moments of cosmic time. Only an omnipresent, cosmic observer, he concludes, who sees the world *sub specie aeternitatis*, can be in the position to draw up a world map.[135] Here the relevance of cosmic time to the theological doctrine of divine eternity becomes explicit.

One final point should be made to avert possible misunderstanding. Padgett intimates that God could not be in cosmic time, or indeed, any measured time, because God would also then have to be spatial as well as temporal.[136] For in order to have cosmic time co-ordinates, God would also have to have spatial co-ordinates as well. Indeed, as we have seen above, He would have to be present at all points on

---

[132] Kerszberg, "Equivalence," p. 376.
[133] Munitz, *Cosmic Understanding*, p. 157.
[134] Kanitscheider, *Kosmologie*, p. 193.
[135] Ibid., p. 194.
[136] Padgett, *God, Eternity and the Nature of Time*, pp. 203-205.

any given hypersurface correlated with a cosmic time t. But then God would be extended physically throughout space, being Himself a sort of divine aether! But such a conception of God's omnipresence contradicts God's transcendence.

Such a conception of omnipresence is, indeed, flawed, but fortunately no such notion is implied by the doctrine of divine eternity which I have assumed.[137] In the first place, the objection does not distinguish between cosmic time as a parameter and as a co-ordinate. Insofar as cosmic time functions as a parameter, God's being in cosmic time does not imply His having spatial co-ordinates. Therefore, God could be in measured time while transcending space. But secondly, and more importantly, the claim is not that God exists in cosmic time as such; rather He exists in metaphysical time as such, but since creation the moments of cosmic time *coincide* with the moments of metaphysical time. It is in virtue of this coincidence that cosmic time measures God's metaphysical time. God thus exists in metaphysical time *per se*, but He exists in cosmic time *per accidens*. God does not exist in cosmic time in virtue of existing at all points on a three-dimensional hypersurface of spacetime, but rather because all the events on a given hypersurface are causally related to God as simultaneous and present. Cosmic time t corresponding to that hypersurface is "now" for Him. In virtue of this coincidence of the moments of cosmic time with the moments of His metaphysical time, God can be said to be in cosmic time in the sense that it is true for any cosmic time t that "God exists at t," and at t God believes "t is now." His omnipresence should be explicated in terms of His being aware of and causally active at every point in space.

---

[137] For a full explication and defense, see my *God, Time and Eternity* (Dordrecht: Kluwer Academic Publishers, forthcoming).

CHAPTER 11

CONCLUSION

The bold claims made on behalf of relativity theory to have destroyed the classical concept of time are seen to be empty. The classical concept of time as enunciated by Newton was founded squarely in a theistic metaphysic. On Newton's view God is an enduring being and His infinite duration constitutes time. Clock-time is but a physical and correctable measure of time itself. Nothing said by Einstein comes even near to refuting Newton's doctrine of metaphysical time. SR and GR were both essentially based on an epistemological positivism which is now universally rejected as untenable and obsolete. The empirical success of relativity theory testifies only to the necessity of revising what Newton called relative time or measured time. But Einstein's theories have zero impact on the classical Newtonian conception of absolute time, God's time.

For Newton absolute time was the seat of temporal becoming and ontological tenses. Although Einstein's original formulation of SR was compatible with the reality of tense, the so-called relativity interpretation of SR is implausible and explanatorily deficient. A spacetime interpretation remedies these defects, but precludes the objectivity of tense and faces other formidable difficulties. A neo-Lorentzian approach to relativity theory, which preserves the classical concept of time, is also free of the defects attending the relativity interpretation but unlike the spacetime interpretation is compatible with the reality of tense. Given an A-Theory of time and God's temporal existence, a neo-Lorentzian interpretation of the mathematical formalism of SR is to be preferred. Moreover, such an approach to relativity theory is actually favored by the discovery of modern equivalents of the aether. Therefore, if time is tensed, the temporality of God and neo-Lorentzian relativity—both plausible hypotheses I should say—follow as a consequence. The A-Theory of time, divine temporality, and neo-Lorentzian relativity thus go together as a coherent package.

BIBLIOGRAPHY

Agazzi, Evandro. "The Universe as a Scientific and Philosophical Problem." In *Philosophy and the Origin and Evolution of the Universe*, pp. 1-51. Edited by Evandro Agazzi and Alberto Cordero. Synthèse Library 217. Dordrecht: Kluwer Academic Publishers, 1991.

Albrecht, A., and Steinhardt, P. I. "Cosmology for Grand Unified Theories with Radioactively Induced Symmetry Breaking." *Physical Review Letters* 48 (1982): 1220-1223.

Angel, Roger. *Relativity: The Theory and its Philosophy*. Foundations and Philosophy of Science and Technology. Oxford: Pergamon Press, 1980.

Arthur, Richard T. W. "Newton's Fluxions and Equably Flowing Time." *Studies in History and Philosophy of Science* 26 (1995): 323-351.

Arzeliès, Henri. *Relativistic Kinematics*. Revised Edition. Oxford: Pergamon Press, 1966.

Aspect, Alain, and Grangier, Philippe. "Experiments on Einstein-Podolsky-Rosen-type Correlations with Pairs of Visible Photons." In *Quantum Concepts in Space and Time*, pp. 1-15. Edited by Roger Penrose and C. J. Isham. Oxford: Clarendon Press, 1986.

Baierlein, Ralph. *Newton to Einstein: The Trail of Light*. Cambridge: Cambridge University Press, 1992.

Balashov, Yuri. "Enduring and Perduring Objects in Minkowki Spacetime." *Philosophical Studies*, forthcoming.

Banks, T. "*TCP*, Quantum Gravity, the Cosmological Constant, and All That...." *Nuclear Physics B* 249 (1985): 332-360.

Barbour, Julian B. *Absolute or Relative Motion?* Volume 1: *The Discovery of Dynamics*. Cambridge: Cambridge University Press, 1989.

_____. "The Timelessness of Quantum Gravity: I." *Classical and Quantum Gravity* 11 (1994): 2853-2873.

_____. "The Timelessness of Quantum Gravity: II." *Classical and Quantum Gravity* 11 (1994): 2875-2897.

Barrow, Isaac. *The Geometrical Lectures of Isaac Barrow*. Translated with Notes by J. M. Child. Chicago: Open Court, 1916.

Barrow, John. D. *The World within the World*. Oxford: Oxford University Press, 1988.

Barrow, John. D., and Tipler, Frank J. *The Anthropic Cosmological Principle*. Oxford: Clarendon Press, 1986.

Bartels, Andreas. *Kausalitätsverletzungen in allgemeinrelativistischen Raumzeiten*. Erfahrung und Denken 68. Berlin: Duncker & Humboldt, 1986.

Bechler, Zev. "Introduction: Some Issues of Newtonian Historiography." In *Contemporary Newtonian Research*, pp. 1-20. Edited by Zev Bechler. Studies in the History of Modern Science 9. Dordrecht: D. Reidel, 1982.

Bell, John S. "How to Teach Special Relativity." In *Speakable and Unspeakable in Quantum Mechanics*, pp. 66-80. Cambridge: Cambridge University Press, 1987.

_____. "John Bell." In *The Ghost in the Atom*, pp. 45-57. Edited by P. C. W. Davies and J. R. Brown. Cambridge: Cambridge University Press, 1986.

_____. "On the Einstein Podolsky Rosen Paradox." *Physics* 1 (1964): 195-200. Reprinted in *Quantum Theory and Measurement*, pp. 403-408. Edited by John Archibald Wheeler and Wojciech Hubert Zurek. Princeton Series in Physics. Princeton: Princeton University Press, 1983.

_____. "On the Impossible Pilot Wave." In *Quantum, Space and Time—The Quest Continues*, pp. 66-76. Edited by Asim O. Barut, Alwyn van der Merwe, and Jean-Pierre Vigier. Cambridge Monographs on Physics. Cambridge: Cambridge University Press, 1984.

_____. "The Paradox of Einstein, Podolsky, and Rosen: Action at a Distance in Quantum Mechanics?" *Speculations in Science and Technology* 10 (1987): 269-285.

Beller, Mara. "Bohm and the 'Inevitability' of Acausality." In *Bohmian Mechanics and Quantum Theory: An Appraisal*, pp. 211-230. Edited by James T. Cushing, Arthur Fine, and Sheldon Goldstein. Boston Studies in the Philosophy of Science 184. Dordrecht: Kluwer Academic Publishers, 1996.

Bergson, Henri. *Duration and Simultaneity*. Translated by Leon Jacobson. With an Introduction by Herbert Dingle. Library of Liberal Arts. Indianapolis: Bobbs-Merrill, 1965.

Bilaniuk, Olexa-Myron, et al. "More about Tachyons." *Physics Today* (December 1969), pp. 47-52.

Black, Max. Review of *The Natural Philosophy of Time*, by G. J. Whitrow. *Scientific American* 206 (April 1962), pp. 179-184.

Bohm, David. *Causality and Chance in Modern Physics*. London: Routledge, Kegan & Paul, 1957.

_____. *The Special Theory of Relativity*. Lecture Notes and Supplements in Physics. New York: W. A. Benjamin, 1965.

Bohr, Niels. "Can Quantum-Mechanical Description of Physical Reality Be Considered Complete?" *Physical Review* 48 (1935): 696-702. Reprinted in *Quantum Theory and Measurement*, pp. 145-151. Edited by John Archibald Wheeler and Wojciech Hubert Zurek. Princeton Series in Physics. Princeton: Princeton University Press, 1983.

Bondi, Hermann. *Cosmology*. Second Edition. Cambridge: Cambridge University Press, 1960.

_____. "Is 'General Relativity' Necessary for Einstein's Theory of Gravitation?" In *Relativity, Quanta, and Cosmology in the Development of the Scientific Thought of Albert Einstein*, pp. 179-186. Edited by Francesco De Finis. 2 Volumes. New York: Johnson Reprint Corp., 1979.

_____. *Relativity and Common Sense*. New York: Anchor Books, 1964.

_____. "The Space Traveller's Youth." *Discovery* 18 (December 1957), pp. 505-510.

Born, Max. *Atomic Physics*. Fifth Edition. Translated by J. Douglas. London: Blackie, 1951.

_____. *Einstein's Theory of Relativity*. Translated by Henry L. Brose. New York: E. P. Dutton, no date.

_____. "Letters to the Editor: Physics: Special Theory of Relativity." *Nature* 197 (30 March 1963), p. 1287.

Bradley, J. *Mach's Philosophy of Science*. London: Athlone Press, 1971.

Brooke, John. H. "The God of Isaac Newton." In *Let Newton Be!*, pp. 169-183. Edited by John Fauvel, Raymond Flood, Michael Shortland, and Robin Wilson. Oxford: Oxford University Press, 1988.

_____. "Science and Religion." In *Companion to the History of Modern Science*, pp. 763-782. Edited by R. C. Olby, G. N. Cantor, J. R. R. Christie, and M. J. S. Lodge. London: Routledge, 1990.

Brown, G. Burniston. "News and Comment." *Physics Bulletin* 19 (1968): 22.

_____. "What Is Wrong with Relativity?" *Bulletin of the Institute of Physics and the Physical Society* 18 (1967): 71-77.

Brown, James Robert. "Einstein's Brand of Verificationism." *International Studies in the Philosophy of Science* 2 (1987): 33-55.

Builder, Geoffery. "The Constancy of the Velocity of Light." *Australian Journal of Physics* 11 (1958): 457-480. Reprinted in *Speculations in Science and Technology* 2 (1979): 421-437.

_____. "Ether and Relativity." *Australian Journal of Physics* 11 (1958): 279-297. Reprinted in *Speculations in Science and Technology* 2 (1979): 230-242.

_____. "The Resolution of the Clock Paradox." *Australian Journal of Physics* 10 (1957): 246-262.

Bunge, Mario. "Physical Time: The Objective and Relational Theory." *Philosophy of Science* 35 (1968): 355-388.

Burge, Tyler. "Philosophy of Language and Mind." *Philosophical Review* 101 (1992): 3-51.

Bush, G. "Note on the History of the FitzGerald-Lorentz Contraction." *Isis* 58 (1967): 230-232.

Callender, Craig. "Is Presentism Worth its Price?" Paper presented at the symposium, "The Prospects for Presentism in Spacetime Theories." Philosophy of Science Association biennial meeting, Kansas City, October, 1998.

_____. "Shedding Light on Time." Preprint.

Callender, Craig, and Weingard, Robert. "The Bohmian Model of Quantum Cosmology." In *PSA 1994*, pp. 218-227. Edited by David Hull, Micky Forbes, and Richard M. Burian. East Lansing, Mich.: Philosophy of Science Association, 1994.

Capek, Milic. "The Inclusion of Becoming in the Physical World." In *The Concepts of Space and Time*, pp. 501-524. Edited by Milic Capek. Boston Studies in the Philosophy of Science 22. Dordrecht: D. Reidel, 1976.

_____. "Introduction." In *The Concepts of Space and Time*, pp. xvii-lvii. Edited by Milic Capek. Boston Studies in the Philosophy of Science 22. Dordrecht: D. Reidel, 1976.

_____. *The Philosophical Impact of Contemporary Physics*. Princeton: D. Van Nostrand, 1961.

_____. "Time-Space Rather than Spacetime." *Diogenes* (Italy) 123 (1983): 324-343.

Capra, Fritjof. *The Tao of Physics*. Second Edition. London: Fontana, 1983.

Carrier, Martin. "Physical Force or Geometrical Curvature?" In *Philosophical Problems of the Internal and External Worlds*, pp. 3-21. Edited by John Earman, Allen I. Janis, Gerald J. Massey, and Nicholas Rescher. Pittsburgh: University of Pittsburgh Press, 1993.

Cassirer, Ernst. *Substance and Function and Einstein's Theory of Relativity* [1923]. Translated by W. C. Swabey and M. C. Swabey. Reprint Edition: New York: Dover, 1953.

Chang, Hasok. "A Misunderstood Rebellion: The Twin Paradox Controversy and Herbert Dingle's Vision of Science." *Studies in the History and Philosophy of Science* 24/5 (1993): 741-790.

Christensen, F. M. *Space-like Time*. Toronto: University of Toronto Press, 1993.

Cleugh, Mary F. *Time and its Importance in Modern Thought*. With a Foreword by L. Susan Stebbing. London: Methuen, 1937.

Clifton, Rob, and Hogarth, Mark. "The Definability of Objective Becoming in Minkowski Spacetime." *Synthèse* 103 (1995): 355-387.

Coburn, Robert C. *The Strangeness of the Ordinary: Problems and Issues in Contemporary Metaphysics*. Savage, Maryland: Rowman & Littlefield, 1990.

Cohen, I. Bernard. *Introduction to Newton's 'Principia'*. Cambridge: Cambridge University Press, 1971.

Cooper, Leon N. *An Introduction to the Meaning and Structure of Physics*. New York: Harper & Row, 1968.

Craig, William Lane. *Divine Foreknowledge and Human Freedom: The Coherence of Theism I: Omniscience*. Brill's Studies in Intellectual History 19. Leiden: E. J. Brill, 1991.

_____. *God, Time and Eternity*. Dordrecht: Kluwer Academic Publishers, forthcoming.

_____. *The* Kalam *Cosmological Argument*. Library of Philosophy and Religion. London: Macmillan, 1979.

_____. "McTaggart's Paradox and the Problem of Temporary Intrinsics." *Analysis* 58 (1998): 122-127.

_____. *The Tensed Theory of Time: a Critical Examination*. Synthèse Library. Dordrecht: Kluwer Academic Publishers, forthcoming.

_____. *The Tenseless Theory of Time: a Critical Examination*. Synthèse Library. Dordrecht: Kluwer Academic Publishers, forthcoming.

Craig, William Lane, and Smith, Quentin. *Theism, Atheism, and Big Bang Cosmology*. Oxford: Clarendon Press, 1993.

Cunningham, E. *Relativity, the Electron Theory, and Gravitation*. Second Edition. Monographs on Physics. London: Longmans, Green, & Co., 1921.

Cushing, James T. "The Causal Quantum Theory Program." In *Bohmian Mechanics and Quantum Theory: An Appraisal*, pp. 1-19. Edited by James T. Cushing, Arthur Fine, and Sheldon Goldstein. Boston Studies in the Philosophy of Science 184. Dordrecht: Kluwer Academic Publishers, 1996.
_____. *Philosophical Concepts in Physics*. Cambridge: Cambridge University Press, 1998.
_____. *Quantum Mechanics: Historical Contingency and the Copenhagen Hegemony*. Science and Its Conceptual Foundations. Chicago: University of Chicago Press, 1994.
_____. "What Measurement Problem?" In *Perspectives on Quantum Reality*, pp. 167-181. Edited by Rob Clifton. University of Western Ontario Series in Philosophy of Science 57. Dordrecht: Kluwer Academic Publishers, 1996.
D'Abro, A. *The Evolution of Scientific Thought* [1927]. Second Revised Edition. N. Y.: Dover Publications, 1950.
D'Inverno, Ray. *Introducing Einstein's Relativity*. Oxford: Clarendon Press, 1992.
Davies, Paul C. W. "The Acceptance Speech of Professor Paul Davies." Westminster Abbey, London, 3 May 1995.
_____. *God and the New Physics*. New York: Simon & Schuster, 1983.
_____. *Space and Time in the Modern Universe*. Cambridge: Cambridge University Press, 1977.
_____. "Spacetime Singularities in Cosmology and Black Hole Evaporations." In *The Study of Time III*, pp. 74-91. Edited by J. T. Fraser, N. Lawrence, and D. Park. Berlin: Springer, 1978.
De Beauregard, Olivier Costa. *Time, the Physical Magnitude*. Boston Studies in the Philosophy of Science 99. Dordrecht: D. Reidel, 1987.
De Sitter, Willem. "On the Relativity of Inertia." In *Koninglijke Nederlandse Akademie van Wetenschappen Amsterdam. Afdeling Wis-en Natuurkundige Wetenschappen. Proceedings of the Section of Science* 19 (1917): 1217-1225.
Di Salle, Robert. "Spacetime Theory as Physical Geometry." *Erkenntnis* 42 (1995): 317-337.
Dicke, R. H. "Mach's Principle and Equivalence." In *Evidence for Gravitational Theories*, pp. 1-50. Proceedings of the International School of Physics, "Enrico Fermi." Course xx. Edited by C. Møller. New York: Academic Press, 1962.
Dieks, Dennis. "Newton's Conception of Time in Modern Physics and Philosophy." In *Newton's Scientific and Philosophical Legacy*, pp. 151-159. Edited by P. B. Scheurer and G. Debrock. Dordrecht: Kluwer Academic Publishers, 1988.
_____. "The 'Reality' of the Lorentz Contraction." *Zeitschrift für allgemeine Wissenschaftstheorie* 15/2 (1984): 330-342.
Dingle, Herbert. "Letters to the Editor: Physics: Special Theory of Relativity." *Nature* 195 (8 September 1962), pp. 985-986.
_____. *Science at the Crossroads*. London: Martin Brian & O'Keefe, 1972.
_____. *The Special Theory of Relativity*. London: Methuen, 1940.
_____. "Special Theory of Relativity." *Nature* 197 (30 March 1963), pp. 1248-1249.

___. "The Time Concept in Restricted Relativity." *American Journal of Physics* 10 (1942): 203-205.

___. "Time in Philosophy and Physics." *Philosophy* 54 (1979): 99-104.

___. "Time in Relativity Theory: Measurement or Coordinate?" In *The Voices of Time*, pp. 455-472. Second Edition. Edited with a new Introduction by J. T. Fraser. Amherst, Mass.: University of Massachusetts Press, 1981.

Dirac, P. A. M. "Discussion." In *Some Strangeness in the Proportion*, pp. 110-111. Edited by Harry Woolf. Reading, Mass.: Addison-Wesley, 1980.

___. "Is There an Aether?" *Nature* 168 (29 November 1951), pp. 906-907.

Dirac, P. A. M., and Infeld, L. "Is There an Aether?" *Nature* 169 (26 April 1952), p. 772.

Dorato, Mauro. *Time and Reality: Spacetime Physics and the Objectivity of Temporal Becoming*. Collana di Studi Epistemologici 11. Bologna: CLUEB, 1995.

Dorling, Jan. "Length Contraction and Clock Synchronization: The Empirical Equivalence of the Einsteinian and Lorentzian Theories." *British Journal for the Philosophy of Science* 19 (1968): 67-69.

Duffy, Michael Ciaran. "The Modified Vortex Sponge: a Classical Analogue for General Relativity." Paper presented at the International Conference of the British Society for the Philosophy of Science, "Physical Interpretations of Relativity Theory." Imperial College of Science and Technology. London, 16-19 September, 1988.

Dugas, René. *A History of Mechanics*. With a Foreword by Louis de Broglie. Translated by J. R. Maddox. New York: Central Book Company, 1955.

Durbin, R.; Loar, H. H.; and Havens, W. W. "The Lifetime of the $\pi+$ and $\pi-$ Mesons." *Physical Review* 88 (1952): 179-183.

Earman, John. "Spacetime, or How to Solve Philosophical Problems and Dissolve Philosophical Muddles without Really Trying." *Journal of Philosophy* 67 (1970): 259-277.

___. "Who's Afraid of Absolute Space?" *Australasian Journal of Philosophy* 48 (1970): 287-319.

___. *World Enough and Spacetime*. Cambridge, Mass.: MIT Press, 1989.

Earman, John, and Friedman, Michael. "The Meaning and Status of Newton's Law of Inertia and the Nature of Gravitational Forces." *Philosophy of Science* 40 (1973): 329-359.

Earman, John, and Mosterin, Jesus. "A Critical Look at Inflationary Cosmology." *Philosophy of Science* 66 (1999): 1-49.

Eberhard, P. H. "Bell's Theorem and the Different Concepts of Locality." *Il Nuovo Cimento* 46B (1978): 392-419.

Eddington, Arthur S. *The Expanding Universe*. Cambridge: Cambridge University Press, 1952.

___. "A Generalization of Weyl's Theory of the Electromagnetic and Gravitational Fields." *Proceedings of the Royal Society of London* A99 (1921): 104-122.

___. *The Mathematical Theory of Relativity*. Second Edition. Cambridge: Cambridge University Press, 1930.

———. *The Nature of the Physical World* [1928]. With an Introductory Note by Sir Edmund Whittaker. Everyman's Library. Reprint Edition: London: I. M. Dent & Sons, 1964.

———. "On the Instability of Einstein's Spherical World." *Monthly Notices of the Royal Astronomical Society* 19 (1930): 668-678.

———. *Space, Time and Gravitation* [1920]. Cambridge Science Classics. Reprint Edition: Cambridge: Cambridge University Press, 1987.

Ehrenfest, Paul. *Zur Krise der Lichthypothese*. Berlin: Springer Verlag, 1913.

Einstein, Albert. "Autobiographical Notes." In *Albert-Einstein: Philosopher-Scientist*, pp. 1-94. Third Edition. Edited by Paul Arthur Schilpp. Library of Living Philosophers 7. La Salle, Ill.: Open Court, 1970.

———. *Äther und Relativitätstheorie*. Berlin: Julius Springer Verlag, 1920.

———. "Bemerkungen zu der Notiz von Hrn. Paul Ehrenfest: 'Die Translation deformierbarer Elektronen und der Flächensatz'." *Annalen der Physik* (4. ser.) 23 (1907): 206-208.

———. *The Collected Papers of Albert Einstein*. Vol. 1: *The Early Years, 1879-1902*. Edited by John Stachel. Princeton, N. J.: Princeton University Press, 1987.

———. "Cosmological Considerations on the General Theory of Relativity." In *The Principle of Relativity*, pp. 177-188. By Albert Einstein, *et al.* With Notes by A. Sommerfeld. Translated by W. Perrett and G. B. Jeffery. Reprint Edition: New York: Dover Publications, 1952.

———. "Dialog über Einwände gegen die Relativitätstheorie." *Naturwissenschaften* (1918): 697-702.

———. "Die formale Grundlage der allgemeinen Relativitätstheorie." *Physikalische Zeitschrift* 14 (1913): 1249-1262.

———. "Die Relativitäts-Theorie." *Vierteljahrsschrift der naturforschenden Gesellschaft in Zürich* 56 (1911): 1-14.

———. "Ernst Mach." *Physikalische Zeitschrift* 17 (1916): 101. Reprinted in Ernst Mach, *Die Mechanik in ihrer Entwicklung historisch-kritisch dargestellt*, pp. 683-689. Edited by Renate Wahsner and Horst-Heino Borzeszkowski. Berlin, DDR: Akademie-Verlag, 1988.

———. "Ether and the Theory of Relativity." In *Sidelights on Relativity*, pp. 2-24. New York: Dover Publications, 1903.

———. "The Foundations of General Relativity Theory." In *General Theory of Relativity*, pp. 141-172. Edited by C. W. Kilmister. Selected Readings in Physics. Oxford: Pergamon Press, 1973.

———. "Fundamental Ideas and Problems of the Theory of Relativity." [1923]. In *Nobel Lectures, Physics: 1901-1921*, pp. 479-490. New York: Elsevier, 1967.

———. *Lettres á Maurice Solovine*. Paris: Gauthiers-Villars, 1956.

———. *The Meaning of Relativity* [1922]. Sixth Edition. Reprint Edition: London: Chapman & Hall, 1967.

———. "On the Electrodynamics of Moving Bodies." Translated by Arthur I. Miller. In Appendix to Arthur Miller, *Albert Einstein's Special Theory of Relativity*, pp. 392-415. Reading, Mass.: Addison-Wesley, 1981.

———. "On the Methods of Theoretical Physics." In *The World as I See It*, pp. 30-40. New York: Covici-Friede, 1934.

———. "The Problem of Space, Ether, and the Field in Physics." [1934]. In *Ideas and Opinions*. pp. 276-285. Translated by Sonja Bargmann. New York: Crown Publishers, 1954.

———. *Relativity: The Special and the General Theory*. Fifteenth Edition. New York: Crown Trade Paperback, 1961.

———. *Relativity: The Special and the General Theory*. Translated by Robert W. Lauren. London: Methuen, 1920.

———. "Remarks on Bertrand Russell's Theory of Knowledge." In *The Philosophy of Bertrand Russell*, pp. 277-291. Third edition. Edited by P. A. Schilpp. Library of Living Philosophers. New York: Tudor Publishing, 1951.

———. "Über das Relativitätsprinzip und die aus demselben gezogenen Folgerungen." *Jahrbuch der Radioaktivität und Elektronik* 4 (1907): 411-462.

———. "Über die Entwicklung unserer Anschauungen über das Wesen und die Konstitution der Strahlung." *Physikalische Zeitschrift* 10 (1909): 817-825.

———. *Über die spezielle und allgemeine Relativitätstheorie* [1917]. Fourth Edition. Reprint Edition: Braunschweig: Friedr. Vieweg & Sohn, 1960.

———. "Wie ich die Welt sehe." In *Mein Weltbild*, pp. 7-45. Edited by Carl Selig. Frankfurt: Ullstein Bücher, 1934.

———. "Zum Ehrenfestschen Paradoxon." *Physikalische Zeitschrift* 12 (1911): 509-510.

Einstein, Albert and Besso, Michele. *Correspondance 1903-1955*. Translated with Notes and an Introduction by Pierre Speziali. Paris: Hermann, 1979.

Einstein, Albert, and Infeld, Leopold. *The Evolution of Physics*. New York: Simon & Schuster, 1938.

Einstein, Albert; Podolsky, B.; and Rosen, N. "Can Quantum Mechanical Description of Physical Reality Be Considered Complete?" *Physical Review* 47 (1935): 777-780. Reprinted in *Quantum Theory and Measurement*, pp. 138-141. Edited by John Archibald Wheeler and Wojciech Hubert Zurek. Princeton Series in Physics. Princeton: Princeton University Press, 1983.

Ellis, B., and Bowman, P. "Conventionality in Distant Simultaneity." *Philosophy of Science* 34 (1967): 116-136.

Epstein, Lewis Carroll. *Relativity Visualized*. San Francisco: Insight Press, 1981.

Epstein, Paul S. "The Time Concept in Restricted Relativity." *American Journal of Physics* 10 (1942): 1-6.

———. "The Time Concept in Restricted Relativity—A Rejoinder." *American Journal of Physics* 10 (1942): 205-208.

Evans, Melbourne G. "On the Falsity of the Fitzgerald-Lorentz Contraction Hypothesis." *Philosophy of Science* 36 (1969): 254-262.

"Far Apart, 2 Particles Respond Faster than Light." By Malcolme W. Browne. *New York Times*, 22 July 1997, p. C1.

Feyerabend, P. K. "Zahar on Einstein." *British Journal for the Philosophy of Science* 25 (1974): 25-28.
Findlay, J. N. "Time and Eternity." *Review of Metaphysics* 32 (1978-79): 3-14.
Fine, Arthur. *The Shaky Game: Einstein, Realism, and the Quantum Theory*. Chicago: University of Chicago Press, 1986.
Fitzgerald, Paul. Critical notice of *Space and Time*, by R. Swinburne. *Philosophy of Science* 43 (1976): 631.
―――. "Relativity Physics and the God of Process Philosophy." *Process Studies* 2 (1972): 251-276.
―――. "The Truth about Tomorrow's Sea Fight." *Journal of Philosophy* 66 (1969): 307-329.
FitzGerald, George Francis. "The Ether and the Earth's Atmosphere." *Science* 13 (1889), p. 390.
Föppl, August. *Einführung in die Maxwell'sche Theorie der Elektrizität*. Leipzig: Teubner, 1894.
Force, James E. "Newton's God of Dominion: The Unity of Newton's Theological, Scientific, and Political Thought." In *Essays on the Context, Nature, and Influence of Isaac Newton's Theology*, pp. 75-102. Edited by James E. Force and Richard H. Popkin. International Archives of the History of Ideas 128. Dordrecht: Kluwer Academic Publishers, 1990.
Frank, Philipp. "The Importance of Ernst Mach's Philosophy of Science for our Times." [1917]. Reprinted in *Ernst Mach: Physicist and Philosopher*, pp. 219-234. Edited by Robert S. Cohen and Raymond J. Seeger. Boston Studies in the Philosophy of Science 6. Dordrecht: D. Reidel, 1970.
―――. *Interpretations and Misinterpretations of Modern Physics*. Actualités Scientifiques et Industrielles 587: Exposés de Philosophie Scientifique 2. Paris: Hermann & Cie, 1938.
―――. *Philosophy of Science*. Englewood Cliffs, N. J.: Prentice-Hall, 1957.
Freundlich, Yehudah. "'Becoming' and the Asymmetries of Time." *Philosophy of Science* 40 (1973): 496-517.
Friedman, Michael. *Foundations of Spacetime Theories*. Princeton: Princeton University Press, 1983.
―――. "Simultaneity in Newtonian Mechanics and Special Relativity." In *Foundations of Spacetime Theories*, pp. 403-432. Edited by John S. Earman, Clark N. Glymour, and John J. Stachel. Minnesota Studies in the Philosophy of Science 8. Minneapolis: University of Minnesota Press, 1977.
Friedman, William. *About Time*. Cambridge, Mass.: MIT Press, 1990.
Friedmann, Aleksandr. "Über die Krümmung des Raumes." *Zeitschrift für Physik* 10 (1922): 377-386.
Frisch, David H., and Smith, James H. "Measurement of the Relativistic Time Dilation Using μ-Mesons." *American Journal of Physics* 31 (1963): 342-355.
Gabbey, Alan. "Newton and Natural Philosophy." In *Companion to the History of Modern Science*, pp. 243-263. Edited by R. C. Olby, G. N. Cantor, J. R. R. Christie, and M. J. S. Lodge. London: Routledge, 1990.

Gale, George. "Cosmos and Conflict: Interactions between Observation, Theory, and Metaphysics in Modern Cosmology." Paper presented at the conference, "The Origin of the Universe." Colorado State University, 22-25 September, 1988.

──────── . "Some Metaphysical Perplexities in Contemporary Physics." Paper presented at the 36th Annual Meeting of the Metaphysical Society of America. Vanderbilt University, 14-16 March, 1985.

Gale, Richard. "Human Time: Introduction." In *The Philosophy of Time: a Collection of Essays*, pp. 293-303. Edited by R. M. Gale. New York: Humanities Press, 1968.

Galilei, Galileo. *Dialogue Concerning the Two Chief World Systems—Ptolemaic and Copernican*. Translated by Stillman Drake. Berkeley: University of California Press, 1962.

Gamow, George. *My World Line*. New York: Viking Press, 1970.

Geroch, Robert. *General Relativity from A to B*. Chicago: University of Chicago Press, 1978.

Ghins, Michel. *L'inertie et l'espace-temps absolu de Newton à Einstein*. Mémoires de la Classe des Lettres 69/2. Brussels: Palais des Académies, 1990.

Glymour, Clark. "Topology, Cosmology, and Convention." In *Space, Time, and Geometry*, pp. 193-216. Edited by Patrick Suppes. Synthèse Library. Dordrecht: D. Reidel, 1973.

Gödel, Kurt. "A Remark about the Relationship between Relativity Theory and Idealistic Philosophy." In *Albert Einstein: Philosopher-Scientist*, pp. 555-562. Edited by Paul Arthur Schilpp. Library of Living Philosophers 7. LaSalle, Ill.: Open Court, 1949.

Goldberg, Stanley. "The Lorentz Theory of Electrons and Einstein's Theory of Relativity." *American Journal of Physics* 37 (1969): 982-994.

──────── . "Putting New Wine in Old Bottles: The Assimilation of Relativity in America." In *The Comparative Reception of Relativity*, pp. 1-26. Edited by Thomas F. Glick. Boston Studies in the Philosophy of Science 103. Dordrecht: D. Reidel, 1987.

──────── . *Understanding Relativity: Origin and Impact of a Scientific Revolution*. Boston: Birkhäuser, 1984.

Grant, Edward. "Medieval and Seventeenth-Century Conceptions of an Infinite Void Space beyond the Cosmos." In *Studies in Medieval and Natural Philosophy*, pp. 39-60. London: Variorum Reprints, 1981.

──────── . *Physical Science in the Middle Ages*. History of Science. New York: John Wiley & Sons, 1971.

Graves, John Cowperthwaite. *The Conceptual Foundations of Contemporary Relativity Theory*. With a Foreword by John Archibald Wheeler. Cambridge, Mass.: MIT Press, 1971.

Greene, Brian. *The Elegant Universe*. New York: W. W. Norton, 1999.

Grieder, Alfons. "Relativity, Causality and the 'Substratum'." *British Journal for the Philosophy of Science* 28 (1977): 35-48.

Griffin, David Ray. "Hartshorne, God, and Relativity Physics." *Process Studies* 21 (1992): 85-112.

_____. "Introduction: Time and the Fallacy of Misplaced Concreteness." In *Physics and the Ultimate Significance of Time*, pp. 1-48. Edited by David R. Griffin. Albany, N. Y.: State University of New York Press, 1986.

Grünbaum, Adolf. "The Denial of Absolute Space and the Hypothesis of a Universal Nocturnal Expansion: A Rejoinder to George Schlesinger." *Australasian Journal of Philosophy* 45 (1967): 61-91.

_____. *Philosophical Problems of Space and Time*. Second Edition. Boston Studies in the Philosophy of Science 12. Dordrecht: D. Reidel, 1973.

_____. "The Pseudo-Problem of Creation in Physical Cosmology." *Philosophy of Science* 56 (1989): 373-394.

_____. "Science and Ideology." *Scientific Monthly* 79 (July 1954): 13-19.

Guth, Alan. "Inflationary Universe: A Possible Solution to the Horizon and Flatness Problems." *Physical Review* D 23 (1981): 347-356.

Gutting, Gary. "Einstein's Discovery of Special Relativity." *Philosophy of Science* 59 (1972): 51-68.

Hafele, J. C., and Keating, R. E. "Around the World Atomic Clocks: Predicted Relativistic Time Gains." *Science* 177 (14 July 1972), pp. 166-170.

Hartle, James, and Hawking, Stephen W. "Wave Function of the Universe." *Physical Review D* 28 (1983): 2960-2975.

Hartshorne, Charles. "Bell's Theorem and Stapp's Revised View of Space-Time." *Process Studies* 7 (1977): 183-191.

Hawking, Stephen W. *A Brief History of Time: From the Big Bang to Black Holes*. With an Introduction by Carl Sagan. New York: Bantam Books, 1988.

_____. "The Existence of Cosmic Time Functions." *Proceedings of the Royal Society of London* A 308 (1968): 433-435.

Hawking, Stephen W., and Moss, I. G. "Supercooled Phase Transitions in the Very Early Universe." *Physics Letters* B 110 (1982): 35-38.

Hay, H. J.; Schiffer, J. P.; Cranshaw, T. E.; and Egelstaff, P. A.. "Measurement of the Red Shift in an Accelerated System Using the Mössbauer Effect in Fe." *Physical Review Letters* 4 (1960): 165.

Healey, Richard. "Introduction." In *Reduction, Time and Reality*, pp. vii-xi. Edited by Richard Healey. Cambridge: Cambridge University Press, 1981.

Heisenberg, Werner. *The Physical Principles of the Quantum Theory*. New York: Dover, 1930.

Heller, Michael; Klimek, Zbigniew; and Rudnicki, Konrad. "Observational Foundations for Assumptions in Cosmology." In *Confrontation of Cosmological Theories with Observational Data*, pp. 1-11. Edited by M. S. Longair. Dordrecht: D. Reidel, 1974.

Herbert, Nick. *Quantum Reality*. Garden City, N. Y.: Doubleday, 1985.

Herneck, Friedrich. "Nochmals über Einstein und Mach." *Physikalische Blätter* 17 (1961): 275-276.

Hirosige, Tetu. "The Ether Problem, the Mechanistic Worldview, and the Origins of the Theory of Relativity." *Historical Studies in the Physical Sciences* 7 (1976): 3-82.

———. "Origins of Lorentz' Theory of Electrons and the Concept of the Electromagnetic Field." *Historical Studies in the Physical Sciences* 1 (1969): 151-209.

Hoefer, Carl, and Cartwright, Nancy. "Substantivalism and the Hole Argument." In *Philosophical Problems of the Internal and External Worlds*, pp. 23-43. Edited by John Earman, Alan I. Janis, Gerald J. Massey, and Nicholas Rescher. Pittsburgh-Konstanz Series in the Philosophy and History of Science. Pittsburgh: University of Pittsburgh Press, 1993.

Hoffmann, Banesh, with Dukas, Helen. *Albert Einstein: Creator and Rebel.* London: Hart-Davis, MacGibbon, 1972.

Holton, Gerald J. "Einstein, Michelson and the 'Crucial' Experiment." In *Thematic Origins of Scientific Thought: Kepler to Einstein*, pp. 261-352. Cambridge, Mass.: Harvard University Press, 1973.

———. "Einstein's Scientific Program: The Formative Years." In *Some Strangeness in the Proportion*, pp. 49-68. Edited by Harry Woolf. Reading, Mass.: Addison-Wesley Publishing Co., 1980.

———. "Mach, Einstein and the Search for Reality." In *Ernst Mach: Physicist and Philosopher*, pp. 165-199. Boston Studies in the Philosophy of Science 6. Dordrecht: D. Reidel, 1970.

———. "On the Origin of the Special Theory of Relativity." In *Thematic Origins of Scientific Thought: Kepler to Einstein*, pp. 165-183. Cambridge, Mass.: Harvard University Press, 1973.

———. "On Trying to Understand Scientific Genius." In *Thematic Origins of Scientific Thought: Kepler to Einstein*, pp. 353-380. Cambridge, Mass.: Harvard University Press, 1973.

———. "Poincaré and Relativity." In *Thematic Origins of Scientific Thought: Kepler to Einstein*, pp. 185-195. Cambridge, Mass.: Harvard University Press, 1973.

———. "Where Is Reality? The Answers of Einstein." In *Science and Synthesis*, pp. 45-69. Edited by UNESCO. Berlin: Springer Verlag, 1971. Reprinted in *Thematic Origins of Scientific Thought: Kepler to Einstein*, pp. 45-69. Cambridge, Mass.: Harvard University Press, 1973.

Hönl, H., and Dehnen, H. "The Aporias of Cosmology and the Attempts at Overcoming Them by Nonstandard Models." In *Old and New Questions in Physics, Cosmology, Philosophy, and Theoretical Biology*, pp. 145-167. Edited by Alwyn van der Merwe. New York: Plenum Press, 1983.

Horwich, Paul G. *Asymmetries in Time.* Cambridge, Mass.: MIT Press, 1987.

———. "On the Existence of Time, Space, and Spacetime." *Noûs* 12 (1978): 397-419.

———. "On the Metric and Topology of Time." Ph. D. Dissertation. Cornell University, 1975.

Hume, David. *An Enquiry concerning Human Understanding.* In *Enquiries concerning Human Understanding and concerning the Principles of Morals* [1777]. Reprinted from the Posthumous Edition. Edited with an Introduction by L. A. Selby-Bigge. Third Edition. Revised with Notes by P. H. Nidditch. Oxford: Clarendon Press, 1975.

──────. *A Treatise of Human Nature* [1739-1740]. Edited by L. A. Selby-Bigge. Revised with Notes by P. H. Nidditch. Oxford: Clarendon Press, 1978.

Illy, Jozsef. "Einstein Teaches Lorentz, Lorentz Teaches Einstein. Their Collaboration in General Relativity, 1913-1920." *Archive for History of Exact Sciences* 39 (1989): 247-289.

Infeld, L. "Clocks, Rigid Bodies, and Relativity Theory." *American Journal of Physics* 11 (1943): 219-222.

Isham, C. J. "Creation of the Universe as a Quantum Process." In *Physics, Philosophy, and Theology: A Common Quest for Understanding*, pp. 375-408. Edited by R. J. Russell, W. R. Stoeger, and G. V. Coyne. Vatican City: Vatican Observatory, 1988.

Ives, Herbert E. "Derivation of the Lorentz Transformations." *Philosophical Magazine* 36 (1945): 392-401. Reprinted in *Speculations in Science and Technology* 2 (1979): 247-257.

──────. "Historical Note on the Rate of a Moving Clock." *Journal of the Optical Society of America* 37 (1947): 810-813.

Ives, Herbert E., and Stilwell, G. R. "An Experimental Study of the Rate of a Moving Atomic Clock." *Journal of the Optical Society of America* 28 (1938): 215-226.

──────. "An Experimental Study of the Rate of a Moving Atomic Clock II." *Journal of the Optical Society of America* 31 (1941): 369-374.

Jammer, Max. *Concepts of Space*. Cambridge, Mass.: Harvard University Press, 1954.

Jánossy, L. *Theory of Relativity Based on Physical Reality*. Budapest: Akadémiai Kiadó, 1971.

Jeans, James. *Physics and Philosophy*. Cambridge: University Press, 1942.

Kaku, Michio. *Hyperspace*. New York: Oxford University Press, 1994.

──────. *Introduction to Superstrings and M-Theory*. Second Edition. Graduate Texts in Contemporary Physics. New York: Springer, 1999.

Kanitscheider, Bernulf. *Kosmologie*. Stuttgart: Philipp Reclam., Jun., 1984.

Kant, Immanuel. *Immanuel Kant's "Critique of Pure Reason."* Translated by Norman Kemp Smith. London: Macmillan, 1970.

Katsumori, Makoto. "The Theories of Relativity and Einstein's Philosophical Turn." *Studies in History and Philosophy of Science* 23 (1992): 557-592.

Kaufmann, W. "Über die Konstitution des Elektrons." *Annalen der Physik* 19 (1906): 487-553.

Kennedy, Roy J., and Thorndike, Edward M. "Experimental Establishment of the Relativity of Time." *Physical Review* 42 (1932): 400ff.

Kerszberg, Pierre. "On the Alleged Equivalence between Newtonian and Relativistic Cosmology." *British Journal for the Philosophy of Science* 38 (1987): 347-380.

Kilmister, C. W. "Why is Relativity Interesting?" *British Journal for the Philosophy of Science* 42 (1991): 413-423.

Kitchener, Richard F. "Introduction: The World View of Contemporary Physics: Does It Need a New Metaphysics?" In *The World View of Contemporary Physics*, pp. 3-24. Edited by Richard F. Kitchener. Albany: State University of New York Press. 1988.

Knox, A. J. "Hendrik Antoon Lorentz, the Ether, and the General Theory of Relativity." *Archive for History of Exact Sciences* 38 (1988): 67-78.

Kostro, Ludwig. "Einstein's New Conception of the Ether." Paper presented at the International Conference of the British Society for the Philosophy of Science, "Physical Interpretations of Relativity Theory." Imperial College of Science and Technology. London, 16-19 September, 1988.

Kripke, Saul A. *Naming and Necessity*. Second Revised Edition. Oxford: Basil Blackwell, 1980.

Kroes, Peter. "The Physical Status of Time Dilation within the Special Theory of Relativity." Paper presented at the International Conference of the British Society for the Philosophy of Science, "Physical Interpretations of Relativity Theory." Imperial College of Science and Technology. London, 16-19 September, 1988.

_____. *Time: Its Structure and Role in Physical Theories*. Synthèse Library 179. Dordrecht: D. Reidel, 1985.

Landsberg, Peter T., and Evans, David A. *Mathematical Cosmology: An Introduction*. Oxford: Clarendon Press, 1977.

Larmor, Joseph. *Aether and Matter*. Cambridge: Cambridge University Press, 1900.

Laue, M. *Das Relativitätsprinzip*. Sammlung Naturwissenschaftlicher und mathematischer Monographien 38. Braunschweig: Friedrich Vieweg & Sohn, 1911.

_____. "Zwei Einwände gegen die Relativitätstheorie und ihre Widerlegung." *Physikalische Zeitschrift* 13 (1912): 118-120.

Le Poidevin, Robin. Critical notice of *Time and Reality*, by Mauro Dorato. *Studies in History and Philosophy of Modern Physics* 28 B (1997): 541-546.

Leclerc, Ivor. "The Relation between Natural Science and Metaphysics." In *The World View of Contemporary Physics*, 25-37. Edited with an Introduction by Richard F. Kitchener. Albany: State University of New York Press, 1988.

Leslie, John. "Risking the World's End." *Canadian Nuclear Society Bulletin* 10 (May/June 1989): 10-15.

Lewis, Delmas. "Persons, Morality, and Tenselessness." *Philosophy and Phenomenological Research* 47 (1986): 305-309.

Linde, A. D. "Chaotic Inflation." *Physics Letters* 8 129 (1983): 177-181.

_____. "Coleman-Weinberg Theory and the New Inflationary Universe Scenario." *Physics Letters* B 114 (1982): 431-435.

_____. "A New Inflationary Universe Scenario: A Possible Solution of the Horizon, Flatness, Homogeneity, Isotropy, and Primordial Monopole Problems." *Physics Letters* B 108 (1982): 389-393.

_____. "The Present Status of the Inflationary Universe Scenario." *Comments on Astrophysics* 10/6 (1985): 229-237.

Lodge, O. J. "Note on Mr. Sutherland's Objection to the Conclusiveness of the Michelson-Morley Aether Experiment." *Philosophical Magazine* 46 (1898): 343-344.

Lorentz, H. A. . "Alte und neue Fragen der Physik." *Physikalische Zeitschrift* 11 (1910): 1234ff. Reprinted in H. A. Lorentz, *Collected Papers*, 7: 205-257. Edited by P. Zeeman and A. D. Fokker. The Hague: Martinus Nijhoff, 1934.

─────── . "Conference on the Michelson-Morley Experiment." *Astrophysical Journal* 68 (1928): 341-402.

─────── . "Considérations élémentaires sur le principe de relativité." *Revue générale des Sciences* 25 (1914): 179ff. Reprinted in H. A. Lorentz, *Collected Papers*, 7: 147-165. Edited by P. Zeeman and A. D. Fokker. The Hague: Martinus Nijhoff, 1934.

─────── . "Deux mémoires de Henri Poincaré sur la physique mathématique." *Acta mathematica* 38 (1914): 293ff. Reprinted in H. A. Lorentz, *Collected Papers*, 7: 258-273. Edited by P. Zeeman and A. D. Fokker. The Hague: Martinus Nijhoff, 1934.

─────── . *The Einstein Theory of Relativity*. New York: Brentano's, 1920.

─────── . "Electromagnetic Phenomena in a System moving with any Velocity Smaller than that of Light." *Proceedings of the Royal Academy of Amsterdam* 6 (1904): 809ff. Reprinted in H. A. Lorentz, *Collected Papers*, 5: 172-197. Edited by P. Zeeman and A. D. Fokker. The Hague: Martinus Nijhoff, 1934-1939.

─────── . "Het relativiteitsbeginsel. Voordrachten in Teyler's Stichtung." *Archives Musée Teyler* 2 (1914): 1-60.

─────── . *Problems of Modern Physics*. Edited by H. Bateman. Boston: Ginn, 1927.

─────── . "The Relative Motion of the Earth and the Ether." *Verslagen Koninklijke Akadamie van Wetenschappen Amsterdam* 1 (1892): 74ff. Reprinted in H. A. Lorentz, *Collected Papers*, 4: 219-223. The Hague: Martinus Nijhoff, 1935-1939.

─────── . *The Theory of Electrons* [1909]. Reprint Edition: New York: Dover, 1952.

─────── . *Versuch einer Theorie der elektrischen und optischen Erscheinungen in bewegten Körpern*, §§89-92. Leiden: E. J. Brill, 1895. Reprinted in *The Principle of Relativity* [1923], 3-7. With Notes by A. Sommerfeld. Translated by W. Perrett and G. B. Jeffery. Reprint Edition: New York: Dover Publications, 1952.

Lorentz, H. A.; Einstein, A.; and Minkowski, H. *Das Relativitätsprinzip*. Fortschritte der mathematischen Wissenschaften 2. Mit Anmerkungen von A. Sommerfeld und Vorwort von O. Blumenthal. Leipzig: B. G. Teubner, 1920.

Lorenz, Dieter. "Über die Realität der FitzGerald-Lorentz Kontraction." *Zeitschrift für allgemeine Wissenschaftstheorie* 13/2 (1982): 294-319.

Lucas, J. R. *A Treatise on Time and Space*. London: Methuen, 1973.

Lucas, J. R., and Hodgson, P. E. *Spacetime and Electromagnetism*. Oxford: Clarendon Press, 1990.

McCormmach, Russell. "H. A. Lorentz and the Electromagnetic View of Nature." *Isis* 61 (1970): 459-497.

McCrea, W. H. "On the Significance of Newtonian Cosmology." *Astronomical Journal* 60 (1955): 271-274.

McGuire, J. E. "Existence, Actuality, and Necessity: Newton on Space and Time." *Annals of Science* 35 (1978): 463-508.

———. "Newton on Place, Time, and God: An Unpublished Source." *British Journal for the History of Science* 11 (1978): 114-129.

———. "Space, Infinity and Indivisibility: Newton on the Creation of Matter." In *Contemporary Newton Research*, pp. 145-190. Edited by Zev Bechler. Studies in the History of Modern Science 9. Dordrecht: D. Reidel, 1982.

Mach, Ernst. *Die Geschichte und die Wurzel des Satzes von der Erhaltung der Arbeit*. In Ernst Mach, *Abhandlungen*. Edited by J. Thiele. Amsterdam: E. J. Bonset, 1969.

———. *Die Mechanik in ihrer Entwicklung historisch-kritisch dargestellt*. Edited by Renate Wahsner and Horst-Heino Borzeszkowski. Berlin, DDR: Akademie-Verlag, 1988.

———. *Erkenntnis und Irrtum*. Second Edition. Leipzig: D. Reidel, 1906.

———. *History and Root of the Principle of the Conservation of Energy*. Translated by Philip E. B. Jourdain. Chicago: Open Court, 1911.

———. *Principles of the Theory of Heat*. Translated by T. J. McCormack. Rev. and compl. by P. E. B. Jourdain and A. E. Heath. With an Introduction by M. J. Klein. Edited by Brian McGuiness. Vienna Circle Collection 17. Dordrecht: D. Reidel, 1986.

———. *Prinzipien der Wärmelehre*. Leipzig: J. A. Barth, 1896.

———. *The Science of Mechanics: A Critical and Historical Account of Its Development*. Translated by Thomas J. McCormack. LaSalle, Ill.: Open Court, 1960.

Maciel, A. K. A., and Tiomno, J. "Analysis of Absolute Spacetime Lorentz Theories." *Foundations of Physics* 19 (1989): 505-519.

Mackie, J. L. "Three Steps toward Absolutism." In *Space, Time and Causality*, pp. 3-22. Edited by Richard Swinburne. Synthèse Library 157. Dordrecht: D. Reidel, 1983.

Malament, David. "Causal Theories of Time and the Conventionality of Simultaneity." *Noûs* 11 (1977): 293-300.

Mansouri, Reza, and Sexl, Roman U. "A Test of the Theory of Special Relativity: I. Simultaneity and Clock Synchronization." *General Relativity and Gravitation* 8 (1977): 497-513.

Marder, L. *Time and the Space-Traveller*. London: George Allen & Unwin, 1971.

Margenau, Henry. "Metaphysical Elements in Physics." *Reviews of Modern Physics* 13 (1941): 176-189.

Margenau, Henry, and Mould, Richard A.. "Relativity: an Epistemological Appraisal." *Philosophy of Science* 24 (1957): 297-307.

Markosian, Ned. "How Fast Does Time Pass?" *Philosophy and Phenomenological Research* 53 (1993): 829-844.

Maudlin, Tim. *Quantum Non-Locality and Relativity.* Aristotelian Society Series 13. Oxford: Blackwell, 1994.

_____. "Space-Time in the Quantum World." In *Bohmian Mechanics and Quantum Theory: An Appraisal,* pp. 285-307. Edited by James T. Cushing, Arthur Fine, and Sheldon Goldstein. Boston Studies in the Philosophy of Science 184. Dordrecht: Kluwer Academic Publishers, 1996.

Mehlberg, Henry. "Philosophical Aspects of Physical Time." *Monist* 53 (1969): 340-384.

_____. *Time, Causality, and the Quantum Theory.* 2 Volumes. Edited by Robert S. Cohen. Boston Studies in the Philosophy of Science 19. Dordrecht: D. Reidel, 1980.

Mellor, D. H. *Real Time.* Cambridge: Cambridge University Press, 1981.

_____. "Special Relativity and Present Truth." *Analysis* 34 (1973-1974): 74-77.

Merricks, Trenton. "Endurance and Indiscernibility." *Journal of Philosophy* 91 (1994): 165-184.

Miller, Arthur I. *Albert Einstein's Special Theory of Relativity.* Reading, Mass.: Addison-Wesley, 1981.

_____. *Imagery in Scientific Thought: Creating 20th Century Physics.* Cambridge, Mass.: MIT Press, 1986.

_____. "On Lorentz's Methodology." *British Journal for the Philosophy of Science* 25 (1974): 29-45.

_____. "On Some Other Approaches to Electrodynamics in 1905." In *Some Strangeness in Proportion,* pp. 66-91. Edited by Harry Woolff. Reading, Mass.: Addison-Wesley, 1980.

Milne, E. A. *Kinematic Relativity.* Oxford: Clarendon Press, 1948.

_____. *Modern Cosmology and the Christian Idea of God.* Oxford: Clarendon Press, 1952.

_____. "A Newtonian Expanding Universe." *Quarterly Journal of Mathematics* 5 (1934): 64-72.

_____. "Presidential Address to the Royal Astronomical Society." *Monthly Notices of the Royal Astronomical Society* 104 (1944): 120-134.

_____. *Relativity, Gravitation and World Structure.* Oxford: Clarendon Press, 1935.

Minkowski, H. "Space and Time." In *The Principle of Relativity,* pp. 75-91. By A. Einstein, *et al.* Translated by W. Perrett and G. B. Jeffery. Reprint Edition: New York: Dover Publications, 1952.

Misner, C.; Thorne, K. S.; and Wheeler, J. A. *Gravitation.* San Francisco: W. H. Freeman, 1973.

Møller, C. *The Theory of Relativity.* Second Edition. Oxford: Clarendon Press, 1972.

More, Henry. "The Immortality of the Soul." In *Philosophical Writings of Henry More,* pp. 57-180. Edited with an Introduction and Notes by Flora Isabel MacKinnon. Wellesley Semi-Centennial Series. New York: Oxford University Press, 1925.

Munitz, Milton K. *Cosmic Understanding*. Princeton: Princeton University Press, 1986.

Nagel, Ernst. "Relativity and Twentieth-Century Intellectual Life." In *Some Strangeness in the Proportion*, pp. 38-45. Edited by Harry Woolf. Reading, Mass.: Addison-Wesley, 1980.

Nanopoulos, D. V. "The Inflationary Universe." *Comments on Astrophysics* 10/6 (1985): 219-227.

Nerlich, Graham. *The Shape of Space*. Cambridge: Cambridge University Press, 1976.

———. "Time as Spacetime." In *Questions of Time and Tense*, pp. 119-134. Edited by R. LePoidevin. Oxford: Oxford University Press, 1998.

———. *What Spacetime Explains*. Cambridge: Cambridge University Press, 1994.

Newton, Isaac. *Isaac Newton's Papers and Letters on Natural Philosophy*. Second Edition. Edited with a General Introduction by I. Bernard Cohen. Cambridge, Mass.: Harvard University Press, 1978.

———. "On the Gravity and Equilibrium of Fluids." In *Unpublished Scientific Papers of Isaac Newton*, pp. 89-156. Edited by A. Rupert Hall and Marie Boas Hall. Cambridge: Cambridge University Press, 1962.

———. *Philosophiae Naturalis Principia Mathematica* [1726]. Third Edition. Edited by Alexandre Koyré and I. Bernard Cohen. 2 Volumes. Cambridge, Mass.: Harvard University Press, 1972.

———. *The Principia*. Translated by I. Bernard Cohen and Anne Whitman. With a guide by I. Bernard Cohen. Berkeley: University of California Press, 1999.

Newton-Smith, W.-H. "Space, Time, and Spacetime: A Philosopher's View." In *The Nature of Time*, pp. 22-35. Edited by Raymond Flood and Michael Lockwood. Oxford: Basil Blackwell, 1986.

Nordenson, Harald. *Relativity, Time, and Reality*. London: George Allen & Unwin, 1969.

Norikov, Igor D. *The River of Time*. Translated by Vitaly Kisin. Cambridge: Cambridge University Press, 1998.

Norton, John D. "Philosophy of Space and Time." In *Introduction to the Philosophy of Science*, pp. 179-232. Edited by Merrilee Salmon. Englewood Cliffs, N. J.: Prentice-Hall, 1992.

———. "The Time Coordinate in Einstein's Restricted Theory of Relativity." *Studium Generale* 23 (1970): 203-223.

Oaklander, L. Nathan. *Temporal Relations and Temporal Becoming*. Lanham, Maryland: University Press of America, 1984.

Ogawa, T. Suyoshi. "Japanese Evidence for Einstein's Knowledge of the Michelson-Morley Experiment." *Japanese Studies in the History of Science* 18 (1979): 73-81.

Ostwald, Wilhelm. *Chemische Energie: Lehrbuch der allgemeinen Chemie*. Second Edition. Leipzig: Verlag von Wilhelm Engelman, 1893.

Ozsváth, I., and Schücking, E. "Finite Rotating Universe." *Nature* 193 (1962), pp. 1168-1169.

Padgett, Alan. *God, Eternity and the Nature of Time.* New York: St. Martin's, 1992.
Pagels, Heinz. *The Cosmic Code.* London: Michael Joseph, 1982.
Pais, Abraham. *'Subtle is the Lord...': The Science and Life of Albert Einstein.* Oxford: Oxford University Press, 1982.
Pap, Arthur. *Introduction to the Philosophy of Science.* London: Eyre & Spottiswoode, 1963.
Park, David. *The Image of Eternity.* Amherst, Mass.: University of Massachusetts Press, 1980.
_____. "What is Time?" In *Time, Creation, and World Order*, pp. 15-22. Acta Jutlandica 74:1. Humanistic Series 72. Aarhus, Denmark: Aarhus University Press, 1999.
Petkov, Vesselin. "Simultaneity, Conventionality, and Existence." *British Journal for the Philosophy of Science* 40 (1989): 69-76.
Petzoldt, Joseph. "Das Verhältnis der Machschen Gedankenwelt zur Relativitätstheorie." In *Die Mechanik in ihrer Entwicklung*, pp. 490-517. By Ernst Mach. Eighth edition. Leipzig: F. A. Brockhaus, 1921.
Plantinga, Alvin. "Reply to Robert Adams." In *Alvin Plantinga*, pp. 371-382. Edited by James Tomberlin and Peter Van Inwagen. Profiles 5. Dordrecht, Holland: D. Reidel, 1985.
Podlaha, M. F. "Length Contraction and Time Dilatation in the Special Theory of Relativity—Real or Apparent Phenomena?" *Indian Journal of Theoretical Physics* 25 (1975): 69-75.
Poincaré, Henri. "Des fondements de la géometrie." *Revue du métaphysique et de morale* 7 (1899): 251-279.
_____. "The Measure of Time." In *The Value of Science* [1905]. Translated by G. B. Halstead. In *The Foundations of Science*, pp. 223-234. Science Press: 1913. Reprint Edition: Washington, D. C.: University Press of America, 1982.
_____. "The Principles of Mathematical Physics." *Monist* 15 (1905): 1-24.
_____. *Science and Hypothesis.* In *The Foundations of Science*, pp. 1-197. Science Press, 1913. Reprint Edition: Washington, D. C.: University Press of America, 1982.
_____. "Space and Time." In *Mathematics and Science: Last Essays* [*Dernières pensées*, 1913], pp. 15-24. Translated by John W. Bolduc. New York: Dover, 1963.
_____. "Sur la dynamique de l'électron." *Rendiconti del Circolo matematico di Palermo* 21 (1906): 129-176. In Henri Poincaré, *Oeuvres de Henri Poincaré*, 9: 494-550. 11 Volumes. Paris: Gauthiers-Villars, 1934-1953.
_____. "Sur les principes de la géometrie." *Revue de métaphysique et de morale* 8 (1900): 73-86.
Popper, Karl R. "A Critical Note on the Greatest Days of Quantum Theory." In *Quantum, Space and Time—The Quest Continues*, pp. 49-54. Edited by Asim O. Barut, Alwyn van der Merwe, and Jean-Pierre Vigier. Cambridge Monographs on Physics. Cambridge: Cambridge University Press, 1984.

_____. "A Note on the Difference between the Lorentz-Fitzgerald Contraction and the Einstein Contraction." *British Journal for the Philosophy of Science* 16 (1966): 332-333.

_____. *Quantum Theory and the Schism in Physics*. Totowa, N. J.: Rowman & Littlefield, 1982.

Powers, Jonathan. *Philosophy and the New Physics*. Ideas. London: Methuen, 1982.

Prokhovnik, S. J. "A Cosmological Basis for Bell's Views on Quantum and Relativistic Physics." Preprint.

_____. "An Introduction to the Neo-Lorentzian Relativity of Builder." *Speculations in Science and Technology* 2 (1979): 225-242.

_____. *Light in Einstein's Universe*. Dordrecht: D. Reidel, 1985.

_____. "The Logic of the Clock Paradox." Paper presented at the International Conference of the British Society for the Philosophy of Science, "Physical Interpretations of Relativity Theory." Imperial College of Science and Technology. London, 16-19 September, 1988.

_____. "The Twin Paradoxes of Special Relativity—Their Resolution and Implications." *Foundations of Physics*. Preprint.

Qadir, Asghar, and Wheeler, John Archibald. "York's Cosmic Time Versus Proper Time as Relevant to Changes in Dimensionless 'Constants,' K-Meson Decay, and the Unity of the Black Hole and Big Crunch." In *From SU(3) to Gravity*, pp. 383-394. Edited by Errol Gotsman and Gerald Tauber. Cambridge: Cambridge University Press, 1985.

Quine, Willard Van Orman. *Word and Object*. Cambridge, Mass.: MIT Press, 1960.

Quinn, Philip L. "Intrinsic Metrics on Continuous Spatial Manifolds." *Philosophy of Science* 43 (1976): 396-414.

Rae, Alastair I. M. *Quantum Physics: Illusion or Reality?* Cambridge: Cambridge University Press, 1986.

Ramige, Eldon Albert. *Contemporary Concepts of Time and the Idea of God*. Boston: Stratford, 1935.

Redhead, Michael. "The Conventionality of Simultaneity." In *Philosophical Problems of the Internal and External Worlds*, pp. 103-128. Edited by John Earman, Alan I. Janis, Gerald J. Massey, and Nicholas Rescher. Pittsburgh-Konstanz Series in the Philosophy and History of Science. Pittsburgh: University of Pittsburgh Press, 1993.

_____. "Nonlocality and Peaceful Coexistence." In *Space, Time and Causality*, pp. 151-189. Edited by Richard Swinburne. Synthèse Library 157. Dordrecht: D. Reidel, 1983.

Rees, Martin J. "The Size and Shape of the Universe." In *Some Strangeness in the Proportion*, pp. 291-301. Edited by Harry Woolf. Reading, Mass.: Addison-Wesley, 1980.

Reichenbach, Hans. "Les fondements logiques de la mécanique des quanta." *Annales de l'Institut Henri Poincaré* 13 (1952): 109-157.

———. *The Philosophy of Space and Time* [1928]. Translated by Maria Reichenbach and John Freund. With an Introduction by Rudolf Carnap. New York: Dover Publications, 1958.
Resnick, Robert. *Introduction to Special Relativity*. New York: John Wiley & Sons, 1968.
Rindler, Wolfgang. "Einstein's Priority in Recognizing Time Dilation Physically." *American Journal of Physics* 38 (1970): 1111-1115.
———. *Introduction to Special Relativity*. Oxford: Clarendon Press, 1982.
Robb, A. A. *A Theory of Time and Space*. Cambridge: Heffer & Sons, 1913.
Rodrigues, Waldyr A. Jr., and Rosa, Marcio A. F. "The Meaning of Time in the Theory of Relativity and 'Einstein's Later View of the Twin Paradox'." *Foundations of Physics* 19 (1989): 705-724.
Rohrlich, Fritz. *From Paradox to Reality*. Cambridge: Cambridge University Press, 1987.
Rompe, R., and Treder, Hans-Jürgen. "Is Physics at the Threshold of a New Stage of Evolution?" In *Quantum, Space and Time—The Quest Continues*, pp. 595-609. Edited by Asim O. Barut, Alwyn van der Merwe, and Jean-Pierre Vigier. Cambridge Monographs on Physics. Cambridge: Cambridge University Press, 1984.
Rosen, Nathan. "Bimetric General Relativity and Cosmology." *General Relativity and Gravitation* 12 (1980): 493-510.
Rosser, W. G. V. *An Introduction to the Theory of Relativity*. London: Butterworths, 1964.
———. *Introductory Relativity*. New York: Plenum Press, 1967.
Rossi, Bruno, and Hall, David B. "Variation of the Rate of Decay of Mesotrons with Momentum." *Physical Review* 59 (1941): 223-228.
Rossi, Bruno, and Hoag, J. Barton. "The Variation of the Hard Component of Cosmic Rays with Height and the Disintegration of Mesotrons." *Physical Review* 57 (1940): 461-469.
Rothman, Tony, and Ellis, George. "Has Astronomy Become Metaphysical?" *Astronomy* (February 1987), pp. 6-22.
Rovelli, Carlo. "Halfway through the Woods: Contemporary Research on Space and Time." In *The Cosmos of Science*, pp. 180-223. Edited by J. Norton and J. Earman. Pittsburgh: University of Pittsburgh Press, 1998.
———. "Quantum Spacetime: What Do We Know?" In *Physics Meets Philosophy at the Planck Scale*. Edited by C. Callender and N. Hugger. Cambridge: Cambridge University Press, forthcoming.
———. "What Does Present Days (*sic*) Physics Tell Us about Time and Space?" Lecture presented at the Annual Series Lectures of the Center for Philosophy of Science of the University of Pittsburgh, 17 September 1993.
Ruderfer, Martin. "Introduction to Ives' 'Derivation of the Lorentz Transformations'." *Speculations in Science and Technology* 2 (1979): 243-246.
Russell, B. "Sur les axiomes de la géométrie." *Revue de métaphysique et de morale* 7 (1899): 684-707.

Ryckman, T. A.. "'P(oint)-C(oincidence) Thinking': The Ironical Attachment of Logical Empiricism to General Relativity (and some lingering Consequences)." *Studies in History and Philosophy of Science* 23 (1992): 471-497.

Salmon, Wesley C. "The Conventionality of Simultaneity." *Philosophy of Science* 36 (1969): 44-63.

──────. "The Philosophical Significance of the One-Way Speed of Light." *Noûs* 11 (1977): 253-292.

──────. *Space, Time, and Motion*. Encino, Calif.: Dickenson Publishing, 1975.

Savitt, Steven F. "There's No Time like the Present (in Minkowski Spacetime)." Paper presented at the symposium, "The Prospects for Presentism in Spacetime Theories." Philosophy of Science Association biennial meeting, Kansas City, October, 1998.

Schaffner, Kenneth F. "Einstein versus Lorentz: Research Programmes and the Logic of Comparative Theory Evaluation." *British Journal for the Philosophy of Science* 25 (1974): 45-78.

──────. "Outlines of a Logic of Comparative Theory Evaluation with Special Attention to Pre- and Post-Relativistic Electrodynamics." In *Historical and Philosophical Perspectives of Science*, pp. 311-354. Edited by Roger H. Stuewer. Minnesota Studies in the Philosophy of Science 5. Minneapolis: University of Minnesota Press, 1970.

Schild, Alfred. "The Clock Paradox in Relativity Theory." *American Mathematical Monthly* 66 (1959): 1-18.

Schlesinger, George N. *Aspects of Time*. Indianapolis: Hackett, 1980.

──────. "What does the Denial of Absolute Space Mean?" *Australasian Journal of Philosophy* 45 (1967): 44-60.

Schücking, E. L. "Newtonian Cosmology." *Texas Quarterly* 10 (1967): 270-274.

Seelig, C. *Albert Einstein: A Documentary Biography*. London: Staples Press, 1956.

Sellars, Wilfrid. "Time and the World Order." In *Scientific Explanation, Space, and Time*, pp. 527-616. Edited by Herbert Feigl and Grover Maxwell. Minnesota Studies in the Philosophy of Science 3. Minneapolis: University of Minnesota Press, 1962.

Selleri, Franco. "Le principe de la relativité et la nature du temps." *Fusion* 66 (1997): 50-60.

Shallis, Michael. "Time and Cosmology." In *The Nature of Time*, pp. 63-79. Edited by Raymond Flood and Michael Lockwood. Oxford: Basil Blackwell, 1986.

Shankland, R. S. "Conversations with Albert Einstein II." *American Journal of Physics* 41 (1973): 895-901.

──────. "Michelson-Morley Experiment." *American Journal of Physics* 32 (1964): 16-35.

Shaw, R. "The Length Contraction Paradox." *American Journal of Physics* 30 (1962): 72.

Sherwin, Chalmers W. "Some Recent Experimental Tests of the 'Clock Paradox'." *Physical Review* 120 (1960): 17-21.

Shimony, Abner. "Metaphysical Problems in the Foundations of Quantum Mechanics." *International Philosophical Quarterly* 18 (1978): 13-17.
Sinks, John D. "On Some Accounts about the Future." *Journal of Critical Analysis* 2 (1971): 8-16.
Sjödin, T. "On the One-Way Velocity of Light and its Possible Measurability." Paper presented at the International Conference of the British Society for the Philosophy of Science, "Physical Interpretations of Relativity Theory." Imperial College of Science and Technology. London, 16-19 September, 1988.
Sklar, Lawrence. "The Conventionality of Simultaneity! Again?" Paper presented at the International Conference of the British Society for the Philosophy of Science, "Physical Interpretations of Relativity Theory." Imperial College of Science and Technology. London, 16-19 September, 1988.
_____. "Modestly Radical Empiricism." In *Observation, Experiment, and Hypothesis in Modern Physical Science*, pp. 1-20. Edited by Peter Achinstein and Owen Hannaway. Studies from the Johns Hopkins Center for the History and Philosophy of Science. Cambridge, Mass.: MIT Press, 1985.
_____. *Philosophy and Spacetime Physics*. Berkeley: University of California Press, 1985.
_____. "Real Quantities and their Sensible Measures." In *Philosophical Perspectives on Newtonian Science*, pp. 57-75. Edited by Phillip Bricker and R. I. G. Hughes. Cambridge, Mass.: MIT Press, 1990.
_____. "Time, Reality and Relativity." In *Reduction, Time and Reality*, pp. 129-142. Edited by Richard Healey. Cambridge: Cambridge University Press, 1981.
Smart, J. J. C. "Quine on Spacetime." In *The Philosophy of W. V. O. Quine*, pp. 495-515. Library of Living Philosophers 18. Edited by Lewis Edwin Hahn and Paul Arthur Schilpp. LaSalle, Ill.: Open Court, 1986.
Smith, Quentin. *Language and Time*. New York: Oxford University Press, 1993.
Smoot, G. F.; Gorenstein, M. Y.; and Muller, R. A. "Detection of Anisotropy in the Cosmic Blackbody Radiation." *Physical Review Letters* 39 (1977): 898-901.
Squires, Euan. *The Mystery of the Quantum World*. Bristol: Adam Hilger, 1986.
Stachel, John. "Einstein and Ether Drift Experiments." *Physics Today* (May 1987), pp. 45-47.
Stapp, Henry Pierce. "Are Faster-Than-Light Influences Necessary?" In *Quantum Mechanics versus Local Realism*, pp. 63-85. Edited by Franco Selleri. Physics of Atoms and Molecules. New York: Plenum Press, 1988.
_____. "Quantum Mechanics, Local Causality, and Process Philosophy." *Process Studies* 7 (1977): 173-182.
Stein, Howard. Critical notice of *The Language of Time*, by Richard M. Gale. *Journal of Philosophy* 66 (1969): 350-355.
_____. "Newtonian Spacetime." *Texas Quarterly* 10 (1967): 174-200.
_____. "On Einstein-Minkowski Space-Time." *Journal of Philosophy* 65 (1968): 5-23.
_____. "On Relativity Theory and the Openness of the Future." *Philosophy of Science* 58 (1991): 146-167.

Stern, Alexander W. "Space, Field, and Ether in Contemporary Physics." *Science* 116 (November 1952), pp. 493-496.
Strong, E. W. "Newton and God." *Journal of the History of Ideas* 13 (1952): 147-167.
Suppe, Frederick. "The Search for Philosophic Understanding of Scientific Theories." In *The Structure of Scientific Theories*, pp. 3-118. Second Edition. Edited by Frederick Suppe. Urbana, Ill.: University of Illinois Press, 1977.
Swinburne, Richard. *Space and Time*. New York: St Martin's Press, 1968.
_____. *Space and Time*. Second Edition. London: Macmillan, 1981.
_____. "Verificationism and Theories of Spacetime." In *Space, Time, and Causality*, pp. 63-76. Edited by Richard Swinburne. Synthèse Library 157. Dordrecht: D. Reidel, 1983.
Taylor, Edwin F., and Wheeler, John Archibald. *Spacetime Physics*. San Francisco: W. H. Freeman, 1966.
Taylor, J. G. *Special Relativity*. Oxford Physics Series. Oxford: Clarendon Press, 1975.
Terletskii, Yakov P. *Paradoxes in the Theory of Relativity*. With a Foreword by Banesh Hoffman. New York: Plenum Press, 1968.
Terrell, James. "Invisibility of the Lorentz Contraction." *Physical Review* 116 (1959): 1041-1045.
Thomas, L. H. "The Kinematics of an Electron with an Axis." *Philosophical Magazine* (1927): 1-21.
Tipler, Frank J. "The Sensorium of God: Newton and Absolute Space." In *Newton and the New Direction of Science*, pp. 215-228. Edited by G. V. Coyne, M. Heller, and J. Zyncinski. Vatican City: Specola Vaticana, 1988.
Tittel, W.; Brendel, J.; Gisin, N.; and Zbinden, H. "Long Distance Bell-type Tests Using Energy-Time Entangled Photons." *Physical Review* A 59/6 (1999): 4150-4163.
Tonnelat, Marie-Antoinette. *Histoire du principe de relativité*. Nouvelle Bibliothèque Scientifique. Paris: Flammarion, 1971.
Valentini, Antony. "Pilot-Wave Theory of Fields, Gravitation and Cosmology." In *Bohmian Mechanics and Quantum Theory: An Appraisal*, pp. 45-66. Edited by James T. Cushing, Arthur Fine, and Sheldon Goldstein. Boston Studies in the Philosophy of Science 184. Dordrecht: Kluwer Academic Publishers, 1996.
Van Inwagen, Peter. "Four Dimensional Objects." *Noûs* 24 (1990): 245-255.
Varicak, V. "Zum Ehrenfestschen Paradoxen." *Physikalische Zeitschrift* 12 (1911): 169-170.
Vigier, Jean-Paul. "Louis de Broglie—Physicist and Thinker." In *Quantum, Space and Time—The Quest Continues*, pp. 3-10. Edited by Asim O. Barut, Alwyn van der Merwe, and Jean-Pierre Vigier. Cambridge Monographs on Physics. Cambridge: Cambridge University Press, 1984.
Weinberg, Steven. *Gravitation and Cosmology: Principles and Explications of the General Theory of Relativity*. New York: John Wiley & Sons, 1972.
Weingard, Robert. "Relativity and the Reality of Past and Future Events." *British Journal for the Philosophy of Science* 23 (1972): 119-121.

Weisskopf, Victor F. "Elementary Particle Physics in the Very Early Universe." In *Astrophysical Cosmology*, pp. 503-528. Edited by H. A. Brück, G. V. Coyne, and M. S. Longair. Pontificiae Academiae Scientiarvm Scripta Varia 48. Vatican City: Pontificia Academia Scientiarvm, 1982.

──────. "The Visual Appearance of Rapidly Moving Objects." *Physics Today* 13 (September 1960), pp. 24-27.

Wenzl, A. "Einstein's Theory of Relativity, Viewed from the Standpoint of Critical Realism, and its Significance for Philosophy." In *Albert Einstein: Philosopher-Scientist*, pp. 583-606. Edited by Paul Arthur Schilpp. Library of Living Philosophers 7. LaSalle, Ill.: Open Court, 1949.

Wertheimer, Max. *Productive Thinking*. Edited by Michael Wertheimer. Enlarged Edition. London: Tavistock, 1961.

Westfall, Richard S. *Never at Rest: A Biography of Isaac Newton*. Cambridge: Cambridge University Press, 1980.

Wheeler, J. A. "From Relativity to Mutability." In *The Physicist's Conception of Nature*, pp. 202-247. Edited by J. Mehra. Dordrecht: D. Reidel, 1973.

──────. *Frontiers of Time*. Amsterdam: North-Holland, 1979.

Whitrow, G. J. *The Natural Philosophy of Time*. Second Edition. Oxford: Clarendon Press, 1980.

Whitt, L. A. "Absolute Space: Did Newton Take Leave of His (Classical) Empirical Senses?" *Canadian Journal of Philosophy* 12 (1982): 709-724.

Whittaker, Edmund. *A History of the Theories of Aether and Electricity*. 2 Volumes. Reprint Edition: New York: Humanities Press, 1973.

Wien, Wilhelm. "Ueber die Fragen, welche die translatorische Bewegung des Lichtäthers betreffen." *Annalen der Physik und Chemie* 65/3 (1898): 1-17.

Will, Clifford M. *Was Einstein Right?* New York: Basic Books, 1986.

Williams, E. J. "The Loss of Energy by B-Particles and its Distribution between Different Kinds of Collisions." *Proceedings of the Royal Society* A 130 (1931): 328-346.

Winnie, John A.. "The Twin-Rod Thought Experiment." *American Journal of Physics* 40 (1972): 1091-1094.

Worrall, John. "How to Remain (Reasonably) Optimistic: Scientific Realism and the 'Luminiferous Ether'." In *PSA 1994*, pp. 334-342. Edited by David Hull, Micky Forbes, and Richard M. Burian. East Lansing, Mich.: PSA, 1994.

Yourgrau, Palle. *The Disappearance of Time*. Cambridge: Cambridge University Press, 1991.

Zahar, Elie. "Why Did Einstein's Programme Supersede Lorentz's? (I)." *British Journal for the Philosophy of Science* 24 (1973): 95-123.

──────. "Why Did Einstein's Programme Supersede Lorentz's? (II)." *British Journal for the Philosophy of Science* 24 (1973): 223-262.

Zeeman, E. C. "Causality Implies the Lorentz Group." *Journal of Mathematical Physics* 5 (1964): 490-493.

Zwart, P. J. "The Flow of Time." *Synthèse* 24 (1972): 133-158.

SUBJECT INDEX

A-Theory of Time, v, 40, 78-83 *pass.*, 103, 112, 150, 155, 157-158, 170, 171, 173, 179, 192, 194, 208, 242
aberration of starlight, 5, 10, 22
absolute motion, 3, 39, 109, 112n, 117-118, 124, 125, 138, 149, 186, 197-199 *pass.*, 202, 223, 233.
absolute time and space, 1-2, 18n, 19, 23-25 *pass.*, 38, 41n-42n, 45-46, 51, 57n, 59, 78, 81, 85, 87, 101, 105-139 *pass.*, 141, 146, 147n, 148, 150, 155, 157, 168, 179, 195, 201, 213, 233-235 *pass.*
ad hoc character of a theory, 15-17 *pass.*, 179, 184
aether drag, 7, 10
aether wind, 7, 22, 65, 220-221, 223
aether, 2-10 *pass.*, 18n, 19, 21-24 *pass.*, 38-42 *pass.*, 66, 79, 85, 88, 92, 96, 97, 122, 125, 132, 137, 138n, 139, 140, 141, 146, 150, 164, 167, 174-177 *pass.*, 179, 180; modern equivalents of: 184, 219-235 *pass.*, 242.
anisotropy of time, 155

B-Theory of time, v, 40-41, 78, 83, 102, 158, 192
beginning of the universe, vn, 157-158, 191, 213, *see also* Big Bang Theory; cosmic time
Bell's Theorem, *see* EPR experiment
Big Bang Theory, 15, 142-143, 154, 156n, 204, 211, 239; *see also* cosmic time

camel and the needle's eye experiment, 89-90
causal theory of time, 151
clock retardation, 14, 17-19n, 47-65 *pass.*, 92, 93, 100, 108, 119, 151, 176, 179, 185; *see also* time dilation
clock synchronization, 28-29, 35, 37-39 *pass.*, 42-47 *pass.*, 74, 81, 88, 96, 130-132 *pass.*, 136, 149, 163, 164, 167, 173, 216, 219
conventionalism, metric, 2, 107-110 *pass.*, 129, 145, 151
conventionality of simultaneity, *see* simultaneity, conventionality of
coordinate vs. parameter time, 207-208, 213, 238, 241
cosmic time, 202-241 *pass.*
creation, vn, 147, 154, 155, 192, 204

direction of time, 155, 159

electrodynamics, 4-10 *pass.*, *see also* quantum electrodynamics
electromagnetic spectrum, 6
electron theory of Lorentz, 11-18 *pass.*, 41, 100, 146, 177-178
emanative causality, 74-115 *pass.*
endurantism, 93n-94n
EPR experiment, 185n, 223-233 *pass.*
eternity, v, 242

fundamental reference frame, 180, 208-209, 218-221n *pass.*, 234, 239

Galilean relativity, 2-7 *pass.*, 27, 109, 117
Galilean spacetime, 186-188 *pass.*,

201
Galilean transformation, 3-4, 17, 122, 163, 183
General Theory of Relativity, 1, 3, 53, 77, 79-81 *pass.*, 99n, 108, 126n, 133-134, 139, 143, 144, 155-156, 159, 174, 178, 186, 189-190, 195-213 *pass.*, 215-216, 218, 232, 234n, 236, 239
gravitation as force vs. spacetime curvature, 189-190
God: as Creator, vn, 143, 154; and conventionalism, 34; and relativity of simultaneity, 49, 117, 119n, 166, 177; as constituting absolute time/space, 110-121 *pass.*, 150, 156n, 171-179 *pass.*, 188, 191, 194, 235-240; as Ultimate Observer, 142, 144; as Designer, 143; and conspiracy of nature, 184-186 *pass.*; and physical time, 195-241 *pass.*

Haefele-Keating experiment, 67-68
hyper-surface, 174, 205-206, 209-210, 214, 216, 236, 238, 240

inertial frame, 2, 174, 200, 205, 208
inflation hypothesis, 15-16
infinite velocity signals, *see* superluminal signals
invisibility of length contraction, 61-62
ionization of charged particles, 65-66
Ives-Stillwell experiments, 66

kalam cosmological argument, vn
Kennedy-Thorndike experiment, 15, 65
Kinematic Relativity, *see* Milne-McCrea cosmology

length contraction, 47, 59-64 *pass.*, 66, 88, 177, 179-180, 182, 185, 186, 193, 209, 219; *see also* Lorentz-FitzGerald contraction
light clock 95-96, 181
light cone structure, 71-73 *pass.*
light, corpuscular vs. wave theory, 5-6, 26; one-way velocity of, 30, 132
Light Postulate, *see* Principle of the Constancy of Light's Velocity
local time, 17, 18n, 19n, 24n-25n, 39, 48, 119, 176-177, 237
Lorentz-FitzGerald contraction, 11-17 *pass.*, 22, 23, 25, 43, 65, 66, 95, 109, 139, 154, 178, 185; *see also* length contraction
Lorentz transformation, 13, 17, 24n, 68, 69, 78, 80, 94, 122, 154, 163, 182, 193

Mach's Principle, 134, 198-199
McTaggart's Paradox, 104
magnet and conductor experiment, 21, 22, 98, 125
meson lifetimes 66-67
metaphysical elements in science, 139-148 *pass.*
Michelson-Morely experiment, 7-10 *pass.*, 14, 15, 22-23, 65, 99, 138, 169, 221, 233-234
Milne-McCrea cosmology, 239-240
Minkowski spacetime, 31, 32, 35, 40-42 *pass.*, 69-80 *pass.*, 83, 85, 87, 88n, 94, 183-184, 193, 200, 207, 230; *see also* spacetime realism
Mössbauer effect, 67

Newtonian time and space, vi, 105-122 *pass.*, 151, 154-162 *pass.*, 166, 169, 170, 171, 179, 188-189, 195, 201, 208, 229, 233, 236, 239-240, 242; *see also* absolute time and space

Subject Index

operational definitions in SR, 28-29, 42, 129-138 *pass.*, 140, 148, 153, 160, 162-169 *pass.*, 176n, 177, 228

perdurantism, 93n-94n, 103, 192
phenomenalism, *see* sensationism
positivism, v, 109, 112-113, 120, 122-161 *pass.*, 165, 169, 174-175, 183, 197, 224, 227-228; *see also* verificationism
Principle of the Constancy of Light's Velocity, 21-22, 24n, 25, 26-27, 36, 38, 42-43, 46, 132, 181-182, 219
Principle of Equivalence, 134, 199-201 *pass.*
Principle of Relativity, 3, 13n, 17n, 18n, 21, 23n, 24n, 25-27, 36, 37, 38, 42-43, 46, 92-93, 100, 126, 132, 133, 137, 146n, 150, 180, 181-182, 195, 198-199, 232
"pure" relativity, 87, 88, 92, 93, 94, 209

quantum electrodynamics, 222-223
quantum mechanics, 142-144 *pass.*, 154-157n, 177, 178, 185n, 221-233 *pass.*

reference frame, 2n, 78, 80, 81, 82, 85, 131-132, 146
relational theories of time/space, 93n, 111-112, 114-115, 139n, 201, 237
relativity of length/time, 42-46 *pass.*, 69, 122, 149
Relativity Postulate, *see* Principle of Relativity
revolving globes experiment, 109, 196-197
rotating bucket experiment, 3, 109, 197

sensationism, 123-125 *pass.*, 127, 130, 134
simultaneity: absolute, 20, 44, 50, 79, 86, 105n, 127, 129, 140, 146, 152, 161-165 *pass.*, 168-169, 172, 174, 176, 187, 210-211, 223, 226-240; conventionality of, 2, 29-43 *pass.*, 81, 83, 129, 149, 151, 162-163, 168; definition of, 27-29 *pass.*, 162-169 *pass.*, 172; relativity of, 29, 40, 43-59 *pass.*, 122, 149, 161, 168, 212, 219, 221n, 230
Special Theory of Relativity: Lorentzian interpretation, 26, 27n, 37, 41, 46, 66, 88, 92, 95, 101, 145-147 *pass.*, 169, 174-194 *pass.*, 213, 226-235 *pass.*, 242; relativity interpretation, 24n, 66, 78-102 *pass.*, 104, 145-147 *pass.*, 178, 182, 189, 190, 192, 229, 233, 242; spacetime interpretation, 24, 79-81 *pass.*, 84, 86, 93-95 *pass.*, 101-102, 145-147 *pass.*, 178, 180, 182-194 *pass.*, 229, 242
spacetime realism, 77, 83n-84n, 85, 87, 93n-94n, 95, 104, 146, 170, 177, 180, 182-183 *pass.*, 202
substantival theories of time/space, 93n, 110-112 *pass.*, 114-116 *pass.*, 139n
superluminal signals, 27, 150, 163-164, 174-176 *pass.*, 184, 227, 229-230n, 233

temporal becoming, 160-161, 170, 192, 208, 242
tensed theory of time, *see* A-Theory of time
tenseless theory of time, *see* B-Theory of time
theory equivalence, 26, 69, 145-147 *pass.*, 153
Thomas precession, 65
time dilation, 18, 47-64 *pass.*, 66, 88,

94, 96, 100, 179-181 *pass.*, 186, 193, 209, 219
time, metaphysical vs. measured, 2, 64, 106-122 *pass.*, 130-131, 133, 149, 150, 152-162 *pass.*, 169, 187, 194, 195, 214-215, 237, 241
time travel, 57, 156-157, 187, 227
timelessness, divine, 116-117
translation, 3, 11, 17, 43
Twin Paradox, 52-59 *pass.*, 67, 71, 93, 94, 95n, 100, 181, 182
twin rod experiment, 88-89

universe, beginning of, *see* beginning of the universe

verificationism, v-vi, 32, 34, 49n, 120, 122, 129, 130, 133, 134, 135n, 136, 138, 139, 140, 147-149 *pass.*, 151, 152-153, 161, 162, 169, 172, 175, 183, 187, 197

world map, 240

Zeeman effect, 13

PROPER NAME INDEX

Achinstein, Peter, 41n
Agazzi, Evandro, 236
Albrecht, A., 16n
Angel, Roger, 94n
Aristotle, 99, 236
Arthur, Richard T. W., 147n
Arzeliès, Henri, 41n–42n, 64n, 78, 93, 193
Aspect, Alain, 226n, 233

Bachelard, Gaston, 138n
Baierlein, Ralph, 152
Balashov, Yuri, 84n, 187n
Banks, T., 155
Barbour, Julian B., vi, 105n, 153, 160
Bargmann, Sonja, 136n, 175n
Barrow, Isaac, 120
Barrow, John D., 144n, 147, 185n, 236
Bartels, Andreas, 236n
Barut, Asim O., 147n, 222n
Bechler, Zev, 113, 116n
Bell, J. S., 86n, 90-91, 94n, 179n, 185n, 223, 225-226, 227n, 229-230, 232, 234
Beller, Mara, 228n
Bentley, Richard, 113
Bergson, Henri, 88
Berkeley, George, 107n
Besso, Michael, 79, 126-127, 128n, 129n
Bilaniuk, Olexa-Myron, 165
Black, Max, 193
Bohm, David, 142, 146n, 192-193n, 225, 227-229, 233
Bohr, Niels, 223-224, 227n, 228n, 229n, 230n
Bolduc, John, 32n
Bondi, Hermann, 57-58, 65, 94n, 105n, 152-153, 196n, 239
Born, Max, 48n, 141-142
Borzeszkowski, Horst-Heino, 123n
Bowman, P., 30n
Bradley, James, 5-6, 124n
Brendel, J., 226
Bricker, Phillip, 108n
Bridgeman, P. W., 138
Brooke, John Hedley, 116n
Brose, Henry L., 141n
Brown, G. Burniston, 48n
Brown, James Robert, 127n, 227n
Brück, H. A., 144n
Builder, Geoffery, 52n, 53n, 99n-100n, 169, 178-182, 220n
Bunge, Mario, 152n
Burge, Tyler, 150
Burian, Richard M., 101n, 229n
Bush, G., 11n

Callender, Craig, 84n, 104n, 168n, 228-229, 233n, 234n
Cantor, G. N., 113n
Capek, Milic, 15n, 20, 38, 57n, 85-86, 120n, 153, 158, 163n
Capra, Fritjof, 105n
Carnap, Rudolf, 34n
Carrier, Martin, 81n, 185
Cassirer, Ernst, 27n, 138-139
Chang, Hasok, 88n
Child, J. M., 120n
Christensen, F. M., 86, 87n, 192
Christie, J. R. R., 113n
Clarke, Samuel, 1, 139
Cleugh, Mary F., 151-152, 162, 166, 193n
Clifton, Rob, 104n, 233n
Coburn, Robert C., 103n
Cohen, I. Bernard, 3n, 106n, 113,

197n
Cohen, Richard S., 152n
Collins, Robin, 185n
Comte, Auguste, 123n-124n
Cooper, Leon N., 168n
Cordero, Alberto, 236n
Coyne, G. V., 144n, 158n, 236n
Craig, William Lane, vn, 40n, 94n, 103n, 121n, 156n, 164n, 165n, 171n, 173n
Cranshaw, T. E., 67n
Cunningham, E., 12n
Cushing, James T., 19n, 69n, 178n, 227n, 228n, 233

D'Abro, A., 105n, 141n, 183-186
D'Inverno, Ray, 54n, 153-154
Darwin, Charles, 143
Davies, Paul C. W., 19n, 57n, 105n, 154, 217, 227n
De Beauregard, Olivier Costa, 184n
De Broglie, Louis, 142, 146n, 194n, 222, 227, 229
De Finis, Francesco, 196n
De Sitter, Willem, 203-204, 213
Debrock, G., 117n
Dehnen, H., 221-222n
Democritus, 222
Di Salle, Robert, 102n
Dicke, R. H., 40
Dieks, Dennis, 95, 97, 100, 117, 146n
Dingle, Herbert, 12n, 24n, 48-49, 52n, 53n, 66n, 87-88, 161n, 162, 193n
Dirac, P. A. M., 221n, 222, 234
Dorato, Mauro, 104n, 208n, 217n, 237n
Dorling, Jan, 49n, 97n
Duffy, Michael Ciaran, 153, 219
Dugas, René, 146, 166n, 193n-194n
Dukas, Helen, 79n
Durbin, R., 67n

Earman, John S., vi, 2n, 16n, 29n, 75n, 80n, 81n, 107n, 108n, 111n, 134, 146n, 147, 165n, 182, 185n, 186-188, 190n
Eberhard, P. H., 232n
Eddington, Sir Arthur, 11-12n, 18n, 61n, 72, 140, 141, 143, 161-162, 171, 203-204, 213-214, 216-217
Egelstaff, P. A., 67n
Ehrenfest, Paul, 24n, 140n
Einstein, Albert, v, vi, 1, 2, 3, 11n, 12, 13, 14, 18, 19-20 pass., 21-46 pass., 48n, 49-52, 55, 59, 61-62, 64, 69, 77-81, 83, 85, 88-89, 91-93, 95-100, 102, 104, 105, 109n, 118, 119, 120, 122-148 pass., 149, 151, 152n, 153n, 154, 156n, 161-164, 166-170, 171, 174-186, 189n, 190, 191n, 193n, 194, 195-204, 207, 208n, 212-214, 216, 218n, 219, 220n, 221-224, 225n, 226n, 227-228, 233, 236, 239, 242
Ellis, B., 30n
Ellis, George, 142
Epstein, Lewis Carroll, 35n, 49n, 51n
Epstein, Paul S., 12n, 48n, 60, 138n
Evans, David A., 239n
Evans, Melbourne G., 15n

Faraday, Michael, 21
Fauvel, John, 116n
Feigl, Herbert, 193n
Fermi, Enrico, 40n
Feyerabend, P. K., 14n
Findlay, J. N., 172-173
Fine, Arthur, 69n, 189
FitzGerald, George Francis, 11, 12, 14, 15, 23n, 25, 43, 65, 109, 154, 178, 180, 185, 226, 227
Fitzgerald, Paul, 105n, 214, 235, 237
Fizeau, Armand, 22
Flood, Raymond, vi, 110n, 116n, 217n
Fokker, A. D., 12n, 175n
Föppl, August, 125
Forbes, Micky, 101n, 229n

PROPER NAME INDEX 275

Force, James E., 116
Frank, Philipp, 26n, 123n, 130n, 136, 140, 161, 169
Franson, James, 230n
Fraser, J. T., 48n, 66n, 105n, 162n, 217n
Fresnel, Augustin Jean, 7, 22
Freud, Sigmund, 127
Freund, John, 29n, 34n
Freundlich, Yehudah, 153
Friedman, Michael, 1, 10-11, 25-26, 29n, 31, 34n, 75n, 85n, 95n, 134n, 137, 164n, 167, 182-183, 186, 198-201, 207n
Friedman, William, 102
Friedmann, Aleksandr, 204-206, 211, 213, 217, 234n, 236, 239
Frisch, David H., 66n

Gabbey, Alan, 113n
Gale, George, 16n, 142-143, 147
Gale, Richard M., 105n, 163n
Galileo Galilei, 2-3, 99, 134, 197
Gamow, George, 204n
Geroch, Robert, 193n
Ghins, Michel, 2n, 6n
Gisin, N., 226
Glymour, Clark, 29n, 32-33n, 75n, 145
Gödel, Kurt, 156, 170n, 187, 200, 211, 214
Goldberg, Stanley, 24n, 181-182n
Goldstein, Sheldon, 69n, 228n
Gorenstein, M. Y., 220
Gotsman, Errol, 217n
Grangier, Philippe, 226n
Grant, Edward, 41n
Graves, John Cowperthwaite, 216n
Greene, Brian, 144n
Grieder, Alfons, 99n
Griffin, David Ray, 110, 227n, 234n
Grossmann, Marcel, 22
Grünbaum, Adolf, 15n, 19, 29, 31, 34-35n, 80n, 99, 102, 112n, 127n, 158, 164, 213n, 235n
Guth, Alan, 15n

Gutting, Gary, 131

Habicht, K., 129
Hafele, J. C., 67-68
Hall, A. Rupert, 114n
Hall, David B., 66n
Hall, Marie Boas, 114n
Halstead, G. B., 13n, 31n, 129n, 172n
Hannaway, Owen, 41n
Hartle, James, 144, 157-158, 186
Hartshorne, Charles, 86n
Havens, W. W., 67n
Hawking, Steven W., 16n, 105n, 144, 157-158, 186, 212
Hay, H. J., 67n
Healey, Richard, 82n, 149n, 150
Heath, A. E., 124n
Hegel, G. F. W., 137
Heisenberg, Werner, 228n, 230n
Heller, Michael, 219-220, 236n
Herneck, Friedrich, 127n
Herz, H., 125
Hilbert, David, 127
Hirosige, Tetu, 13n, 122n, 128n, 175n
Hoag, J. Barton, 66n
Hodgson, P. E., 49n, 56n, 108n, 174, 184n
Hoffmann, Banesh, 56n, 79n
Hogarth, Mark, 104n
Holton, Gerald, 13, 16, 17, 21n, 22, 24n, 120, 122n, 123, 125, 126n, 127n, 130, 133n, 136, 137, 166, 171, 181-182n
Hönl, H., 221-222
Horwich, Paul Gordon, 34n, 93n, 103n, 156n
Hubble, Edwin, 204, 211, 212, 218, 221n
Hugger, N., 168n
Hughes, R. I. G., 108n
Hull, David, 101n, 229n
Hume, David, 127-129, 135, 143

Illy, Jozsef, vin, 175n, 176n, 177n
Infeld, Leopold, 48n, 79, 222n

Isham, C. J., 158, 226n
Ives, Hubert E., 52n, 65-66, 153n, 178-179

Jacobson, Leon, 88n
Jammer, Max, 108-109, 112-113n
Janis, Allen I., 81n, 147n, 165n, 185n
Jánossy, L., 65n, 91
Jeans, Sir James, 140n, 152
Jeffery, G. B., 11n, 69n, 191n, 202n
Jourdain, Philip E. B., 124n, 126n

Kaku, Michio, 144n
Kanitscheider, Bernulf, 203, 206n, 211, 220-221, 240
Kant, Immanuel, 191
Katsumori, Makoto, 131n
Kaufmann, Walter, 137
Keating, R. E., 67-68
Kennedy, Roy, 15, 65
Kerszberg, Pierre, 239-240
Kilmister, C. W., 133n, 145-146n, 195n
Kitchener, Richard F., 116n, 150
Klein, M. J., 124n
Klimek, Zbigniew, 219-220
Knox, A. J., 122, 177n
Kostro, Ludwik, 23n-24n, 79
Koyré, Alexandre, 106n
Kripke, Saul A., 34n
Kroes, Peter, 59n, 63n, 95-97, 100, 107-108, 160, 207-208, 215-218, 238

Lanczos, C., 127n
Landsberg, Peter T., 239n
Larmor, Joseph, 12n, 18, 65, 119, 122, 227
Lawrence, N., 105n, 217n
Le Poidevin, Robin, 80n, 104n
Leclerc, Ivor, 116n
Leibniz, G. W., 1, 107n, 111n, 120, 139
Lenzen, Victor F., 136
Leslie, John, 16, 143

Lewis, Delmas, 103n
Linde, A. D., 16n
Loar, H. H., 67n
Lockwood, Michael, vi, 110n, 217n
Lodge, M. J. S., 113n
Lodge, Sir Oliver J., 122n
Longair, M. S., 144n, 220n
Lorentz, H. A., vi, 2, 10-19, 22, 23, 25, 26, 27n, 39, 40n, 41, 43, 48, 52n, 61, 62, 65, 66, 69, 78, 80, 86, 88-95, 97n, 99-100, 109, 119, 122, 125n, 132, 137, 139, 140, 145, 146, 154, 163n, 166n, 168n, 169n, 173-194 *pass.*, 205, 213, 221n, 226, 227, 228n, 231-234, 237, 239
Lorenz, Dieter, 89n
Lucas, J. R., 49n, 56n, 107, 108n, 119-120, 174, 184n, 192

Mach, Ernst, 107n, 120, 122n, 123-131, 133, 134-137, 147n, 166, 171n, 182n, 196, 198-199
Maciel, A. K. A., 178n
Mackie, J. L., 2n, 49n
MacKinnon, Flora Isabel, 114n
Maddox, J. R., 146n, 194n
Malament, David, 30-31, 80n
Mansouri, Reza, 221
Marder, L., 49n, 50n, 52n, 57, 60n, 94n, 182n
Margenau, Henry, 139-140, 161, 168n
Maric, Mileva, 22
Markosian, Ned, 1n
Massey, Gerald J., 81n, 147n, 165n, 185n
Maudlin, Tim, 165-166, 223n, 228n, 229n, 230-233, 235
Maxwell, Grover, 193n
Maxwell, James Clerk, 6, 7, 14, 179, 222
McCormack, Thomas J., 123n, 124n
McCormmach, Russell, 13, 178n
McCrea, W. H., 239

Proper Name Index

McGuiness, Brian, 124n
McGuire, J. E., 114-116n, 117n
McTaggart, John McTaggart Ellis, v, 104
Mehlberg, Henry, 152n, 155
Mehra, J., 154n
Mellor, D. H., 81-83, 93n-94n, 103-104, 170
Merricks, Trenton, 103n
Michelson, A. A., 7, 8, 10, 11, 14, 15, 16n, 18n, 22-23n, 65, 92, 99, 122n, 138, 169, 184n, 221
Miller, Arthur I., 2n, 11n, 14, 19n, 23n, 27, 29n, 48n, 66, 92-93n, 99, 125n, 126n, 221n
Miller, D. C., 184n
Milne, E. A., 143, 151, 205, 235n, 239-240
Minkowski, Hermann, 11n, 24, 25n, 30-32, 35, 40-41, 69-70, 77-80, 83, 85-87, 94, 103n, 104n, 137, 145, 146, 175n, 176-178, 183, 191-194, 200, 207, 230
Misner, Charles W., 120, 143n, 144n, 154n, 157, 193n, 205-206, 207, 210n
Møller, C., 12n, 40n, 153
More, Henry, 114-115
Morley, E. W., 8, 10, 11, 14, 15, 18, 22-23n, 65, 99, 122n, 169, 221n
Moss, I. G., 16n
Mössbauer, R. L., 67
Mosterin, Jesus, 16n
Mould, Richard A., 168n
Muller, R. A., 220
Munitz, Milton K., 210, 238, 240

Nagel, Ernst, 136n
Nanopoulos, D. V., 16n
Nerlich, Graham, 33, 35n, 78-80n, 87, 95, 103n, 110n, 145, 165n, 189
Newton, Isaac, v, vi, 2, 3, 5, 14, 41n, 42, 85, 99, 106-121 *pass.*, 122-125, 127-128, 130-131, 137, 139, 147, 150, 151, 154, 155, 161, 162, 166, 171, 177, 179, 182, 186, 188-189, 194, 195, 197, 201-202, 207, 213-215, 220, 233, 235-240, 242
Newton-Smith, W.-H., vin, 110n
Nidditch, P. H., 128n
Nordenson, Harald, 20, 27n
North, J. D., 193n
Norton, John D., 109n, 138n, 146n, 151, 190n
Novikov, Igor D., 223

Oaklander, L. Nathan, 102n
Ogawa, T. Suyoshi, 23n
Olby, R. C., 113n
Ostwald, Wilhelm, 125, 136
Ozsváth, I., 200, 211

Padgett, Alan, 237n, 240
Pagels, Heinz, 131, 153
Pais, Abraham, 19n, 24n, 129, 175n, 184n
Pap, Arthur, 158
Park, David, 105n, 108n, 217n, 236
Pauli, W., 228n
Penrose, R., 226n
Perrett, W., 11n, 69n, 191n, 202n
Petkov, Veselin, 82n, 216n
Petzoldt, Joseph, 127, 137
Planck, Max, 126, 137, 143, 147, 154-156, 168, 177
Plantinga, Alvin, 193
Podlaha, M. F., 92
Podolsky, B., 224, 225n, 226n, 227n
Poincaré, Henri, 2, 13, 17, 18-19n, 31-34, 39, 85-86, 91-92, 122, 129, 145, 167, 171-174, 180, 183, 188, 213, 214, 221n, 227, 237
Popkin, Richard H., 116n
Popper, Karl R., 97n, 233
Powers, Jonathan, 160
Prokhovnik, Simon J., vii, 52n, 138n, 178, 179n, 180-181, 208n, 218-220, 234, 236n
Putnam, Hilary, 103-104, 216n

Qadir, Asghar, 217n
Quine, Willard Van Orman, 34n, 105n
Quinn, Philip L., 35n

Rae, Alastair I. M., 226
Ramige, Eldon Albert, 191-192n
Redhead, Michael, 165n
Rees, Martin J., 211-212
Reichenbach, Hans, 29, 31, 34, 80n, 138, 145, 191n
Reichenbach, Maria, 29n, 34n
Rescher, Nicholas, 81n, 147n, 165n, 185n
Resnick, Robert, 51
Rindler, Wolfgang, 17-18n, 57n, 94n, 105
Robb, A. A., 30-31, 80
Robertson, H. P., 153n, 206, 211, 212, 217, 218, 235, 238, 239
Rodrigues, Waldyr A., Jr., 59n
Rohrlich, Fritz, 57n, 109n
Rompe, R., 147, 222
Rosa, Marcio A. F., 59n
Rosen, Nathan, 221n, 224, 225n, 226n, 227n
Rosser, W. G. V., 64n, 105n
Rossi, Bruno, 66n
Rothman, Tony, 142
Rovelli, Carlo, 146n, 159-160, 168n, 190n
Ruderfer, Martin, 179
Rudnicki, Konrad, 219-220
Russell, Bertrand, 33, 128n
Russell, R. J., 158n
Ryckman, T. A., 130n

Sagan, Carl, 105n, 158n
Salmon, Merrilee, 109n, 151n
Salmon, Wesley C., 29-30n, 80n, 165n
Savitt, Steven, F., 83n-84n
Schaffner, Kenneth F., 14-16, 18n, 26n, 125

Schelling, Friedrich, 137
Scheurer, P. B., 117n
Schiffer, J. P., 67n
Schild, Alfred, 49n, 56
Schilpp, Paul Arthur, 25n, 34n, 127n, 128n, 156n, 191n
Schlesinger, George N., 103n, 112n
Schücking, E., 200, 211, 239
Schwarzschild, K., 186
Seelig, Carl, 126n, 137n
Selby-Bigge, L. A., 128n
Selig, Carl, 171n
Sellars, Wilfrid, 193
Selleri, Franco, 219n, 230n
Sexl, Roman U., 221
Shallis, Michael, 216-217, 237n
Shankland, R. S., 22
Shaw, R., 89n
Sherwin, Charles W., 67n
Shimony, Abner, 233n
Shortland, Michael, 116n
Sinks, John D., 105n
Sjödin, T., 163-164, 166-167
Sklar, Lawrence, 31-32, 39-42, 82, 108n, 109, 144, 145, 149, 169-170
Smart, J. J. C., 34
Smith, James H., 66n
Smith, Norman Kemp, 191n
Smith, Quentin, vn, vii, 103n, 121n, 171n, 173n
Smoot, G. F., 220
Sommerfeld, A., 11n, 191n, 202n
Speziali, Pierre, 79n, 126n
Squires, Euan, 142, 226n
Stachel, John J., 22n, 23, 29n, 75n
Stapp, Henry Pierce, 221, 227, 230n
Stebbing, L. Susan, 152n, 193n
Stein, Howard, 30n, 41-42, 103-104, 105n, 113
Steinhardt, P. I., 16n
Stern, Alexander W., 222n
Stilwell, G. R., 66
Stoeger, W. R., 158n
Strong, E. W., 112

Proper Name Index

Sudarshan, E. C. G., 165
Suppe, Frederick, 150n, 167n, 168n
Suppes, Patrick, 33n, 145n
Swabey, M. C., 27n, 139n
Swabey, W. C., 27n, 139n
Swinburne, Richard, vi, 2n, 86, 107n, 151, 165n, 213n, 218n, 235n

Tauber, Gerald, 217n
Taylor, Edwin F., 2n, 60n, 84-85, 94, 105n, 182n
Taylor, J. G., 2n, 11n, 65, 105n
Terrell, James, 61-62
Thiele, J., 126n
Thomas, L. H., 65
Thorndike, E. M., 15, 65
Thorne, Kip S., 120, 143n, 144n, 154n, 157n, 193n, 205-206, 207n, 210n
Tiomno, J., 178n
Tipler, Frank J., 185n, 236
Tittel, W., 226
Tiwari, S. C., 162n
Tomberlin, James, 193n
Tonnelat, Marie-Antoinette, 131n
Treder, H.-J., 147, 222

Valentini, Antony, 228n, 234n
Van der Merwe, Alwyn, 147n, 222n
Van Fraassen, Bas, 80n
Van Inwagen, Peter, 103n, 193n
Varicak, V., 88
Vigier, Jean-Pierre, 147n, 222n, 227
Von Laue, Max, 137, 150
Von Neuman, John, 225

Wahsner, Renate, 123n
Walker, A. G., 206, 211, 212, 217, 218, 235, 238, 239
Weinberg, Steven, 16n, 144n, 189-190
Weiner, Armin, 127n
Weingard, Robert, 82, 229, 234n
Weisskopf, Victor F., 61n, 144
Wells, H. G., 57
Wenzl, A., 191
Wertheimer, Michael, 135n
Westfall, Richard S., 114n, 116n
Wheeler, John Archibald, 2n, 60n, 84-85, 94, 105n, 120, 143n, 144, 154, 155n-156n, 157, 182n, 193n, 205-206, 207n, 210n, 216n, 217n, 224n
Whitman, Anne, 3n, 106n, 197n
Whitrow, G. J., 193n, 203n, 210-211, 212, 214, 237n
Whitt, L. A., 113n
Whittaker, Sir Edmund, 12n, 72n, 140n, 143, 162n, 222-223
Wien, Wilhelm, 22, 137
Will, Clifford M., 153
Williams, E. J., 66n
Wilson, Robin, 116n
Winnie, John A., 80n, 89
Woolf, Harry, 93n, 136n, 212n
Wolters, Gerion, 127n
Worrall, John, 101n

Young, Thomas, 6
Yourgrau, Palle, 170n

Zahar, Elie, 12-15, 17, 177-178n, 221n
Zbinden, H., 226
Zeeman, E. C., 80n
Zeeman, P., 12n, 13, 175n
Zurek, Wojciech Hubert, 224n
Zwart, P. J., 151
Zyncinski, J., 236n

PHILOSOPHICAL STUDIES SERIES

1. Jay F. Rosenberg: *Linguistic Representation.* 1974 ISBN 90-277-0533-X
2. Wilfrid Sellars: *Essays in Philosophy and Its History.* 1974 ISBN 90-277-0526-7
3. Dickinson S. Miller: *Philosophical Analysis and Human Welfare.* Selected Essays and Chapters from Six Decades. Edited with an Introduction by Lloyd D. Easton. 1975
   ISBN 90-277-0566-6
4. Keith Lehrer (ed.): *Analysis and Metaphysics.* Essays in Honor of R. M Chisholm. 1975
   ISBN 90-277-0571-2
5. Carl Ginet: *Knowledge, Perception, and Memory.* 1975 ISBN 90-277-0574-7
6. Peter H. Hare and Edward H. Madden: *Causing, Perceiving and Believing.* An Examination of the Philosophy of C. J. Ducasse. 1975 ISBN 90-277-0563-1
7. Hector-Neri Castañeda: *Thinking and Doing.* The Philosophical Foundations of Institutions. 1975 ISBN 90-277-0610-7
8. John L. Pollock: *Subjunctive Reasoning.* 1976 ISBN 90-277-0701-4
9. Bruce Aune: *Reason and Action.* 1977 ISBN 90-277-0805-3
10. George Schlesinger: *Religion and Scientific Method.* 1977 ISBN 90-277-0815-0
11. Yirmiahu Yovel (ed.): *Philosophy of History and Action.* Papers presented at the First Jerusalem Philosophical Encounter (December 1974). 1978 ISBN 90-277-0890-8
12. Joseph C. Pitt (ed.): *The Philosophy of Wilfrid Sellars: Queries and Extensions.* 1978
    ISBN 90-277-0903-3
13. Alvin I. Goldman and Jaegwon Kim (eds.): *Values and Morals.* Essays in Honor of William Frankena, Charles Stevenson, and Richard Brandt. 1978 ISBN 90-277-0914-9
14. Michael J. Loux: *Substance and Attribute.* A Study in Ontology. 1978 ISBN 90-277-0926-2
15. Ernest Sosa (ed.): *The Philosophy of Nicholas Rescher.* Discussion and Replies. 1979
    ISBN 90-277-0962-9
16. Jeffrie G. Murphy: *Retribution, Justice, and Therapy.* Essays in the Philosophy of Law. 1979
    ISBN 90-277-0998-X
17. George S. Pappas (ed.): *Justification and Knowledge.* New Studies in Epistemology. 1979
    ISBN 90-277-1023-6
18. James W. Cornman: *Skepticism, Justification, and Explanation.* With a Bibliographic Essay by Walter N. Gregory. 1980 ISBN 90-277-1041-4
19. Peter van Inwagen (ed.): *Time and Cause.* Essays presented to Richard Taylor. 1980
    ISBN 90-277-1048-1
20. Donald Nute: *Topics in Conditional Logic.* 1980 ISBN 90-277-1049-X
21. Risto Hilpinen (ed.): *Rationality in Science.* Studies in the Foundations of Science and Ethics. 1980 ISBN 90-277-1112-7
22. Georges Dicker: *Perceptual Knowledge.* An Analytical and Historical Study. 1980
    ISBN 90-277-1130-5
23. Jay F. Rosenberg: *One World and Our Knowledge of It.* The Problematic of Realism in Post-Kantian Perspective. 1980 ISBN 90-277-1136-4
24. Keith Lehrer and Carl Wagner: *Rational Consensus in Science and Society.* A Philosophical and Mathematical Study. 1981 ISBN 90-277-1306-5
25. David O'Connor: *The Metaphysics of G. E. Moore.* 1982 ISBN 90-277-1352-9

PHILOSOPHICAL STUDIES SERIES

26. John D. Hodson: *The Ethics of Legal Coercion.* 1983　　ISBN 90-277-1494-0
27. Robert J. Richman: *God, Free Will, and Morality.* Prolegomena to a Theory of Practical Reasoning. 1983　　ISBN 90-277-1548-3
28. Terence Penelhum: *God and Skepticism.* A Study in Skepticism and Fideism. 1983
　　ISBN 90-277-1550-5
29. James Bogen and James E. McGuire (eds.): *How Things Are.* Studies in Predication and the History of Philosophy of Science. 1985　　ISBN 90-277-1583-1
30. Clement Dore: *Theism.* 1984　　ISBN 90-277-1683-8
31. Thomas L. Carson: *The Status of Morality.* 1984　　ISBN 90-277-1619-9
32. Michael J. White: *Agency and Integrality.* Philosophical Themes in the Ancient Discussions of Determinism and Responsibility. 1985　　ISBN 90-277-1968-3
33. Donald F. Gustafson: *Intention and Agency.* 1986　　ISBN 90-277-2009-6
34. Paul K. Moser: *Empirical Justification.* 1985　　ISBN 90-277-2041-X
35. Fred Feldman: *Doing the Best We Can.* An Essay in Informal Deontic Logic. 1986
　　ISBN 90-277-2164-5
36. G. W. Fitch: *Naming and Believing.* 1987　　ISBN 90-277-2349-4
37. Terry Penner: *The Ascent from Nominalism.* Some Existence Arguments in Plato's Middle Dialogues. 1987　　ISBN 90-277-2427-X
38. Robert G. Meyers: *The Likelihood of Knowledge.* 1988　　ISBN 90-277-2671-X
39. David F. Austin (ed.): *Philosophical Analysis.* A Defense by Example. 1988
　　ISBN 90-277-2674-4
40. Stuart Silvers (ed.): *Rerepresentation.* Essays in the Philosophy of Mental Representation. 1988　　ISBN 0-7923-0045-9
41. Michael P. Levine: *Hume and the Problem of Miracles.* A Solution. 1989　ISBN 0-7923-0043-2
42. Melvin Dalgarno and Eric Matthews (eds.): *The Philosophy of Thomas Reid.* 1989
　　ISBN 0-7923-0190-0
43. Kenneth R. Westphal: *Hegel's Epistemological Realism.* A Study of the Aim and Method of Hegel's *Phenomenology of Spirit.* 1989　　ISBN 0-7923-0193-5
44. John W. Bender (ed.): *The Current State of the Coherence Theory.* Critical Essays on the Epistemic Theories of Keith Lehrer and Laurence BonJour, with Replies. 1989
　　ISBN 0-7923-0220-6
45. Roger D. Gallie: *Thomas Reid and 'The Way of Ideas'.* 1989　　ISBN 0-7923-0390-3
46. J-C. Smith (ed.): *Historical Foundations of Cognitive Science.* 1990　ISBN 0-7923-0451-9
47. John Heil (ed.): *Cause, Mind, and Reality.* Essays Honoring C. B. Martin. 1989
　　ISBN 0-7923-0462-4
48. Michael D. Roth and Glenn Ross (eds.): *Doubting.* Contemporary Perspectives on Skepticism. 1990　　ISBN 0-7923-0576-0
49. Rod Bertolet: *What is Said.* A Theory of Indirect Speech Reports. 1990
　　ISBN 0-7923-0792-5
50. Bruce Russell (ed.): *Freedom, Rights and Pornography.* A Collection of Papers by Fred R. Berger. 1991　　ISBN 0-7923-1034-9
51. Kevin Mulligan (ed.): *Language, Truth and Ontology.* 1992　　ISBN 0-7923-1509-X

PHILOSOPHICAL STUDIES SERIES

52. Jesús Ezquerro and Jesús M. Larrazabal (eds.): *Cognition, Semantics and Philosophy.* Proceedings of the First International Colloquium on Cognitive Science. 1992 ISBN 0-7923-1538-3
53. O.H. Green: *The Emotions.* A Philosophical Theory. 1992 ISBN 0-7923-1549-9
54. Jeffrie G. Murphy: *Retribution Reconsidered.* More Essays in the Philosophy of Law. 1992 ISBN 0-7923-1815-3
55. Phillip Montague: *In the Interests of Others.* An Essay in Moral Philosophy. 1992 ISBN 0-7923-1856-0
56. Jacques-Paul Dubucs (ed.): *Philosophy of Probability.* 1993 ISBN 0-7923-2385-8
57. Gary S. Rosenkrantz: *Haecceity.* An Ontological Essay. 1993 ISBN 0-7923-2438-2
58. Charles Landesman: *The Eye and the Mind.* Reflections on Perception and the Problem of Knowledge. 1994 ISBN 0-7923-2586-9
59. Paul Weingartner (ed.): *Scientific and Religious Belief.* 1994 ISBN 0-7923-2595-8
60. Michaelis Michael and John O'Leary-Hawthorne (eds.): *Philosophy in Mind.* The Place of Philosophy in the Study of Mind. 1994 ISBN 0-7923-3143-5
61. William H. Shaw: *Moore on Right and Wrong.* The Normative Ethics of G.E. Moore. 1995 ISBN 0-7923-3223-7
62. T.A. Blackson: *Inquiry, Forms, and Substances.* A Study in Plato's Metaphysics and Epistemology. 1995 ISBN 0-7923-3275-X
63. Debra Nails: *Agora, Academy, and the Conduct of Philosophy.* 1995 ISBN 0-7923-3543-0
64. Warren Shibles: *Emotion in Aesthetics.* 1995 ISBN 0-7923-3618-6
65. John Biro and Petr Kotatko (eds.): *Frege: Sense and Reference One Hundred Years Later.* 1995 ISBN 0-7923-3795-6
66. Mary Gore Forrester: *Persons, Animals, and Fetuses.* An Essay in Practical Ethics. 1996 ISBN 0-7923-3918-5
67. K. Lehrer, B.J. Lum, B.A. Slichta and N.D. Smith (eds.): *Knowledge, Teaching and Wisdom.* 1996 ISBN 0-7923-3980-0
68. Herbert Granger: *Aristotle's Idea of the Soul.* 1996 ISBN 0-7923-4033-7
69. Andy Clark, Jesús Ezquerro and Jesús M. Larrazabal (eds.): *Philosophy and Cognitive Science: Categories, Consciousness, and Reasoning.* Proceedings of the Second International Colloquium on Cogitive Science. 1996 ISBN 0-7923-4068-X
70. J. Mendola: *Human Thought.* 1997 ISBN 0-7923-4401-4
71. J. Wright: *Realism and Explanatory Priority.* 1997 ISBN 0-7923-4484-7
72. X. Arrazola, K. Korta and F.J. Pelletier (eds.): *Discourse, Interaction and Communication.* Proceedings of the Fourth International Colloquium on Cognitive Science. 1998 ISBN 0-7923-4952-0
73. E. Morscher, O. Neumaier and P. Simons (eds.): *Applied Ethics in a Troubled World.* 1998 ISBN 0-7923-4965-2
74. R.O. Savage: *Real Alternatives, Leibniz's Metaphysics of Choice.* 1998 ISBN 0-7923-5057-X
75. Q. Gibson: *The Existence Principle.* 1998 ISBN 0-7923-5188-6
76. F. Orilia and W.J. Rapaport (eds.): *Thought, Language, and Ontology.* 1998 ISBN 0-7923-5197-5

PHILOSOPHICAL STUDIES SERIES

77. J. Bransen and S.E. Cuypers (eds.): *Human Action, Deliberation and Causation.* 1998
    ISBN 0-7923-5204-1
78. R.D. Gallie: *Thomas Reid: Ethics, Aesthetics and the Anatomy of the Self.* 1998
    ISBN 0-7923-5241-6
79. K. Korta, E. Sosa and X. Arrazola (eds.): *Cognition, Agency and Rationality.* Proceedings of the Fifth International Colloquium on Cognitive Science. 1999   ISBN 0-7923-5973-9
80. M. Paul: *Success in Referential Communication.* 1999   ISBN 0-7923-5974-7
81. E. Fischer: *Linguistic Creativity.* Exercises in 'Philosophical Therapy'. 2000
    ISBN 0-7923-6124-5
82. R. Tuomela: *Cooperation.* A Philosophical Study. 2000   ISBN 0-7923-6201-2
83. P. Engel (ed.): *Believing and Accepting.* 2000   ISBN 0-7923-6238-1
84. W.L. Craig: *Time and the Metaphysics of Relativity .* 2000   *ISBN 0-7923-6668-9*

KLUWER ACADEMIC PUBLISHERS – DORDRECHT / BOSTON / LONDON

CPSIA information can be obtained at www.ICGtesting.com
Printed in the USA
LVOW10*1719081214

417825LV00010B/86/P

9 780792 366683